T0177190

Theoretical Foundations of Nanoscale Quantum Devices

Nanooptics, which describes the interaction of light with matter at the nanoscale, is a topic of great fundamental interest to physicists and engineers and allows the direct observation of quantum mechanical phenomena in action. This self-contained and extensively referenced text describes the underlying theory behind nanodevices operating in the quantum regime for use both in advanced courses and as a reference for researchers in physics, chemistry, electrical engineering, and materials science. Presenting an extensive theoretical toolset for design and analysis of nanodevices, the authors demonstrate the art of developing approximate quantum models of real nanodevices. The rudimentary mathematical knowledge required to master the material is carefully introduced, with detailed derivations and frequent worked examples allowing readers to gain a thorough understanding of the material. More advanced applications are gradually introduced alongside analytical approximations and simplifying assumptions often used to make such problems tractable while representative of the observed features.

Malin Premaratne is Vice President of the Academic Board of Monash University, Australia. He is a Fellow of the Optical Society of America and a Fellow of the Institute of Engineers, Australia. His industrial experience includes consultancy roles for Cisco, Lucent Technologies, Ericsson, Siemens, VPIsystems, Telcordia Technologies, Ciena, and Tellium. He is also a visiting researcher at the Jet Propulsion Laboratory at Caltech, Oxford University, the Institute of Optics, University of Rochester, Australian National University, and the University of Melbourne. He has served as Associate Editor for *IEEE Photonics Technology Letters*, *IEEE Photonics Journal*, and *OSA Advances in Optics and Photonics Journal*.

Govind P. Agrawal is James C. Wyant Professor of Optics at the University of Rochester. He is a Fellow of IEEE and The Optical Society of America, and a Life Fellow of the Optical Society of India. He has been awarded the IEEE Photonics Society Quantum Electronics Award, the Riker University Award for Excellence in Graduate Teaching, the Esther Hoffman Beller Medal, the Max Born Award of the Optical Society, and the Quantum Electronics Prize of the European Physical Society. He has also served as Editor-in-Chief for the *OSA Advances in Optics and Photonics Journal*.

Theoretical Foundations
of Nanoscale Quantum Devices

Malin Premaratne

Monash University, Clayton Victoria, Australia

Govind P. Agrawal

University of Rochester, New York, USA

CAMBRIDGE
UNIVERSITY PRESS

University Printing House, Cambridge CB2 8BS, United Kingdom

One Liberty Plaza, 20th Floor, New York, NY 10006, USA

477 Williamstown Road, Port Melbourne, VIC 3207, Australia

314–321, 3rd Floor, Plot 3, Splendor Forum, Jasola District Centre, New Delhi – 110025, India

79 Anson Road, #06–04/06, Singapore 079906

Cambridge University Press is part of the University of Cambridge.

It furthers the University's mission by disseminating knowledge in the pursuit of education, learning, and research at the highest international levels of excellence.

www.cambridge.org
Information on this title: www.cambridge.org/9781108475662
DOI: 10.1017/9781108634472

© Malin Premaratne and Govind P. Agrawal 2021

This publication is in copyright. Subject to statutory exception and to the provisions of relevant collective licensing agreements, no reproduction of any part may take place without the written permission of Cambridge University Press.

First published 2021

Printed in the United Kingdom by TJ Books Ltd, Padstow Cornwall

A catalogue record for this publication is available from the British Library.

ISBN 978-1-108-47566-2 Hardback

Cambridge University Press has no responsibility for the persistence or accuracy of URLs for external or third-party internet websites referred to in this publication and does not guarantee that any content on such websites is, or will remain, accurate or appropriate.

For Erosha, Gehan, Sayumi,
Ginger, Pepper, and Sugar
—Malin Premaratne

For Anne, Sipra, Caroline, and Claire
—Govind P. Agrawal

Contents

Preface

> I regarded as quite useless the reading of large treatises of pure analysis: too large a number of methods pass at once before the eyes. It is in the works of application that one must study them; one judges their utility there and appraises the manner of making use of them.
>
> Joseph Louis Lagrange

The use of the prefix "nano," standing for a billionth (10^{-9}) of a base unit, has become quite prevalent in recent years. This is mainly due to the enormous success of the microelectronics industry in reducing the size of transistors inside a computer chip (the smallest feature size has reached below 5 nm), leading to the advent of smartphones in 2007. The words nanoscience and nanotechnology have also become common over the last 20 years. It is commonly understood that nanoscience refers to the science of nanoscale devices, while nanotechnology makes use of such devices for practical applications. Contrary to our day-to-day experience, nanoscale devices behave quite differently. In this book, we build a tool kit useful for designing, controlling, and understanding the operation of such devices.

Understanding of nature is based on simplified models with varying degrees of complexity that can be easily manipulated and observed. Our motivation behind this book is to show that such models of nanoscale quantum devices can be used to understand their operation and to optimize their performance. We achieve this goal by adopting quantum mechanics for the description of nanoscale devices. The quantum effects often dominate in such devices and must be included for their realistic description. Quantum mechanics is extraordinarily successful, and its predictions have been confirmed to an astonishing level of precision in a broad spectrum of experiments. Even though it has a mystique axiomatic foundation, it provides an accurate mapping between the real-world devices and theoretical models used for them. As one delves deeper, it becomes clear that mathematics plays an integral bridging role. Thus, we do not shy away from introducing sophisticated mathematical tools, but provide examples and sufficient details for understanding the material in a manner that we have not found in other texts. It is our firm belief that the complexity of underlying theoretical tools should not be a barrier for graduate students and scientists to model nanoscale devices accurately.

This book is an attempt to bring widely spread diverse theoretical tools under a single volume that should prove useful for both graduate students and scientists or engineers interested in using nanoscale quantum devices for various applications. Many books cover distinct areas of nanoscience and nanotechnology separately. This book is intended not to replace them but to complement them. The book content is the tool kit that authors use in their day-to-day research work as active researchers in the field of nanophotonics. We hope that our exposition to quantum systems will illuminate and inspire the reader to become proficient in designing and analyzing novel nanoscale devices with diverse applications.

The book is organized as follows. Its first chapter provides historical introduction to the field of nanoscience and introduces features such as quantum confinement, quantum interference, and quantum transport that become relevant for nanoscale devices. We review in Chapter 2 the quantum-mechanical concepts and mathematical tools needed for later chapters. Linear response theory is discussed in Chapter 3, where we also calculate the generalized susceptibility and introduce the fluctuation-dissipation theorem. The effects of the environment that lead to dissipation and decoherence in a quantum device are covered in Chapter 4, where we use a master-equation approach to derive the Lindblad and Redfield equations. The focus of Chapter 5 is on the flow of current inside quantum devices. We first use a simple approach based on scattering matrices and later apply the nonequilibrium Green's function method to calculate the current. The phenomenon of quantum tunneling is covered in Chapter 6, where we calculate the tunneling current using two different methods. Chapter 7 is devoted to quantum noise. After discussing important noise sources leading to thermal noise, shot noise, and Brownian motion, we introduce the quantum Langevin equations and apply them to lasers to calculate the noise spectra associated with the intensity and phase fluctuations. The concept of a squeezed state is also covered in this chapter.

This monograph should serve well the needs of the nanoscience community interested in modeling and understanding the operation of nanoscale quantum devices. The potential readership is likely to consist of graduate students enrolled in MS and PhD programs and scientists working in fields such as nanophotonics and plasmonics. The book may also be useful for a high-level graduate course devoted to nanoscale devices. The extensive bibliography at the end of this book should also be helpful to readers.

It is a pleasure to acknowledge the help we have received from various sources in writing this book. Many individuals have contributed to the completion of this book, either directly or indirectly. We are thankful to all of them, especially to our graduate students and postdoctoral fellows, whose curiosity helped us in understanding better the material presented in this book. Particular mention is due to Nicholas Gibbons of Cambridge University Press, who helped us steer this project to completion. No authors could wish for a more supportive editor, and we thank him, Sarah Lambert, Henry Cockburn, and rest of the CUP team. Last but not least, we thank Erosha and Anne for their love, encouragement, and support.

1 Introduction

The principles of physics, as far as I can see, do not speak against the possibility of maneuvering things atom by atom. It is not an attempt to violate any laws; it is something, in principle, that can be done; but in practice, it has not been done because we are too big.

Richard Feynman

1.1 Nanoscience and Nanotechnology

The widely used International System of Units (the SI, short for Système International) was adopted in 1889. It is based on seven base units – second (s), meter (m), kilogram (kg), ampere (A), kelvin (K), mole (mol), and candela (cd) – for measuring time, length, mass, electric current, temperature, amount of substance, and luminous intensity, respectively. Prefixes representing integer powers of 10 are added to these base units to produce multiples and submultiples of the original unit. The SI system also specifies that Latin terms should be used for negative powers of 10 – e.g., milli (m), micro (μ), nano (n) – and Greek terms should be used for positive powers of 10 – e.g., kilo (k), mega (M), giga (G). The prefix nano was adopted in 1958 to precisely mean 10^{-9} SI units. According to the Oxford English Dictionary, the word *nano* originates from the classical Latin *nanus*, or its ancient Greek etymon *nanos* ($\nu\acute{\alpha}\nu o\varsigma$), meaning dwarf [1].

In 1974, Norio Taniguchi introduced the term *nanotechnology* to describe his work on ultrafine machining and its potential for engineering devices at a submicrometer scale [2]. The modern usage of this term extends well beyond this simple machine metaphor and corresponds to a transformational technology capable of assembling, manipulating, and controlling individual atoms, molecules, or their interactions on a nanometer scale (1 to 100 nm). Even though this usage captures the essence of present-day nanotechnology, it is based on the size of objects involved and thus has many deficiencies. International Organization for Standardization (ISO), for example, has proposed to broaden the scope by including materials having at least one internal or surface feature, where the onset of size-dependent phenomena differs from the properties of individual atoms and molecules. Such structures enable novel applications and lead to improved materials, devices, and systems by exploiting nanoscale properties.

Nanoscience can be described as the science of nanoscale devices. Essentially, one can consider nanoscience as the bridge between classical and quantum physics – it is a scale

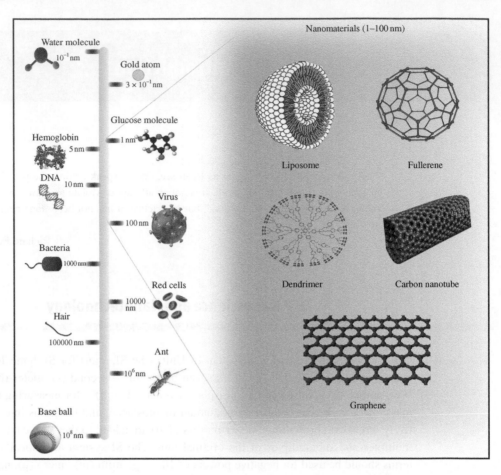

Fig. 1.1 Making sense of length scales involved in nanotechnology. Typical bond lengths are 0.12 to 0.15 nm, and a DNA double-helix has a diameter of 2 nm. The smallest cellular life-form, bacteria of the genus Mycoplasma, is 200 nm in length, and the diameter of a human hair is about 10^5 nm (or 0.1 mm). *Source:* https://commons.wikimedia.org/w/index.php?curid=32395880 ©Sureshbup. (The color version of this figure is available online.)

where we can make use of both aspects to harness collective rather than the individual properties of atoms and molecules. As we shall see later, these collective properties of individual building blocks predominantly define the novel aspects of nanostructures. Figure 1.1 shows a variety of objects covering length scales ranging from 0.1 nm to 1 cm. An expanded view of a few nanoscale (1 to 100 nm) objects involved in the development of nanotechnology is shown on the right side of this figure.

1.1.1 Historical Perspective

Historically, James Clerk Maxwell proposed in 1867 the use of tiny machines to violate the second law of thermodynamics, which states that entropy of a closed system cannot decrease. According to this law, heat must flow from hot to cold, and a perpetual motion machine cannot be built. The gedanken experiment he proposed is known as Maxwell's demon and involves a machine (or demon) guarding a tiny hole between two gas reservoirs at the same temperature. The demon can measure the speed of individual molecules and let through only the fast ones, which would create a temperature difference between the two reservoirs without doing any work. As the second law of thermodynamics has stood the test of time, it is unlikely that Maxwell's demon would succeed, but it is fascinating to see that molecular-level sensing and manipulation ideas were conceived more than 150 years ago.

More recently, in a 1959 lecture titled "There's plenty of room at the bottom" and given to the American Physical Society, the physicist Richard Feynman alluded to the possibility of having miniaturized devices, made of a small number of atoms and working in compact spaces, for exploiting specific effects unique to their size and shape to control synthetic chemical reactions and to produce useful devices or substances.

Historical evidence exists showing that humans have exploited the interaction of light with nanoparticles, without understanding the physics behind it. An intriguing example is provided by the Lycurgus Cup shown in Figure 1.2. It is thought to have been made by Roman craftsmen during the fourth century. The cup contains gold and silver nanoparticles embedded in the glass and exhibits a color-changing property that makes its glass take on different hues, depending on the light source. It appears jade-green when observed in reflected light. However, when light is shone into the cup, it appears translucent-red from the outside. The second object in Figure 1.2, a stained glass window at Lancaster Cathedral showing Edmund and Thomas of Canterbury, uses trapped gold and silver nanoparticles in the glass to generate the ruby-red and deep-yellow colors, respectively. These visual effects can be explained using modern theories on plasmon generation, but it is still a puzzle how

Fig. 1.2 Historical evidence of the use of plasmonic effects. The Lycurgus Cup (left), thought to have been made during the fourth century, exhibits a color-changing property. It appears jade-green when looked at in reflected light but translucent-red when light is shone into the cup because it is made of glass containing gold and silver nanoparticles. The stained glass window (right) at Lancaster Cathedral in which gold and silver nanoparticles were used to generate the ruby red and deep yellow colors. (The color version of this figure is available online.)

ancient blacksmiths knew the precise material properties and compositions to realize them in practice.

Regardless of the current advances that allow humans to harness the power of nanotechnology, the striking reality is that natural processes have cleverly utilized nanotechnology effects for billions of years. Examples include harvesting of solar energy though photosynthesis, accurate replication of the DNA structure, and repair of any damage to DNA incurred because of endogenous or exogenous factors. Discovery of such effects unique to the nanoscale is the prime task of *nanoscience*. Theoretical know-how and understanding developed through nanoscience is used for *nanotechnology* that benefits the society through its specific applications such as longer lasting tennis balls, more efficient solar cells, and cleaner diesel engines. However, from prehistoric times to now, there are numerous cases where the use of a technology preceded the underlying science; practitioners were unaware of the reasons for peculiar behavior they found in materials and devices much different from familiar individual atoms, molecules, and bulk matter, yet proceeded to use them in applications – a model that modern engineers and scientists appear to emulate even now!

1.1.2 New Features Appearing at the Nanoscale

Materials interacting with electromagnetic and other fields exhibit phenomena on a broad range of spatial and temporal scales. A basic postulate in physics is the independence of any observation with respect to the choice of time, place, and units. It requires that physical quantities rescale by the same amount throughout space-time, yet it does not imply physics to be scale invariant. It is very clear that, at the smallest scale, physics demands a quantized treatment, and Planck's constant h sets the smallest observable limit. The standard model of elementary particles identifies four fundamental forces that govern our universe: gravity, the weak force, the strong force, and the electromagnetic force. Each of these forces has a specific coupling strength and a specific distance dependence. The gravitational and electromagnetic forces scale as $1/r^2$ (called the inverse-square law) and they can act over long distances, but the weak and strong forces act only over short distances. At distances greater than 10^{-14} m, the strong force is practically unobservable, and the weak force has no influence over distances greater than 10^{-18} m. All this suggests that we need to pay attention to the scale and units used for measuring different quantities.

A common feature of all forces is that they fade away as one moves from the source. Quantum field theory explains any force between two objects using exchange particles, which are virtual particles emitted from one object (source) and absorbed at the other (sink). Four types of exchange particles – photons, gluons, weak bosons, and gravitons – give rise to four forces; they all have a spin of 1 in units of $\hbar \equiv h/(2\pi)$ and transfer momentum between the two interacting objects. The rate at which momentum is exchanged is equal to the force created between the two objects. Quantum field theory shows that this force weakens as the distance between objects increases. For example, electromagnetic force between two charge particles decreases as $1/r^2$, but this dependence becomes

$1/r^4$ for dipole–dipole interactions. As the origin of most physical or chemical properties can be traced back to the interactions among the atomic or molecular constituents, all such properties tend to carry remnants of the inverse-distance dependence and manifest as size-dependent features for the nanoscale objects. For example, the following material properties become size dependent (to various degrees) when at least one of the dimensions goes below 100 nm:

• Mechanical properties [3]: elastic moduli, adhesion, friction, capillary forces;
• Thermal properties [4, 5]: melting point, thermal conductivity;
• Chemical properties [6, 7]: reactivity, catalysis;
• Electrical properties [8]: quantized conductance, Coulomb blockade;
• Magnetic properties [9]: spin-dependent transport, giant magnetoresistance;
• Optical properties [10]: band structure, band-gap energy, nonlinear response.

A practical and useful aspect of this size dependence is that it gives engineers the ability to tune one or more properties of bulk materials by resizing them to the nano-regime (1 nm to 100 nm). It is this feature that lies behind the concept of metamaterials – artificially designed materials that allow one to use nanotechnology for practical applications.

1.1.3 Surface-to-Volume Ratio

In nanosize objects, surface atoms behave somewhat differently from their bulk atoms. A simple way to judge whether the surface or bulk effects dominate is to consider the ratio of surface area A and volume V of a nanostructure. Table 1.1 compares the ratio A/V for three solids in the shape of a sphere, a cube, and a right-square pyramid. It shows that this ratio scales as $1/r$, where r is a measure of linear size. This scaling is found to hold for all regular, simple structures. Even for a complicated structure, if we can identify a single size parameter (e.g., by enclosing the structure inside a sphere of radius r), the same scaling holds approximately.

Physically, the $1/r$ scaling implies that, when the size of a three-dimensional structure shrinks, the ratio A/V increases. The drastic effect on surface area can be seen in the example shown in Figure 1.3. The cube A has 1 m sides with a surface area of $6 \, \mathrm{m}^2$. If it is divided into smaller 1 cm-size cubes (part B), each cube has an area of 6 cm^2 but there are 10^6 such cubes, resulting in a total surface area of $600 \, \mathrm{m}^2$. If 1 nm-size cubes are made of cube A (part C), the total surface area would become $6000 \, \mathrm{km}^2$. Even though the total volume remains the same in all three cases, the collective surface area is greatly increased with a reduced size of each cube.

A drastic increase in the area of surfaces (or interfaces) can lead to entirely new electronic and vibrational states associated with each surface. Indeed, surface effects are responsible for melting initiated at a surface (premelting) and a lower melting temperature of a compact object (compared to its bulk counterpart) [11, 12, 13]. Also, considerable variations in the thermal conductivity of nanostructures can be partially attributed to an enhanced surface area [14]. For example, thermal conductivity of a nanowire can be much

Table 1.1	Surface-to-volume ratio of three regular solids.		
	A	V	A/V
Cube of side r	$6r^2$	r^3	$\frac{6}{r}$
Sphere of radius r	$4\pi r^2$	$\frac{4}{3}\pi r^3$	$\frac{3}{r}$
Pyramid with side r	$3r^2$	$\frac{1}{3\sqrt{2}}r^3$	$\frac{9\sqrt{2}}{r}$

Fig. 1.3   Dramatic increase in total surface area of nanostructures for a given volume. The cube in part A has 1-m sides, smaller cubes in part B have 1-cm sides, and the smallest cubes in part C have 1-nm sides. The total surface area of all cubes first increases from 6 m^2 to 600 m^2 and then to 6000 km^2 for the same volume.

lower than that of the corresponding bulk material [15]; similarly, carbon nanotubes exhibit much higher thermal conductivity compared to diamonds [16].

As an another example of the surface effects, even though bulk gold is relatively inert chemically, it shows high chemical reactivity in the form of a nanosize cluster. This can be partially attributed to the abundance of surface atoms in a gold nanocluster that behave like individual atoms [17]. Similarly, bulk silver tends not to react with hydrochloric acid. However, high reactivity of silver nanoparticles with hydrochloric acid has been observed and is attributed to the electronic structure of the surface states [18].

In addition to the thermal and chemical properties, mechanical and electrical properties are also affected by an increased surface area. For example, indium arsenide (InAs) nanowires exhibit a monotonic decrease in mobility as their radius is reduced to below 10 nm. The low-temperature transport data show clearly that it is surface-roughness scattering that leads to mobility degradation [19]. Moreover, a reduced coordination of surface atoms and the presence of surface charges can exert a relatively high stress that is well beyond the elastic regime [20, 21]. Peculiar enough, charges present on the polar surfaces of thin zinc-oxide (ZnO) nanobelts can cause spontaneous formation of rings and coils [22]. A large surface area and the resulting surface effects can also explain the occurrence of unexpected phenomena at the nanoscale such as shape-based memory and pseudoelasticity. For example, Young's modulus of films that are thinner than 10 atomic layers is found to be 30% smaller than the bulk value. All these observations indicate that the enhancement of surface-to-volume ratio plays an important role for nanosize objects.

1.2 Characteristic Length Scales

To understand the physics at the nanoscale, it is useful to have a clear understanding of a few length scales that have fundamental meaning associated with them. Whenever a characteristic dimension of a nanosize object becomes comparable to one of these length scales, we expect to see effects peculiar to the nanoscale, normally absent from its bulk counterpart. If we control the movement of elementary particles (electrons, holes, excitons, …) by shaping a nanosize object, the quantum effects appear especially in those regions that have dimensions comparable to a specific length scale.

Quantum mechanics shows that moving particles of matter, whether large or small, can be described both as waves and as particles (wave–particle duality). Early work in this area concentrated on demonstrating the wavelike properties of fundamental particles such as electrons, protons, and neutrons. More recently, massive particles such as C_{60} fullerene (a molecule with 60 carbon atoms, size 1.1 nm), tetraphenylporphyrin (a biodye molecule, size 2 nm), and $C_{60}F_{48}$ (a fluorinated buckyball of 108 atoms) have been shown to comply with the wave–particle duality description.

When dealing with nanosize objects, it is often not clear whether a classical or quantum model should be used to describe their behavior, and there are situations when both are needed. One good example is the photoelectric effect of light, for which both wave and particle aspects of light are required for a complete description. Another example is diffraction of electrons from crystals. The length scale that governs the wave nature of particles is known as the de Broglie wavelength, denoted as λ_{DB} and defined in Aside 1.1. This length scale plays an important role in nanoscience. We will see later that when motion of a nanosize object is restricted to dimensions smaller than its de Broglie wavelength, not only quantum-size effects appear but the associated density of states is also modified.

Further insight could be gained by looking at the analogous situation in optics. The geometrical-optics approximation of light does not take into account its wave nature. It provides us with a strong clue when we could discard the wave effects. Geometrical optics works well for analyzing interaction of light with objects whose physical dimensions are larger than the wavelength λ of light. For example, if light passes through a slit whose width is much larger than λ, we can use geometrical optics. However, when the slit width becomes comparable to λ, the wave nature of light cannot be ignored, and the diffraction effects will be pronounced and noticeable. If we replace the wavelength of light by the de Broglie wavelength of a particle, the same reasoning can be used to decide when the wave nature of the particle becomes relevant. We shall occasionally use the term *quantum particle* to remind the reader that we intend to use both properties judiciously as needed to model certain effects.

Aside 1.1   de Broglie Wavelength

The de Broglie wavelength, associated with any moving particle with momentum p, is defined as

$$\lambda_{DB} = h/p, \tag{1.1}$$

where the Planck constant has a value $h = 6.626 \ 10^{-34}$ J-s. It is useful to express p in terms of relativistic energy E and mass m of the moving particle:

$$E^2 = p^2c^2 + m^2c^4, \tag{1.2}$$

where c is the speed of light in vacuum. This relation implies that λ_{DB} can be written as

$$\lambda_{DB} = \frac{hc}{\sqrt{E^2 - m^2c^4}}. \tag{1.3}$$

When a particle is traveling close to the speed of light, its rest-mass energy mc^2 can be discarded compared to E to obtain

$$\lambda_{DB} \approx hc/E. \tag{1.4}$$

If the particle is travelling much slower than the speed of light ($pc \ll mc^2$), we can use the approximation $E \approx mc^2 + p^2/2m$. Such a particle has a kinetic energy given by $E_k = p^2/2m$, and its de Broglie wavelength is given by

$$\lambda_{DB} \approx \frac{h}{\sqrt{2mE_k}} = \frac{h}{mv}, \tag{1.5}$$

where v is the particle's speed.

The magnitude of λ_{DB} is immeasurably small for macroscopic objects such as humans, but it is of the same order (0.1 nm) as chemical bonds for subatomic particles. As its value depends on the momentum, λ_{DB} can vary considerably for any particle, depending on its mass and speed (or equivalently its kinetic energy). For example, consider a situation where electrons have a kinetic energy of about 100 eV, resulting in a de Broglie wavelength of about 0.1 nm. This value is comparable to typical spacing between atoms in a crystal. As a result, electrons behave as waves and are diffracted by a crystal. However, if the kinetic energy of electrons is increased to 100 MeV, λ_{DB} becomes so small that electrons behave as particles, showing no diffraction effects in the same crystal.

Depending on the context, de Broglie wavelength has different names. The *thermal length* λ_T, for example, is a length scale related to the de Broglie wavelength of a gas consisting of noninteracting, slowly moving particles in equilibrium at temperature T. The kinetic energy E_k for such particles is of the order of $\pi k_B T$, where k_B is the Boltzmann constant. Using this value in Eq. (1.5), thermal length of this gas is obtained from

$$\lambda_T = \frac{h}{\sqrt{2\pi m k_B T}}. \tag{1.6}$$

Consider a gas in thermal equilibrium. The characteristic interparticle distance d for such a gas can be estimated from the packing density of cubes of volume d^3 filling a unit volume, giving $d = n^{1/3}$, where n is the number density of particles. As an example, oxygen at room temperature has d around 3 nm, whereas λ_T is estimated to be around 0.02 nm from Eq. (1.6). Since $\lambda_T \ll d$, it follows that this gas does not require a quantum-mechanical treatment (except for collisions that do require a quantum treatment).

As a second example, consider electron–hole pairs inside a semiconductor. Since the effective mass of electrons and holes can vary from 1% to 100% of a free electron's mass, the corresponding de Broglie wavelength at 300 K is in the range of 73 nm to 7.3 nm. As the temperature approaches 3 K, these values increase by a factor of 10, as seen clearly from Eq. (1.6). Apart from temperature, the de Broglie wavelength can also be manipulated in semiconductors through impurity doping. The main point to stress is that λ_{DB} of charged carriers in semiconductors can vary from 7 nm to 1 μm, depending on the experimental conditions, and a quantum treatment may be needed in some cases.

The situation becomes somewhat different for metals, which are characterized by nearly free-moving electrons whose density can be controlled thermally, chemically, or optically. Metals exhibit high electrical conductivity as well as high thermal conductivity compared to other materials. Metals also generally obey the Wiedemann–Franze law at high temperatures (which states that the ratio of thermal to electrical conductivity is proportional to the temperature). Owing to the abundance of electrons with little interaction among them, the simplest model assumes that metals are a collection of positive ions in a sea of non-interacting electrons and treats metals as a free-electron gas. The positive ions are assumed to form a regular lattice and provide a periodic potential in the Schrödinger equation. As a consequence of the Bloch theorem, the electron's wave function is proportional to $\exp(i\mathbf{k} \cdot \mathbf{r})$, where $\hbar\mathbf{k}$ represents the electron's momentum and $\mathbf{r}$ is its position vector. The Fermi energy for a free-electron gas can be written as $E_F = \hbar^2 k_F^2 / 2m$, where m is the effective mass of electrons in the metal and k_F is the wave number for this energy.

The energy distribution for a free-electron gas is governed by the Fermi–Dirac distribution; it is similar in nature to the Maxwell–Boltzmann distribution of an ideal gas (or Planck's blackbody radiation). The Fermi–Dirac distribution depends on the chemical potential μ of the material and its temperature, both of which can be experimentally measured. The significance of chemical potential is most readily seen in the limiting case of $T = 0$ K, for which the ground state is obtained by placing all electrons into the lowest available energy levels up to the energy E_F. As a result of this filling, there is a sharp boundary in the three-dimensional $\mathbf{k}$-space between the filled and empty states. This surface is known as the *Fermi surface* and its shape is spherical for a free-electron gas.

The Fermi wavelength λ_F is a length scale defined as the de Broglie wavelength of electrons at the Fermi energy E_F. If the momentum of electrons at this energy is $p_F\hat{\mathbf{p}}$, where $\hat{\mathbf{p}}$ is a unit vector, then $E_F = p_F^2/2m$ is the kinetic energy of each electron. Thus, the Fermi wavelength is given by

$$\lambda_F = \frac{h}{p_F} = \frac{h}{\sqrt{2mE_F}}. \tag{1.7}$$

This relation provides good agreement with experimentally observed behavior because only electrons in the vicinity of the Fermi surface participate in most physical processes. For this reason, the Fermi energy and the shape of the Fermi surface are important in practice. Typically, λ_F for metals is around $0.6\,\text{nm}$, and this value can be controlled by changing the density of free electrons. This dependence is widely exploited in semiconductors by changing the type and density of dopants. The Fermi wavelength as a length scale plays an important role in both the design and analysis of metal-based nanoscale devices. For example, as we will see later, λ_F is an important parameter for understanding quantum conductance of nanostructures.

Sometimes a characteristic temperature, called *Fermi temperature* and denoted by T_F, is also defined as $T_F = E_F/k_B$. For most metals, its value lies in the range $10^4\,\text{K}$ to $10^5\,\text{K}$ and is much higher than room temperature or typical temperatures at which metals are used. Thus, under typical operation, temperature changes do not affect the Fermi momentum p_F much. Suppose the perturbation to p_F is given by $\delta\mathbf{p}$. Then, the dispersion relation providing the electron's energy E_e as a function of its momentum can be written as

$$E_e = \frac{1}{2m}(p_F\hat{\mathbf{p}} + \delta\mathbf{p}) \cdot (p_F\hat{\mathbf{p}} + \delta\mathbf{p}) = E_F + \mathbf{v}_F \cdot \delta\mathbf{p} + o(\delta p^2), \tag{1.8}$$

where $\mathbf{v}_F$ is the *Fermi velocity* with the magnitude p_F/m. Since $\delta\mathbf{p}$ is relatively small, we retain only the first-order term. This approximation is appropriately called semiclassical approximation and it enables us to interpret the electron's motion analogous to a classical particle moving with the Fermi velocity. Because of a relatively large value of the Fermi temperature for most metals, they can be described with sufficient accuracy assuming zero-temperature conditions.

Another widely used length scale is the electron's *mean free path* l_e. Its value is calculated by averaging the distance different electrons travel between two scattering events. Clearly, this value is not precise unless the type of scattering (elastic, inelastic, ...) is specified. For example, one can calculate this quantity for elastic scattering, occurring when electrons come near impurities or collide with solid boundaries. Noting that only electrons close to the Fermi surface participate in such scattering, the average time τ_e between two elastic scattering events, known as the *mean collision time*, can be used to find the electron mean free path as $l_e = \tau_e v_F$. The collision time τ_e is also called the relaxation time for elastic collisions of electrons.

However, inelastic collisions of electrons are more prevalent in solids. These can occur when electrons interact with other electrons, phonons, plasmons, or change their energy through interband transitions. The fraction of electrons subjected to such inelastic processes depends on the energy of electrons. Inelastic scattering affects phase coherence of electrons and, if it occurs multiple times, all coherence is lost. In this situation, electrons obey a diffusion law with a diffusion coefficient D_{ie}, and it becomes possible to define another length scale, the *inelastic scattering length*, as

$$l_{ie} = \sqrt{D_{ie}\tau_{ie}}, \tag{1.9}$$

where τ_{ie} is the mean time between two inelastic scattering events.

Coherence plays a central role in all scattering events because it is related to the phase of a wave function in quantum mechanics. The degree of coherence is set by the extent of correlation between the phases of a wave at different locations (spatial coherence) or at different times (temporal coherence). Temporal coherence measures how monochromatic a wave is. For example, if an electromagnetic wave has a spectral bandwidth $\Delta\nu$, its coherence time is defined as $t_c = 1/\Delta\nu$. In practice, $\Delta\nu$ is taken to be the full width at half maximum (FWHM) of the optical spectrum. However, more precise measures can be found in Ref. [23]. The coherence time is the average time over which phase information is preserved during propagation. Since light travels a distance ct_c during this time, a length scale known as the coherence length is defined as

$$l_c = ct_c = c/\Delta\nu. \tag{1.10}$$

One may wonder whether the preceding approach used for photons and electrons can be extended to phonons, which are responsible for thermal effects. The main quantity of interest is temperature, and how it affects the coherence of phonons. By considering the coherence of blackbody radiation [24] and analyzing thermal conductivity of superlattices [25], it has been deduced that the coherence time of phonons is approximately $t_c - h/k_BT$. This can be understood by relating thermal spreading of energy levels (h/t_c) to thermal energy k_BT. As before, if we combine the coherence time with the diffusion coefficient characterizing many scattering events, a new length scale, *thermal diffusion length l_t*, can be defined as

$$l_t = \sqrt{\frac{D_{ie}\hbar}{k_BT}}. \tag{1.11}$$

As physical mechanisms responsible for the two length scales, l_{ie} and l_t, occur simultaneously, they are not independent of each other. In practice, the smaller of the two lengths sets an upper limit on the phase coherence of any particle requiring a quantum description.

Three other length scales become important when we consider the interaction of charged particles, such as electrons and holes. Even though these length scales have a semiclassical origin, they provide considerable insight, without requiring a full quantum-mechanical treatment. The first length scale is the *Debye length l_D*, defined in Aside 1.2 and applicable to plasmas such as a free-electron gas. Its physical meaning can be understood as follows. Even though charges in a plasma interact with each other, the plasma itself can be considered electrically neutral, until an external charge is introduced in the middle of it. The electrical potential of such a localized charge falls with distance exponentially, and the Debye length is equal to the length where it has fallen to $1/e$ of its initial value. The importance of the Debye length in nanotechnology has been discussed in Refs. [26, 27].

Aside 1.2   Debye Length

Consider a plasma of free electrons inside an electrically neutral metal. It differs from a normal gas because of Coulomb interactions among its charged particles, even though it has a tendency to remain electrically neutral. If the neutrality condition is disturbed by introducing an external charge within the plasma, it quickly reacts to smear out deviations from the charge neutrality. The restoration of charge neutrality happens in a spherical volume centered at the external charge, and radius of this sphere is called the Debye length.

To find the Debye length, let $+Q$ be the magnitude of external charge (assumed to be located at the origin). As positive ions are much heavier than electrons, they barely move but the charge density of surrounding electrons changes due to their high mobility. In thermal equilibrium, the number density of electrons changes from its initial value N_0 to $N_e(\mathbf{r}) = N_0 \exp[-q_e\phi(\mathbf{r})/k_BT]$, where $\phi(\mathbf{r})$ is the electrostatic potential at a distance $\mathbf{r}$ from the charge location. There are many assumptions built into this expression including electron density obeys the Boltzmann distribution, the potential $\phi(\mathbf{r})$ is isotropic around the point test-charge, and $q_e\phi(\mathbf{r}) \ll k_BT$.

We can find $\phi(\mathbf{r})$ by solving the Poisson equation, $\nabla^2\phi(\mathbf{r}) = \rho/\epsilon$, with the charge density $\rho = -q_e[N_e(\mathbf{r}) - N_0]$. Considering the spherical symmetry of the situation, we need to solve

$$\frac{1}{r^2}\frac{d}{dr}\left(r^2\frac{d\phi}{dr}\right) \approx \frac{\phi(r)}{l_D^2}, \tag{1.12}$$

where we replaced N_e by the first two terms in its Taylor-series expansion and introduced the Debye length as

$$l_D = \sqrt{\frac{\epsilon k_BT}{N_0 q_e^2}}. \tag{1.13}$$

We solve Eq. (1.12) with the substitution $\phi = u/r$. The resulting equation, $d^2u/dr^2 = u/l_D^2$, provides the solution as $\phi(r) = (C/r)\exp(-r/l_D)$, where we retained only the exponentially decaying part. The constant C can be related to the charge Q by integrating the Poisson equation over the volume of a sphere of radius r. Using the result $C = Q/(4\pi\epsilon)$, we obtain

$$\phi(r) = \frac{1}{4\pi\epsilon}\frac{Q}{r}\exp\left(-\frac{r}{l_D}\right). \tag{1.14}$$

This equation shows that the electrostatic potential $\phi(r)$ decreases faster than $1/r$ because of the finite Debye length.

The Landau length l_L is the mean distance between electrons and ions in a plasma for which recombination does not take place. It is defined as the distance from a charge at which the electrostatic energy is equal to the thermal energy $k_B T$, where k_B is the Boltzmann constant. It follows from Coulomb's law that

$$l_L = \frac{q_e^2}{4\pi \epsilon_0 k_B T},$$ (1.15)

where q_e is the charge of an electron and ϵ_0 is the vacuum permittivity.

The final length scale is the Bohr radius a_0, defined to be the distance between the nucleus and the electron in a hydrogen atom in its ground state. Its derivation uses classical physics and equates the centripetal force, mv^2/a_0, of an electron circling the proton to the Coulomb force, $q_e^2/(4\pi \epsilon_0 a_0^2)$, between the two particles. It also assumes that angular momentum of the electron is quantized such that $mva_0 = \hbar$ in the ground state of the hydrogen atom. It follows that

$$a_0 = \frac{4\pi \epsilon_0 \hbar^2}{m_e q_e^2}.$$ (1.16)

The concept of the Bohr radius can be applied to excitons, which are electron–hole pairs bound together through the attractive Coulomb force. The role of the proton is played by the positively charged hole in this case.

However, the situation is more complicated for excitons for two reasons. First, the mass appearing in Eq. (1.16) may need to be replaced with an effective mass of the electron, when an electron orbits a hole. If a hole orbits an electron, m_e in Eq. (1.16) should be replaced with the hole's effective mass. As the two effective masses are often different, it becomes necessary to indicate this difference by using a_{0e} and a_{0h} for electrons and holes, respectively.

Second, electrons and holes have comparable masses. As both particles can move within the exciton during its finite lifetime, the model used earlier for finding the Bohr radius of a hydrogen atom is not applicable. Classical mechanics suggests that their combined motion can be analyzed by taking into account the reduced mass of the exciton given by $1/m_{eh} = 1/m_e + 1/m_h$. Using this value in Eq. (1.16), the Bohr radius of the exciton, denoted by a_{0eh}, becomes

$$a_{0eh} = \frac{4\pi \epsilon_0 \hbar^2}{m_e q_e^2} + \frac{4\pi \epsilon_0 \hbar^2}{m_h q_e^2} = a_{0e} + a_{0h}.$$ (1.17)

Excitons are called Frenkel or Wannier types depending on whether the Bohr radius a_{0eh} is smaller or larger than the interatomic spacing of a material. The Bohr model is not a very accurate representation of either of these excitons, but it provides physical insight that is not easy to gain from a detailed quantum-mechanical analysis.

1.3 Quantum Confinement

As mentioned earlier, quantum effects become important when at least one dimension of an object becomes comparable to one or more length scales discussed in Section 1.2. One of such effects is known as the *quantum confinement*. A quantum particle can take only discrete energy values. Moreover, its discrete energy levels depend not only on its properties (such as its mass and speed) but also on the environment in which it is placed. If we restrict the movement of the particle in any dimension, the energies that particle could attain are set by the Schrödinger equation,

$$i\hbar(\partial \,|\Psi\rangle \,/\partial t) = H\,|\Psi\rangle,\tag{1.18}$$

where the Hamiltonian H of the quantum particle is given by

$$H = \frac{p^2}{2m_e} + V(\mathbf{r}) = \frac{1}{2m_e}(-i\hbar\nabla)^2 + V(\mathbf{r}).\tag{1.19}$$

For example, if free electrons in a metal nanostructure are confined to a region whose size is smaller than the Fermi wavelength (de Broglie wavelength of electrons at the Fermi energy), electrons can acquire only discrete energy values found by solving Eq. (1.18) with a potential $V(\mathbf{r})$ that vanishes inside the region but is infinite outside of it. The appearance of $+i$ in Eq. (1.18) and $-i$ in Eq. (1.19) implies a specific sign convention discussed in Aside 1.3.

Aside 1.3   Complex Numbers and Sign of the Imaginary Part

Although all physical quantities are real under measurements, their mathematical description often employs complex notation with equations containing either $+i$ or $-i$, defined as the two square roots of -1 [28, 29, 30]. It is obvious that the choice of this sign should not change any experimentally observed quantity even though complex quantities are used to analyze it. Examples include the complex refractive index of materials and the way Schrödinger's equation is written in Eq. (1.18).

It is important to choose between $+i$ and $-i$ in a consistent manner and ensure that the same choice is made for all equations in this book. We make this choice by specifying the electric field of a forward propagating plane along the positive z direction in the form $E(z,t) = E_0 \exp(-\alpha z)\cos(\beta z - \omega t + \phi)$, where ω is the angular frequency, β is the propagation constant, α is the attenuation coefficient, and ϕ is the phase. We represent this wave as a phasor using the complex notation and write the electric field as

$$E(z,t) = A\exp[i(\beta + i\alpha)z - i\omega t], \qquad A = E_0 e^{i\phi},\tag{1.20}$$

where A is the complex amplitude and it is understood that the real part of E must be taken to obtain the actual electric field. Our notation establishes the rotation direction of the phasor as clockwise with increasing time. With this sign convention, the energy and momentum operators, $i\hbar\frac{\partial}{\partial t}$ and $-i\hbar\frac{\partial}{\partial z}$ respectively, agree with their conventionally accepted forms in quantum mechanics. We use this convention throughout the book.

We also need to make a choice for the sign of the exponential term in the Fourier transform used often in this book. Consider an arbitrary electric field $E(z,t)$. We define its Fourier transform with respect to both z and t as

$$F\{E\}(k,\omega) = \iint_{-\infty}^{\infty} E(z,t)\exp(-ikz + i\omega t)\,dz\,dt,\tag{1.21}$$

where k and ω are the Fourier variables corresponding to z and t, respectively. The inverse Fourier transform is then defined as

$$E(z,t) = \left(\frac{1}{2\pi}\right)^2 \iint_{-\infty}^{\infty} \mathcal{F}\{E\}(k,\omega) \exp(ikz - i\omega t)\, dk\, d\omega, \qquad (1.22)$$

where the factor of 2π ensures $\mathcal{F}^{-1}\mathcal{F} = 1$. It is important to note that opposite signs are used for the spatial and temporal Fourier transforms.

Even though the fundamentals remain the same, the situation becomes more complex for semiconductor materials. This is because both holes and electrons coexist inside a semiconductor. We need to consider not only the de Broglie wavelengths of electrons and holes, but also their Bohr radii to determine the size at which the quantum-confinement effects become important. The smallest of these quantities is $\sim$10 nm in typical semiconductors, enabling one to observe quantum effects more easily compared to metallic structures.

As a concrete example, suppose the critical dimension of interest in a semiconductor has a length d_c. Depending on how d_c compares to the exciton's Bohr radius a_{0eh}, three different regimes should be considered. In the *strongest-confinement* regime ($d_c \ll a_{0eh}$), electrostatic energy resulting from the Coulomb force between an electron and a hole is much smaller than quantum-confinement energies. As a result, the electron–hole pair forming the exciton can be analyzed independently of its constituents. In the *intermediate confinement* regime ($d_c \sim a_{0eh}$), motions of the electron and the hole are strongly correlated through the Coulomb attraction. As a result, energy levels of the exciton depend on both the Coulomb interaction and the boundary conditions applicable to the wave function of the exciton. In the *weak-confinement* regime ($d_c \gg a_{0eh}$), the motion of the exciton's center of mass determines the energy spectra owing to a much stronger electron–hole Coulomb interaction.

1.3.1 Dimensionality of Nanostructures

An object may not be nanoscale in all three dimensions. When an object is smaller than a critical length scale along a specific dimension, its motion is constrained only along that dimension. Conventionally, this dimension is not counted when dimensionality of that object is calculated. In other words, dimensionality of an object refers to the dimensions in which its motion is not restricted. Clearly, all bulk materials are three dimensional (3D) according to this convention used to classify nanoscale structures.

We can now understand what is meant by zero-dimensional (0D), one-dimensional (1D), and two-dimensional (2D) materials [31]. The 2D, 1D, and 0D nanostructures are also referred to in literature as quantum wells (QWs), quantum wires (QWRs), and quantum dots (QDs), respectively. However, caution must be exercised because quantum wells often confine electrons only partially, and their behavior may differ from that of a true 2D structure. For this reason, QWs, QWRs, and QDs with partial confinement are called Q2D, Q1D, and Q0D structures, respectively, with Q standing for their quasi-dimensional nature.

Often, a higher-dimensional structure can be built from nanostructures of lower dimensionality. Moreover, multiple copies of nanosize structures could be assembled in 1D, 2D, or 3D to form composite materials, known as nanostructured materials. An example is provided by QWs of two different types stacked on top of each other. If this assembly is periodic, the resulting structure is called a superlattice.

The simplest nanosize device for observing quantum confinement in all three dimensions is a quantum dot. It can be made using metal oxides (such as ZnO), semiconductor materials (such as CdSe), or large molecules (such as fullerene C60). Among these, semiconductor QDs have prominence in applications because, when pumped electrically or optically, they emit light whose wavelength depends on the QD's size. To understand the origin of this size dependence, we review briefly the physics of semiconductors.

The electronic structure of a bulk semiconductor is characterized through its energy bands, characterized mathematically by $E_n(\mathbf{k})$ and separated by forbidden energy gaps. Here, E_n is the energy of an electron of momentum $\hbar\mathbf{k}$ in the nth band. The valence band is the last band full of electrons. The band above it is called the conduction band; it contains empty delocalized electronic states. Figure 1.4 shows these two bands, whose parabolic shape results from the relation $E_n = \hbar^2 k^2/(2m_e)$. The minimum energy required to transfer an electron from the valence band to the conduction band is called the band-gap energy E_g of the semiconductor. Whenever a photon having energy $E_{ph} > E_g$ is absorbed, the excited electron moves to an empty state in the conduction band, leaving a hole (an unoccupied electronic state) in the valence band. Often, the excited electron and its associated hole attract each other through the Coulomb force and form an exciton. Thermal effects in bulk semiconductors make it difficult to observe such excitons until the temperature is reduced considerably below room temperature.

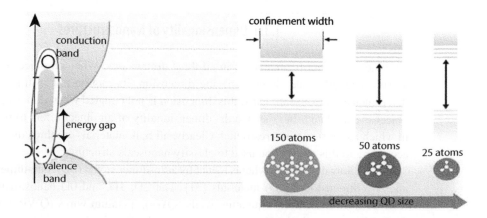

Fig. 1.4   A continuum of energy levels in the valence and conduction bands of a bulk semiconductor (left) transforms into discrete states in QDs (right). Energy of photons (arrow length) emitted by a QD increases as its size shrinks because discrete energy levels of a QD change with its size.

If QDs are made using a semiconductor material, electrons can only move within their volume. The resulting confinement in all three dimensions modifies electronic states in each band such that energy can only take discrete values. A second consequence of this confinement is an increase in the binding energy of excitons, enabling one to observe the excitonic effects at room temperature and to exploit the nonlinear effects in optoelectronic devices. Under suitable conditions, an exciton can be forced to emit light, if its electron jumps back to its original state inside the valence band (recombines with the hole), releasing the energy difference in the form of a photon. Such light emission enables one to observe the size dependence of allowed energy levels in QDs.

As an example, cadmium selenide (CdSe) in its bulk form is a semiconductor with a band gap, $E_g = hc/\lambda = 1.74$ eV, that corresponds to a wavelength $\lambda = 690$ nm. However, one can make CdSe QDs with a size as small as 0.4 nm (containing about 1500 atoms). Such QDs are found to emit light at 530 nm when pumped optically [32]. The reason behind this change in the wavelength is that the lowest energy level of the QDs has an energy larger than that the edge of the conduction band, resulting in a larger photon energy or shorter emitted wavelength, when an electron occupying this state combines with the hole. Indeed, it is possible to control the emission wavelength by varying the diameter of QDs, as shown schematically in Figure 1.4.

Figure 1.5 shows 36 configurations illustrating how the nanoscale dimensioning of materials leads to different arrangements [31]. The notation $kDlmn\ldots$ is used to differentiate different structures, where the integer k denotes the dimensionality of the whole nanostructure, while integers $l, m, n, \ldots$ denote the dimensionality (0, 1, 2 or 3) of its distinct build units; the number of integers equals the number of such units. Using this classification scheme, it is possible to count the number of possible nanostructures in different dimensions; 3 of type kD (top row), 9 of type kDl using one build unit, 19 of type $kDlm$ using two build units, and 5 of type $kDlmn$ using three build units. Adding these, we obtain 36 $(3 + 9 + 19 + 5)$ classes of nanostructures shown in Figure 1.5.

Even though all energy levels are discrete in 0D QDs made of a semiconductor material, they form mini-bands in 1D (QWRs) and 2D (QWs) structures. These features open up many novel opportunities for engineering material properties. Even though metals look dull because they lack a band gap of semiconductors, exotic behavior can be seen in metal nanoclusters because they exhibit resonant oscillations of electrons at certain optical frequencies. When such clusters are subjected to an external electromagnetic field, dynamics of the electron gas is characterized by the presence of plasmonic oscillations. To the lowest order, the linear response of an electron gas is represented by the plasma frequency scaling with the number density N_0 of electrons as $\omega_p \propto \sqrt{N_0}$. More specifically, the plasma frequency is independent of the size of the object containing the electron gas [33].

From a physics standpoint, plasma frequency marks the limiting frequency beyond which a metal can no longer screen electric fields. The oscillations arise owing to the appearance of a restoring Coulomb force when electrons are displaced from their charge-neutral configuration (thus creating a net positive charge). Owing to their inertia, electrons do not simply replenish the positive region, but travel further away, thus re-creating an excess positive charge. This effect gives rise to coherent oscillations of an electron gas at the plasma frequency. The coherence of these oscillations is progressively destroyed by

Elementary building units :

Fig. 1.5 Classification of nanostructures based on dimensionality. Using the notation *kDlmn* and limiting to 3 built units, we obtain 36 possible classes Ref. [31].

collisions of an electron with other electrons or phonons [34, 35, 36]. Although plasmons can also decay radiatively (an energy-losing process), this kind of decay is negligible under adiabatic conditions. In contrast, radiationless decay occurs mainly via dephasing (known as Landau damping). This kind of decay leaves the energy of electrons unchanged but destroys coherence of the collective oscillations.

It is interesting to ask what would happen if one continues to decrease the size of a metal cluster to the level of a few nanometers. In this situation, the Fermi wavelength of

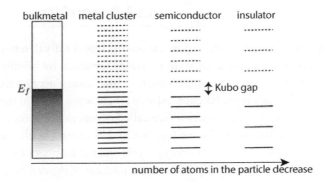

number of atoms in the particle decrease

Fig. 1.6   Increase in the Kubo gap as the size of a metal cluster is reduced from very large (bulk) to a level containing only a few atoms. As a result, a metal first behaves as a semiconductor and then as an insulator [37].

electrons becomes comparable to the dimensions of the nanostructure, kicking in quantum-confinement effects. Even though a large metal structure has closely spaced energy levels, Kubo used the free-electron model to show that average spacing between energy levels increases as its volume decreases [38, 39]. As a result, a gap – known as the Kubo gap – opens up between the Fermi energy (the highest occupied state) and the lowest unoccupied state, as shown in Figure 1.6. This Kubo gap, essentially being the average spacing between consecutive energy levels in a nanostructure, depends inversely on the number of atoms N_v in the nanocluster as

$$\delta_K = \frac{4E_F}{3N_v}. \tag{1.23}$$

When thermal energy $k_B T$ of electrons becomes significantly smaller than the energy of the Kubo gap, the metallic cluster displays properties different from those expected from a metal. As seen in Figure 1.6. a sufficiently small metal cluster shows semiconducting or insulator behavior, depending on its size.

So far, we have considered nanosize objects in which quantum confinement is built into the structure during fabrication. It is also possible to change the confinement of a structure using external fields. For example, when a strong magnetic field B_z is applied in the z direction of a bulk metal, energy levels of the electrons change as

$$E_n = \left(n + \frac{1}{2}\right)\hbar\omega_c + \frac{\hbar^2 k_z^2}{2m_e}, \tag{1.24}$$

where n is a positive integer, $\omega_c = q_e B_z/m_e$ is the cyclotron frequency, and $\hbar k_z$ is the momentum of electrons in the z direction before the magnetic field is applied [40]. This result shows that the electron's energy has two contributions. The kinetic energy part, parallel to the magnetic field, appears to not change with the magnetic field. However, the motion perpendicular to the magnetic field is quantized in multiples of $\hbar\omega_c$. This situation is similar to the discrete energy levels of a Q1D nanostructure considered earlier. If a strong magnetic field is applied to a 2D structure (e.g., bilayer graphene [41]), it develops discrete energy levels similar to a Q0D system. These energy levels are known as *Landau levels*.

1.3.2 Density of States

In quantum-mechanical terms, a nanostructure is fully described if one knows its all energy levels and the associated wave functions, obtained by solving the underlying Schrödinger equation. To complicate the situation, the same energy value could be associated with more than one wave function. Also, a single wave function may correspond to two or more energy values. Such situations are called degeneracies of the system. Even though it is hard to find each individual wave function and its energy, applications often require knowledge of the number of distinct energy levels and the number of wave functions associated with each energy. The remedy is to use the concept of the *density of states* (DOS), $D_{os}(E)$. This quantity provides the number of available quantum states per differential energy interval (dE) per unit volume (area or length for 2D and 1D systems) in the vicinity of energy E. The DOS can be viewed as a distribution function for the energy states in a nanostructure. We can integrate it over a finite range to find the total number of states in a given energy interval. For example, the number of states that fall within the energy range $[E_i, E_f]$ in a unit volume can be found using the integral $\int_{E_i}^{E_f} D_{os}(E)\, dE$.

We can use the DOS concept even for a nanostructure with discrete energy levels E_n with n ranging from 1 to $\mathbb{N}$. If its volume is V, the DOS can be written as

$$D_{os}(E) = \frac{1}{V} \sum_{n \in \mathbb{N}} \delta(E - E_n). \tag{1.25}$$

The degeneracy of an energy level can be included by multiplying the corresponding delta function with its degeneracy value. The validity of this definition is easily seen by integrating Eq. (1.25) over the energy range $[E_i, E_f]$:

$$\int_{E_i}^{E_f} D_{os}(E)\, dE = \frac{1}{V} \sum_{n \in \mathbb{N}} \int_{E_i}^{E_f} \delta(E - E_n)\, dE. \tag{1.26}$$

As the integral reduces to 1 whenever an energy level is within the range $[E_i, E_f]$, the sum over n simply counts the number of energy level in the interval $[E_i, E_f]$.

The preceding definition is not convenient in some specific situations. In those cases, it is useful to employ an auxiliary function representing the number of states having energy less than E in a unit volume (or area in 2D systems). This function is known as the cumulative DOS and is defined as $CD_{os}(E) = \int D_{os}(E)\, dE$. We can find $D_{os}(E)$ by differentiating $CD_{os}(E)$ with respect to E.

The main deficiency of the definition of DOS in Eq. (1.25) is that it does not depend on the location $\mathbf{r}$ within a nanosize particle. It is entirely possible for the wave function to vanish or become so small at certain locations that it is unlikely that the particle will ever be there. To make the DOS definition meaningful in such cases, the *local DOS* (LDOS) is introduced as

$$LD_{os}(E, \mathbf{r}) = \frac{1}{V} \sum_{n \in \mathbb{N}} \|\Psi_n(\mathbf{r})\| \delta(E - E_n), \tag{1.27}$$

where $\Psi_n(\mathbf{r})$ is the wave function for the energy eigenvalue E_n. An illuminating discussion of LDOS in nanoplasmonic systems in the frequency range dominated by a localized surface plasmons can be found in Ref. [42].

We should mention that several different definitions for DOS are used in literature, based on the context of specific applications. Sometimes, normalization of the DOS in Eq. (1.25) by volume is not carried out. This normalization does make sense for large structures such as a bulk crystal. However, for small structures, one may simply count the total number of states without normalization. In some situations, it makes sense to use a variable other than energy as the integrating variable. For example, when the DOS of electromagnetic modes in vacuum is calculated for blackbody radiation (Planck's distribution), energy is considered to have the form $E = n\hbar\omega$, where n is an integer and ω is the frequency of the radiation. In this case, it is common to use ω as the integrating variable and consider the DOS a function of frequency such that $D_{os}(E)\,dE = D_{os}(\omega)\,d\omega$ [see Eq. (1.29) in Aside 1.4]. It is also important to consider other aspects of a quantum state. For example, an electron can have its spin up or down according to Pauli's exclusion principle. Similarly, a photon could be in two orthogonal polarization states. The DOS associated with quantum states is often multiplied by an occupancy number to find the relevant DOS for the particle of interest.

Aside 1.4   Density of States for Photons

Consider a rectangular cavity with its sides d_1, d_2, d_3 aligned with the three Cartesian axes. Maxwell's equations show that such a cavity supports plane waves of the form $\exp(\pm i\mathbf{k} \cdot \mathbf{r})$, where the wave vector $\mathbf{k}$ indicates the direction of propagation of a specific plane wave. Using the boundary condition that the electric field must vanish at all surfaces of this cavity, the three components of $\mathbf{k}$ are found to be quantized such that they take discrete values $k_i = n_i(2\pi/d_i)$, where n_i is an integer (positive or negative). The frequency of an allowed mode is $\omega = ck$, where k is the magnitude of the $\mathbf{k}$ vector. Noting that each photon has an energy $E = \hbar\omega$, we want to calculate the DOS as a function of E.

In the k-space, the volume associated with each state is $(2\pi)^3/V$, where $V = d_1 d_2 d_3$ is the volume of the cavity. The number of quantum states having energy less than $E = \hbar ck$ per unit volume of the cavity, representing the cumulative DOS, is found by calculating how many states are contained in a sphere of radius k in the k space:

$$\mathrm{CD}_{os}(E) = \frac{2}{V}\frac{4\pi k^3/3}{(2\pi)^3/V} = \frac{k^3}{3\pi^2} = \frac{E^3}{3\pi^2\hbar^3 c^3},\tag{1.28}$$

where the factor of 2 accounts for the possibility that each photon can have two orthogonal polarization states. Differentiating the cumulative DOS with respect to E, we obtain the DOS for photons:

$$D_{os}(E) = \frac{E^2}{\pi^2\hbar^3 c^3} \quad \rightarrow \quad D_{os}(\omega) = \frac{\omega^2}{\pi^2 c^3},\tag{1.29}$$

where we used the relation $D_{os}(E)\,dE = D_{os}(\omega)\,d\omega$ with $E = \hbar\omega$. This result is for a 3D cavity. The same argument can be used in fewer dimensions. Noting that the sphere becomes a circle of area πk^2 in the 2D case and just a line of length k in the 1D case, the DOS for photons in these two cases is found to be

$$D_{os}^{(2D)}(E) = \frac{E}{\pi\hbar^2 c^2}, \qquad D_{os}^{(1D)}(E) = \frac{1}{\pi\hbar c}. \tag{1.30}$$

The DOS of a homogeneous structure containing electrons depends on its dispersion relation $E(\mathbf{k})$ showing how an electron's energy depends its momentum $\hbar\mathbf{k}$. Since this relation depends on the dimensionality of that structure, we consider the general dispersion relation $E = E_d(\mathbf{k})$, where d denotes the dimensionality of the structure. This relation describes a surface of constant energy in the k space. The unit vector normal to this surface is related to $\nabla_\mathbf{k} E_d$ as

$$\hat{\mathbf{k}}_\perp = \nabla_\mathbf{k} E_d(\mathbf{k})/|\nabla_\mathbf{k} E_D(\mathbf{k})|. \tag{1.31}$$

Using it, the perpendicular and parallel components of the vector $d\mathbf{k}$ are found to be $d\mathbf{k}_\perp = d\mathbf{k} \cdot \hat{\mathbf{k}}_\perp$ and $d\mathbf{k}_\parallel = d\mathbf{k} - d\mathbf{k}_\perp$, respectively. The differential volume in the d-dimensional k space can be written as $d^d\mathbf{k} = d^{d-1}\mathbf{k}_\parallel d\mathbf{k}_\perp$.

We saw in Aside 1.4 that each dimension with length d_i has a k-space projection of $2\pi/d_i$. The number of states in the k space per unit volume is given by $d^d\mathbf{k}/(2\pi)^d$, assuming a d-dimensional hypersphere in the k space of the structure. Differential geometry tells us that $d^{d-1}\mathbf{k}_\parallel = 2k^{d-1}\pi^{d/2}/\Gamma(d/2)$, where $\Gamma(x)$ is the gamma function. Using this result, the DOS of the structure is given by

$$D_{os}(E) = \frac{\pi^{d/2}}{2^{d-1}\pi^d \Gamma(d/2)} \frac{k^{d-1}}{|\nabla_\mathbf{k} E_D(\mathbf{k})|}, \tag{1.32}$$

where we have not yet included the degeneracy of the energy states.

As a simple example, consider free electrons inside a conductor in d dimensions large enough that no quantum confinement occurs. The dispersion relation in this case is parabolic and is given by $E = \hbar^2 k^2/2m_e$. Considering that each state can be occupied by two electrons, we multiply the preceding result by 2 and use $k = \sqrt{2m_e E}/\hbar$ to obtain

$$D_{os}(E) = 2\left(\frac{m_e}{2\pi}\right)^{d/2} \frac{E^{(d-2)/2}}{\Gamma(d/2)\hbar^d}. \tag{1.33}$$

Table 1.2 lists the DOS of both free electrons and photons for $d = 0, 1, 2$, and 3. It is clear that the dimensionality of the structure affects the functional dependence of DOS on energy and plays a critical role.

Table 1.2 Density of states $D_{os}(E)$ in 0D, 1D, 2D, and 3D structures.

	0D	1D	2D	3D
Electrons	$2\delta(E)$	$(2m_e/\pi^2\hbar^2)^{1/2}E^{-1/2}$	$m_e/(\pi\hbar^2)$	$\frac{1}{2\pi^2}(2m_e/\hbar^2)^{3/2}E^{1/2}$
Photons	0	$(\pi\hbar c)^{-1}$	$(\pi\hbar^2 c^2)^{-1}E$	$(\pi^2\hbar^3 c^3)^{-1}E^2$

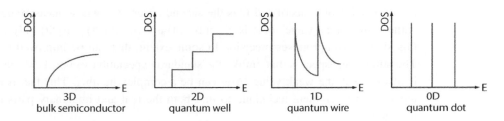

Fig. 1.7   Parabolic form of the DOS in a bulk semiconductor and changes in its form when electrons are confined in more and more directions using QWs, QWRs, and QDs.

Consider how the DOS change happens when a bulk structure is confined in one or more dimensions. Table 1.2 shows that the DOS of a 3D structure varies with E as $E^{1/2}$. Figure 1.7 shows how this DOS is modified from its parabolic form in a bulk medium as electrons are confined in one, two, and three directions using first QWs, then QWRs, and finally QDs.

It is interesting to ask what happens when a structure, fully localized in d dimensions, is replaced with its quasi-partner where localization is not total. For example, as seen in Table 1.2, the DOS of a strictly 1D free-electron gas varies with E as $E^{-1/2}$. In a quasi-1D structure such as a quantum wire, multiple branches display the same $E^{-1/2}$ dependence starting at the discrete energy states of the QWR. The results given in Table 1.2 and Figure 1.7 are for idealized isotropic systems with relatively simple geometries. In practice, many complications arise owing to nonideal conditions such as a finite size of samples, lattice defects, and surface effects. To include them, one must resort to numerical schemes to predict the DOS of such structures.

1.4 Quantum Interference

The wave function of a quantum particle governs all of its properties. This wave function $\Psi(\mathbf{r}, t)$, obtained by solving Eq. (1.18), is a complex quantity. The "Born rule" states that $|\Psi(\mathbf{r}, t)|^2$ represents the probability of finding the quantum particle at the location $\mathbf{r}$ at time t. The phase of the wave function describes relative positions of the peaks and valleys of the wave associated with this particle [40]. Although no technique exists that can determine the wave function completely [43], the so-called weak measurements can be used to measure it approximately [44]. The fundamental idea behind such a method is to measure sequentially two complementary variables of the system, while minimizing the disturbance induced by the first measurement. Recall that a measurement is carried out by coupling a measuring apparatus to the quantum system.

The weak-measurement technique is based on reducing the coupling between the measuring apparatus and the quantum system to minimize the disturbance of the quantum state [45]. A characteristic feature of a weak measurement is that it does not affect a subsequent measurement of the same or another observable in the limit of negligible coupling. If A is the observable of concern and C is the second observable whose measurement gives the value c, the "weak value" of A is given by $\langle A \rangle_W = \langle c | A | \Psi \rangle / \langle c | \Psi \rangle$ [44]. Even though this strategy compromises precision to some extent, that can be improved by averaging. The surprising aspect is that, unlike the standard expectation value $\langle A \rangle = \langle \Psi | A | \Psi \rangle$ that is always real, the weak value $\langle A \rangle_W$ can be a complex number. This fact is exploited in the weak-measurement technique to find both the real and imaginary parts of the wave function.

To illustrate the versatility of this method, Lundeen *et al.* [44] consider an example where a weak measurement of the position ($A = \pi_x = |x\rangle \langle x|$) is followed by a strong (or normal) measurement of momentum, giving the value p. The weak measurement of the position is then obtained from

$$\langle \pi_x \rangle_W = \langle p | x \rangle \langle x | \Psi \rangle / \langle p | \Psi \rangle. \tag{1.34}$$

Noting that $\langle p | x \rangle = \exp(ipx/\hbar)$ and $\Psi(x) = \langle x | \Psi \rangle$, we see that $\langle \pi_x \rangle_W$ is proportional to the wave function $\Psi(x)$ of the particle when $p = 0$. In other words, at each position x, the observed position and momentum values are proportional to the real and imaginary parts of $\Psi(x)$, and thus they can be used to construct the wave function of a quantum particle. Normalization of the wave function is needed to remove the proportionality constant. As normalization involves only the magnitude of the wave function, this technique does not measure the absolute phase of a wave function.

We briefly discuss the physical meaning of a wave function's phase. Consider two quantum states, $|\Psi_a\rangle$ and $|\Psi_b\rangle$, describing the same physical state of a system. This is possible if and only if they are linearly dependent (i.e., $|\Psi_a\rangle = \eta |\Psi_b\rangle$, where η may be a complex number). If $|\Psi_a\rangle$ is normalized such that $\langle \Psi_a | \Psi_a \rangle = 1$, the linear relationship between the two states requires that $|\eta|^2 = 1$. Thus, η is in the form $\eta = e^{i\phi}$, where ϕ is a constant phase. As the two states represent the same physical state, one can argue that the phase of a quantum state has no physical meaning.

However, it is possible to differentiate the two states by letting them interfere with each other to produce an interference pattern. Mathematically,

$$\langle (\Psi_a + \Psi_b) | (\Psi_a + \Psi_b) \rangle = 2 + \langle \Psi_a | \Psi_b \rangle + \langle \Psi_b | \Psi_a \rangle = 2[1 + \cos(2\phi_r)], \tag{1.35}$$

where $\phi_r = \phi_a - \phi_b$ is the relative phase of two wave functions normalized such that $\langle \Psi_a | \Psi_a \rangle = 1$ and $\langle \Psi_b | \Psi_b \rangle = 1$. This result shows clearly that it is possible to measure the relative phase of two quantum states. The relative phase ϕ_r does not change if we multiply both quantum states by the same phase factor $\exp(i\phi)$, making it clear that the absolute phase of a wave function is not a relevant quantity.

The relative phase of two quantum states can be measured through quantum interference, whose effects are analogous to those familiar with classical interference in optics. Indeed, Young's double-slit experiment can be repeated for quantum particles (such as electrons or neutrons) to produce similar fringe patterns. However, there are important differences between the classical and quantum interferences. The most important one is that the amplitude of an electromagnetic field varies in space and time through a measurable quantity such as an electric field that has a measurable phase, in contrast to that of a quantum wave function. As only the relative phase between two quantum states can be observed, it is important to know how it can be controlled in practice for exploitation of quantum effects.

The mapping $\Psi_a(\mathbf{r}, t) \rightarrow \exp(i\zeta)\Psi_a(\mathbf{r}, t)$ is called a *phase transformation* in quantum mechanics. If ζ is a constant everywhere both in space and time, the transformation is referred to as a *global phase transformation*. However, if ζ varies with $\mathbf{r}$ and t, the transformation is referred to as a *local phase transformation*. Although only some expectation values of a physical observable can be invariant under local phase transformations, all expectation values for the same observable are invariant under global phase transformations.

As our focus in this book is mainly on the interaction of charges with electromagnetic fields inside a nanoscale device, it is useful to analyze how the phase of a quantum state evolves in an electromagnetic environment. An electromagnetic wave is described by its electric field $\mathbf{E}$ and magnetic flux density $\mathbf{B}$. The dynamics of these vectors and material charges are governed by the Maxwell equations and the Lorentz force equation. As is well known in electromagnetic theory, it is useful to introduce the scalar potential $V(\mathbf{r}, t)$ and the vector potential $\mathbf{A}(\mathbf{r}, t)$. The electric and magnetic fields can be written in terms of them as [30]

$$\mathbf{E} = -\nabla V - \frac{\partial \mathbf{A}}{\partial t}, \qquad \mathbf{B} = \nabla \times \mathbf{A}. \tag{1.36}$$

However, the two potentials, $V(\mathbf{r}, t)$ and $\mathbf{A}(\mathbf{r}, t)$, are not uniquely defined. The same values of the electric and magnetic fields are obtained if we modify them as

$$V'(\mathbf{r}, t) = V'(\mathbf{r}, t) - \frac{\partial \xi}{\partial t}, \qquad \mathbf{A}'(\mathbf{r}, t) = \mathbf{A}(\mathbf{r}, t) + \nabla \xi, \tag{1.37}$$

where $\xi = \xi(\mathbf{r}, t)$ is an arbitrary scalar function. As $\mathbf{E}$ and $\mathbf{B}$ are the real physically measurable quantities, the transformation of the two potentials by an arbitrary ξ function should have no observable consequence. Transformations that leave the observable physical quantities unchanged are known as *gauge transformations*, and the function $\xi(\mathbf{r}, t)$ is referred to as a gauge function.

Let us consider how the Hamiltonian of a freely moving electron changes in the presence of an electromagnetic field. Because of the Lorentz force acting on the electron, its momentum p is affected by the electromagnetic field. In the Coulomb gauge, the Hamiltonian H given in Eq. (1.19) changes such that [46]

$$H = \frac{1}{2m_e} \left(-i\hbar\nabla - q_e\mathbf{A} \right)^2 + q_eV. \tag{1.38}$$

If a gauge transformation is carried out on the fields interacting with the electron, we need to use new potentials as indicated in Eq. (1.37). The resulting Hamiltonian has a time-dependent term $q_e(\partial\xi/\partial t)$ when V in Eq. (1.38) is replaced with V'. Its presence modifies the phase of the wave function as $|\Psi'\rangle = |\Psi\rangle \exp(iq_e\xi/\hbar)$. Clearly, the gauge transformation of an electromagnetic field results in a local-phase transformation on the wave function of the electron.

The phase of a wave function can also depend on the vector potential. Suppose $|\Psi_0\rangle$ is the quantum state in the absence of any electric or magnetic field. If a magnetic field

Fig. 1.8 Wave function of a charged particle, moving along the loop C in the presence of a magnetic field B, acquires a phase shift that is proportional to the flux passing through the area enclosed by that loop.

turns on adiabatically (i.e., Hamiltonian changes slowly with time), an electron will follow a circular trajectory C, as seen in Figure 1.8. It follows from Eq. (1.38) that the quantum state of this electron at any point $\mathbf{r}$ along this loop can be written as [46]

$$|\Psi_A\rangle = \exp\left(\frac{iq_e}{\hbar}\int_{\mathbf{r}_0}^{\mathbf{r}}\mathbf{A}\cdot d\mathbf{r}\right)|\Psi_0\rangle, \tag{1.39}$$

where $|\Psi_0\rangle$ is the state in the absence of the magnetic field and $\mathbf{r}_0$ is an arbitrary reference point on the loop C. Clearly, the phase acquired by the electron on completing the loop C is given by $\phi_c = (q_e/\hbar)\oint\mathbf{A}\cdot d\mathbf{r}$. Using Stoke's theorem, we can transform this closed-loop integral to the following surface integral:

$$\phi_c = \frac{q_e}{\hbar}\oint\mathbf{A}\cdot d\mathbf{r} = \frac{q_e}{\hbar}\iint_S\mathbf{B}\cdot d\mathbf{s}, \tag{1.40}$$

where we used the relation $\mathbf{B} = \nabla\times\mathbf{A}$ and the vector $d\mathbf{s}$ is normal to the surface area S enclosed by the loop C. This phase is gauge invariant because the surface integral depends on the applied magnetic field through $\mathbf{B}$, which does not change regardless of the gauge. Note also that this integral does not change if the flux through the loop remains constant while the trajectory changes. Even in the extreme case where the magnetic field is shielded such that it is finite near the center but vanishes close to electron's trajectory, the electron will still acquire the same phase shift. This phenomenon is exploited in the Aharonov–Bohm effect described in Aside 1.5.

Aside 1.5   Aharonov–Bohm Effect

Aharonov and Bohm [47] discovered in 1959 that the behavior of a quantum system can be altered by a magnetic field in a nonlocal manner (i.e., behavior changes even when the magnetic field is zero everywhere in the vicinity of all charges). Figure 1.9 shows a setup where an electron beam is split into two, and the two beams are made to interfere after taking different paths (γ_1 and γ_2) around a shielded cylinder containing a magnetic field. Suppose the quantum state of an electron just before the beam's splitting is $|\Psi_0\rangle$. The electron must take either the path γ_1 or γ_2. Based on Eq. (1.39), we can write the quantum state of that electron after traversing one of these paths as

$$|\Psi_{\gamma_1}\rangle = \exp\left(\frac{iq_e}{\hbar}\int_{\gamma_1}\mathbf{A}\cdot d\mathbf{r}\right)|\Psi_0\rangle, \qquad |\Psi_{\gamma_2}\rangle = \exp\left(\frac{iq_e}{\hbar}\int_{\gamma_2}\mathbf{A}\cdot d\mathbf{r}\right)|\Psi_0\rangle. \tag{1.41}$$

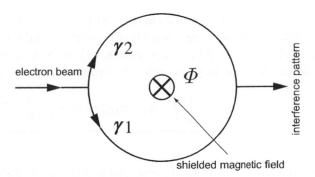

Fig. 1.9 Illustration of the Aharonov–Bohm effect. An interference pattern is observed when an electron beam takes two different paths around a shielded magnetic field.

Since it is not possible to know which path is taken by an electron, its state $\langle\Psi\rangle$ after the two beams are combined is a superposition of the two states (i.e., $|\Psi\rangle = |\Psi_{\gamma_1}\rangle + |\Psi_{\gamma_2}\rangle$). As a result, the probability of finding the electron at a location $\mathbf{r}$ is given by

$$|\langle\mathbf{r}|\Psi\rangle|^2 = 2\,|\langle\mathbf{r}|\Psi_0\rangle|^2 \left[1 + \cos^2(\phi_r)\right], \qquad (1.42)$$

where ϕ_r is the relative phase difference. Using Stoke's theorem, ϕ_r is found to be

$$\phi_r = \frac{q_e}{\hbar} \oint \mathbf{A} \cdot d\mathbf{r} = \iint_S \mathbf{B} \cdot d\mathbf{s}, \qquad (1.43)$$

where the first integral is over the entire loop (after the direction of the γ_2 path is reversed) and S is the area enclosed by this loop.

It is important to realize that, even when the magnetic field is switched off ($\mathbf{B} = 0$), an interference pattern forms commensurate with the de Broglie wavelength of electrons. The effect of the Aharonov–Bohm phase induced by an external magnetic field is to shift that interference pattern by an amount that depends on the magnetic field through ϕ_r. The remarkable feature is that even a shielded magnetic field far from the electron beam affects the quantum state of electrons, and thus can be detected through an interferometric measurement. One may ask how the electron knows that such a magnetic field exists. The answer is that the vector potential has a nonzero value near electron's path, even though $\mathbf{B}$ itself vanishes owing to magnetic shielding. This has led to philosophical discussions whether the vector potential is more fundamental than the field that creates it.

It has been shown that Eqs. (1.39) and (1.40) can be generalized to nonadiabatic cyclic systems [48, 49]. In Figure 1.8, the quantum state $|\Psi(t)\rangle$ is cyclic in time with a period τ because the electron returns to its original location after each round trip over the loop. Because this quantum state accumulates phase as it propagates along the loop, the cyclic feature implies that there is a net phase shift of $\Delta\phi$ after each round trip such that $|\Psi(\tau)\rangle = \exp(i\Delta\phi)\,|\Psi(0)\rangle$.

We can introduce a transformation, $|\Psi(t)\rangle = \exp[if(t)]\,|\widehat{\Psi}(t)\rangle$, such that the new state does not change after one round trip (i.e., $|\widehat{\Psi}(\tau)\rangle = |\widehat{\Psi}(0)\rangle$). This is possible if the function $f(t)$ satisfies $f(\tau) - f(0) = \Delta\phi$ (modulo 2π). With the help of the Schrödinger equation, one can show that the phase shift $\Delta\phi$ consists of two parts [50]:

$$\Delta\phi = \underbrace{-\frac{1}{\hbar}\int_0^\tau \langle\Psi(t)|H(t)|\Psi(t)\rangle\,dt}_{\Delta\phi_d} + \underbrace{i\int_0^\tau \langle\widehat{\Psi}(t)|\frac{d}{dt}|\widehat{\Psi}(t)\rangle\,dt}_{\Delta\phi_g}, \qquad (1.44)$$

where $\Delta\phi_d$ is called the *dynamic phase* and $\Delta\phi_g$ is called the *geometric phase*. The latter is responsible for the Aharonov–Bohm effect. As $\Delta\phi_g$ is independent of the specific Hamiltonian $H(t)$ that produces motion along the curve C, it is only related to the geometry of the motion and does not depend on the nature of interactions in the quantum system. It is also known in literature as Pancharatnam–Berry phase (or just as the Berry phase).

One may wonder whether the scalar potential can also give rise to a geometric phase. This is indeed the case, and the associated effect is called the Aharonov–Casher effect [51]. It was predicted in 1984 and involves the interaction of a static electric field with a moving magnetic dipole. It is commonly identified as a dual effect because moving electrical charges in the Aharonov–Bohm effect are replaced by moving magnetic dipoles, and the vector potential induced by a static magnetic field is replaced by the scalar potential induced by static electric field. The reason that a magnetic dipole is needed is related to the fundamental absence of magnetic monopoles.

The original Aharonov–Bohm effect considered electrons propagating in a vacuum. It is interesting to ask what would happen if the vacuum path is replaced by a metallic path. As electrons in metals diffuse through a lattice consisting of positive ions, it is possible that phase coherence may be lost during collisions of an electron with the lattice. Consider first the scenario where electrons undergo inelastic scattering such that each collision destroys phase information. Owing to the change in energy of electrons when subjected to inelastic scattering, the relative phase in Eq. (1.42) acquires a time-dependent term such that

$$\phi_r = \frac{q_e}{\hbar}\oint \mathbf{A}\cdot d\mathbf{r} + (\phi_{\gamma_1} - \phi_{\gamma_2}) + \frac{E_{\gamma_1} - E_{\gamma_2}}{\hbar}t, \qquad (1.45)$$

where ϕ_{γ_1} and ϕ_{γ_2} are the phase shifts along the two paths, and E_{γ_1} and E_{γ_2} are the new energies of the electrons after inelastic collisions along the paths γ_1 and γ_2, respectively. As this phase oscillates with time at a frequency $(E_{\gamma_1} - E_{\gamma_2})/\hbar$, it vanishes when averaged over time. Thus, the quantum-interference effects cannot be observed in the case of inelastic scattering.

The situation changes if elastic scattering dominates because it not only preserves phase coherence but also does not change the energy of scattered electrons. The resulting relative phase is similar to that in Eq. (1.45), but without the last time-dependent term. The only constraint is that the metal path must be shorter than $10\,\mu m$, a typical elastic scattering length for metals. Clearly, the interference effects can be observed in this situation. Indeed, the effect is observable as conductance fluctuations in metals: constructive interference gives high conductance and destructive interference gives low conductance. An experimental demonstration of this effect was carried out in 1985 using a gold ring (diameter 8 μm, thickness 0.4 μm) with a strong shielded magnetic field at its center [52]. A striking feature of this experiment was that sometimes electrons circulated two or more times in the loop before interference happened. This was deduced from the frequency of oscillations that depends on the total path length before interference occurs. This effect is known as the Altshuler–Aronov–Spivak effect in literature and is an interesting example of weak localization theory of interfering electrons [53]. Another interesting example of quantum interference is related to the persistent currents in small isolated metal rings; it is discussed in Aside 1.6.

We have seen that phase changes in the quantum state of a nanosize object can lead to quantum-interference effects, which act as a fingerprint of the quantum character of the underlying dynamics. Such effects have no classical counterparts. In addition to the phenomena discussed in this section, there are other aspects of quantum interference that manifest often in a disguised form. Examples include quantized conductance, ballistic transport, quantum Hall effect, universal fluctuations of conductance, Anderson localization in disordered wires, and resonant tunneling. We consider some of these in the next section on quantum transport.

Aside 1.6   Persistent Currents in Small Isolated Metal Rings

Owing to inelastic scattering of an electron with phonons and other electrons, currents in metals eventually cease to exist after all fields are turned off. However, in metal rings that are smaller than the electron's dephasing length – the typical distance an electron travels before it and loses its phase information through inelastic scattering [54] – it is possible to induce a perpetual current flow simply by threading the center of the ring with a magnetic flux [55]. The manifestation of this effect is not only a signature of phase coherence of electrons in metals but also an example of the impact of vector potential seen in the Aharonov–Bohm effect.

Consider a metal ring of radius r as shown in Figure 1.10, subjected to a homogeneous magnetic field oriented perpendicular to the plane of the ring. The phase shift induced by the magnetic field can be calculated with the help of Eq. (1.40). Using the polar coordinates (ρ, θ) with the origin at the center of the ring and noting that $\mathbf{A}$ points in the $\hat{\theta}$ direction

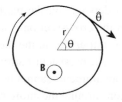

Fig. 1.10   Illustration of a persistent current in a small metallic ring subjected to a magnetic field.

as indicated in Figure 1.10, we can carry out the integration and obtain $\phi_c = (q_e/\hbar)\Phi_R$, where $\Phi_R = 2\pi r A$ is the total flux through the ring.

We seek quantum states that are stationary on the ring, that is, we look for solutions of the Schrödinger equation (1.18) in the form $\Psi = \psi \exp(iEt/\hbar)$, where E is the energy of the state. This form leads to the time-independent Schrödinger equation, $H\psi = E\psi$. We solve it with the Hamiltonian in Eq. (1.38) with $V = 0$ and $A = \Phi_R/(2\pi r)$. Writing this equation in polar coordinates and noting that $\rho = r$ is a constant for the rotating electron, we need to solve

$$\frac{1}{2m_e} \left(\frac{i\hbar}{r} \frac{d}{d\theta} + \frac{q_e \Phi_r}{2\pi r} \right)^2 \psi(\theta) = E\psi(\theta). \tag{1.46}$$

We look for a plane-wave solution in the form $\psi(\theta) = Ce^{ik\theta}$, where C is a constant. When we impose the condition $\psi(\theta + 2\pi) = \psi(\theta)$ for the wave function to be periodic on the ring, the constant k is quantized and assumes discrete values $k_n = n$, where n can be any integer ($n = 0, \pm 1, \ldots$). For each value of n, we obtain a quantum state $\psi_n(\theta)$ with the energy eigenvalue E_n such that

$$\psi_n(\theta) = \frac{1}{\sqrt{2\pi}} \exp(in\theta), \qquad E_n = \frac{\hbar^2}{2m_e r^2} \left(n - \frac{q_e}{h} \Phi_R \right)^2, \tag{1.47}$$

where the constant C was found through normalization.

To get an estimate of the persistent current, we assume that the temperature of the system is close to absolute zero (which was indeed the case in Ref. [55]). Recalling that the energy levels are filled up to the Fermi level in metals, the total energy of the system is approximately $E = 2\sum_n E_n$, where the factor of 2 accounts for the electron's spin and the sum is restricted such that $E_n < E_F$. Noting that E changes if Φ_R changes with time and using Faraday's law, the current I is related to E as

$$I = -\frac{\partial E}{\partial \Phi_R} = \frac{\hbar q_e}{m_e r^2} \sum_{E_n < E_F} \left(n - \frac{q_e}{h} \Phi_R \right). \tag{1.48}$$

Clearly, the current depends on the number of electrons in the ring and its radius but it does not depend on the cross-sectional area of the ring [56, 57]. Even though I is affected by defects and other impurities in the metal ring, it may not entirely vanish if such disorder is moderate. Hence, persistent current may exist in rings with a finite resistance. An instrument such as an ammeter cannot be used to measure this current, as it requires phase coherence of electrons around the entire ring. In experiments, the presence of such currents was deduced indirectly by measuring a tiny magnetic moment induced by the circulating current.

1.5 Quantum Transport

Many quantum devices depend on charge transport for their operation. Ohm's law was the cornerstone of charge transport in the twentieth century. It states that the current through a conducting material is linearly proportional to the voltage across it ($I = GV$); the constant of proportionality is called the conductance.

Historically, charge transport in conductors was systematically modeled by Drude and Lorentz. The Drude theory predicts that the conductivity σ of a metal is given by $\sigma = N_e q_e^2 \tau_e / m_e$, where the mean collision time τ_e of electrons is related to their mean free path l_e as $\tau_e = l_e / v_F$ (see Section 1.2). The underlying assumption in this theory is that the scattering processes involved are incoherent, resulting in charge transport that is essentially diffusive in nature. It is also assumed that different scattering processes (scattering from electrons, phonons, impurities, vacancies, and dislocations) are independent. It follows from Matthiessen's rule that their scattering rates can be simply added [58].

When the temperature of the metal transporting current is reduced to the extent that the quantum nature of electrons dominates over their thermal motion, Matthiessen's rule breaks down. It has also been found that Ohm's law itself fails when the drift velocity of electrons cannot increase indefinitely with increasing electric field and saturates to a constant value. It was eventually realized that the assumption that electrons behave like classical particles inside a metal leads to wrong quantitative predictions, especially in nanostructures.

1.5.1 Charge Transport in Nanostructures

According to a semiclassical theory, there are no fundamental restrictions on the conductivity of any material, and it can have all continuous values in a range that depends on the intrinsic properties of a specific material. However, following the discovery of the integer quantum Hall effect, it became clear that conductance is not a continuous variable but changes in steps of a basic quantum unit, the so-called *conductance quantum* given by

$$G_Q = \frac{q_e^2}{h}.$$ (1.49)

Existence of the conductance quantum was first observed in a 2D-electron gas formed between GaAs and AlGaAs semiconducting layers [59, 60]. This discovery opened up a new area of research on quantum localization.

The subsequent discovery of the *fractional quantum Hall effect* propelled research on charge transport in new directions. A very useful relationship, due to Landauer (generalized later by Büttiker [61, 62]), states that conductance of a quantum material can be calculated by multiplying the conductance quantum G_Q with the quantum-mechanical transmission coefficient of that material. Details of this relation are given in Aside 1.7 (see also Section 5.2). The important question that seems like a paradox is how dissipation occurs as a result of current flowing through a quantum conductor. Whenever a current I flows through a material with conductance G, energy is also dissipated at a rate I^2/G. As elastic scattering is the only process responsible for the appearance of the conductance, there is no dynamical mechanism that can cause energy dissipation. The answer comes from the appearance of a contact resistance, primarily due to mismatch of available conduction channels between the quantum conductor and its reservoir (see Aside 5.2). The reasoning is based on the fluctuation-dissipation theorem, which predicts the behavior of systems obeying detailed balance; electrical resistances in quantum systems are covered by this theorem (see also Section 3.3).

Aside 1.7   Landauer–Büttiker Formula

Consider an ideal quantum conductor connected to two reservoirs at different chemical potentials, as shown in Figure 1.11. We assume that this conductor is connected to reservoirs through quasi-1D leads. When an electron enters these leads, it immediately loses any phase information. Both contacts are assumed to be in thermodynamic equilibrium so that the occupation probabilities for electrons are given by the Fermi–Dirac distributions,

$$f_{FD}(E - \mu_j) = \left[1 + \exp\left(\frac{E - \mu_j}{k_B T} \right) \right]^{-1} \quad (j = 1, 2), \tag{1.50}$$

where μ_1 and μ_2 are the chemical potentials of reservoirs 1 and 2, respectively, and T is the temperature. If V is the voltage applied between the two contacts, the chemical potentials relate to each other through $\mu_1 - \mu_2 = q_e V$.

We can relate the current to the transmission probability $T_{21}(E)$ of an electron of energy E as

$$I = \frac{2q_e}{h} \int_0^\infty T_{21}(E) \left[f_{FD}(E - \mu_1) - f_{FD}(E - \mu_2) \right] dE. \tag{1.51}$$

This equation is known as the *Tsu–Esaki equation* and is widely used for quantum devices. For sufficiently small voltages, we can expand $f_{FD}(E - \mu_j)$ in a Taylor series around the energy E_F, retaining only the first two terms in the expansion. The term in the square brackets inside the integral is then reduced to $-q_e V(\partial f_{FD}/\partial E)$, where the derivative is

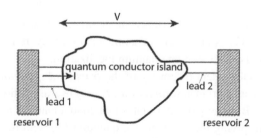

Fig. 1.11 Schematic of a quantum conductor connected to two reservoirs at different chemical potentials via two leads; current flows because of the voltage difference V between the two leads.

evaluated at the energy $E - E_F$. In the limit $T \to 0$, this derivative becomes the delta function $\delta(E - E_F)$, and the conductance, $G = \frac{I}{V}$, of the quantum conductor becomes

$$G = 2G_Q T_{21}(E_F),\tag{1.52}$$

where the factor 2 accounts for the electron spin. This relationship is sometimes called the Landauer formula in literature [63, 64]. It can be extended to a quantum conductor with multiple terminals, resulting in the so-called Landauer–Büttiker formula [61, 62].

In quantum devices, charge transport cannot be studied by counting the flow of electrons because the classical picture of an electron being a point charge loses its validity. For this reason, electric current through a quantum device is quantified by calculating total charge transferred through the conductor per unit time. The paradox of electrical current is that even though charges are quantized, transferred charge can have practically any value, even a fraction of the charge of a single electron (i.e., the electrical current unlike charges is not quantized). This can be understood by noting that electric current through a conductor is merely a displacement of the electron cloud against the lattice of positive cores. As this displacement can be by any amount, the associated electrical current is a continuously varying quantity. To calculate the current, we use the concept of current density in quantum mechanics. If the wave function of the electron is Ψ, the current density is given by

$$\mathbf{J} = -\frac{iq_e\hbar}{2m_e}\left(\Psi^*\nabla\Psi - \Psi\nabla\Psi^*\right) = \frac{q_e\hbar}{m_e}\Im\left[\Psi^*\nabla\Psi*\right],\tag{1.53}$$

where $\Im[\ldots]$ stands for the imaginary part.

As an electron moves inside a material, it is scattered from its path through collisions (elastic or inelastic) with other electrons, photons, and ions. Three different scenarios shown schematically in Figure 1.12 are possible for a conductor of length L and width W. If all collisions are elastic such that the phase of Ψ is preserved during each collision, charge transport maintains its coherent nature and is classified as being *ballistic transport*. Ballistic transport happens when both L and W are much smaller than the electron's mean free path l_e. It is important to realize that the mean free path is unrelated to the phase coherence of electrons, which is determined by the phase coherence length l_ϕ, defined as the distance an electron travels before its phase is randomized. This quantity is closely related to dephasing time τ_ϕ, defined as the time after which it is not possible to trace back

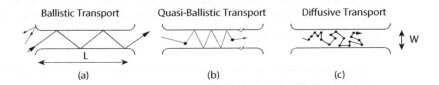

Fig. 1.12 Illustration of three charge-transport regimes in a conductor of length L and width W.

Table 1.3 Transport mechanisms for a system of length L.	
Quantum Systems	Classical Systems
ballistic when $L \ll l_e$	diffusive when $L \gg l_e$
phase coherent when $L \ll l_\phi$	incoherent when $L \gg l_\phi$

the phase information of an electron and determine electrons' prior locations [54]. Information about several such time and length scales can be found in Ref. [65]. Note that the conductance in the case of ballistic transport is not infinite but must be an integer multiple of the conductance quantum given in Eq. (1.49).

In the other extreme where inelastic scattering dominates and phase coherence is not preserved after each collision, charge transport is classified as being *diffusive transport*. As seen in Figure 1.12(c), diffusive transport happens when channel dimensions are larger than l_e but smaller than the localization length ξ. The quasi-ballistic transport regime lies in between these two regimes. It occurs when scattering from the surface of a nanostructure is as important as internal scattering. The three regimes in Figure 1.12 can be classified by comparing the conductance G of a device with the conductance quantum G_Q: $G \ll G_Q$ in regime (a), $G \approx G_Q$ in regime (b), and $G \gg G_Q$ in regime (c). In the first regime ($G \ll G_Q$), transport takes place in rare discrete events as a result of single-electron tunneling.

The device length L plays the critical role in determining whether the quantum nature of charge transport must be considered for any device. Table 1.3 lists the conditions that must be satisfied in the classical and quantum regimes. The mean free path l_e of electrons is the length that divides these two regimes. The second length scale is the phase coherence length l_ϕ. Charge transport remains phase coherent when $L \ll l_\phi$. On the other hand, if the device length L exceeds both l_e and l_ϕ, charge transport becomes incoherent and can be treated classically.

1.5.2 Tunneling in Nanostructures

Tunneling is an excellent example of the wave nature of quantum particles. Although not possible classically, a quantum particle trapped by a potential-energy barrier (owing to it not having enough kinetic energy) can still cross the barrier with a finite probability through a process referred to as tunneling. The reason behind tunneling is that the particle's wave function decays exponentially inside the barrier. If this barrier is thin enough that amplitude of the wave function does not decline to zero at the other side of the barrier, the particle has a finite probability of being found on that side. See Aside 1.8 where the operation of a *tunneling junction* is discussed.

Aside 1.8 Tunneling Junction

One basic, yet useful device is the tunneling junction, built by placing two electrodes of the same or different materials across a barrier layer. As there are many choices for the

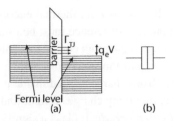

Fig. 1.13 (a) Schematic of a tunneling junction and (b) the symbol used to represent it.

electrodes and barrier-layer materials (including semiconductors, magnetic and nonmagnetic metals, or superconductors), a wide variety of tunneling junctions exists. All of them operate on the same quantum-mechanical principle, shown schematically in Figure 1.13. Electrons in each metallic electrode fill the quantum states up to the level of Fermi energy. When no voltage is applied, the Fermi levels for two electrodes coincide, and no current passes through the tunneling junction. When a bias voltage V is applied, the Fermi levels of the two electrodes are separated by an amount $q_e V$. When this voltage exceeds a threshold value, electrons can tunnel electrons through the gap, resulting in a finite current. Based on quantum theory, the tunneling rate is given by [66]

$$\Gamma_{TJ}(\Delta W) = \frac{1}{q_e}\left[1 - \exp\left(-\frac{\Delta W}{k_B T}\right)\right]^{-1} I\left(\frac{\Delta W}{q_e}\right), \tag{1.54}$$

where ΔW is the change in the free energy of the device because of tunneling and $I(V)$ is the current through the barrier at voltage V in the absence of tunneling effects. It can be approximated by $I(V) = V/R_{TJ}$, where R_{TJ} is the resistance of the tunneling barrier. Frequently, the approximation $\Delta W = \frac{q_e}{2}(V_i + V_f)$ is used, where V_i and V_f are the voltage drops across the barrier before and after the tunneling event, respectively.

Charge transport in a quantum device is affected by the electric capacitance of that device. When such a device is isolated from neighboring devices by a barrier with much higher resistance (conductance G_B of the barrier satisfies $G_B \ll G_Q$), it is possible not only to transport quantized charges in such devices but also to study interactions among these charges. Such a barrier is normally established by connecting the quantum conductor to neighboring devices via a tunneling junction.

Consider a quantum device containing N electrons and possessing a small capacitance C_{QC}. If E_S is the stored electrostatic energy in this capacitor, it is easy to conclude that it depends on N as $E_S(N) = (q_e N)^2/2C_{QC}$. If another electron is added to this device, additional energy needed is

$$E_S(N+1) - E_S(N) = (2N+1)E_{CE}, \qquad E_{CE} = q_e^2/(2C_{QC}). \tag{1.55}$$

The quantity E_{CE} is referred to as the charging energy of a quantum conductor, as this is the amount of energy needed to charge a neutral quantum conductor with one electron. For many nanoscale devices, the capacitance C_{QC} is so small that this charging energy becomes relatively large. If $E_{CE} \gg k_B T$, thermal energy cannot overcome the energy barrier. As a result, a large potential difference must be established across the tunnel junction to transport charges through a quantum conductor. This process becomes harder as we pump more and more charges into the quantum conductor, owing to repulsion experienced by the incoming charge from charges stored inside it. The result is that, if applied voltage is not sufficiently high, incoming charges cannot find a path to the quantum conductor, a process known as *Coulomb blockade*; it plays a significant role in nanoscale electronic devices.

Let us consider a simple example to illustrate the concept of Coulomb blockade. The electrostatic energy stored in an isolated capacitor is given by $Q^2/2C$, where C is the capacitance and Q is the charge on the positive electrode (and $-Q$ is the charge on the negative electrode). If an electron tunnels from the negative electrode to the positive one, total charge on the positive electrode becomes $Q - q_e$, while total charge on the negative electrode becomes $-(Q - q_e)$. Thus, energy stored in the capacitor becomes $(Q - q_e)^2/2C$. As no additional energy was supplied to the capacitor from an external source, such a charge transfer should never change the energy of the system. This condition can only be met if the initial charge is larger than $q_e/2$, or the initial voltage is greater than $q_e/2C$. Thus, if the initial voltage is below $q_e/2C$ (but nonzero), electron transfers are forbidden for such a system. This is a classic example of the manifestation of the Coulomb-blockade phenomenon.

To observe Coulomb blockade, one should be able to locate electrons inside a quantum conductor. However, if electrons can move in and out of this device freely, it is not possible to localize them. The remedy is to close off any conductive channel through which electrons can escape to the surrounding environment (i.e., electrons must tunnel in and out of the quantum conductor whenever charge exchanges happen). The Landauer formula given in Eq. (1.52) shows that each channel contributes at most $2G_Q$ to the conductance. Thus, if the conductance of the quantum conductor is smaller than this value, electrons must tunnel in and out of the conductor.

The simplest circuit where Coulomb blockade can be observed is the single-electron box shown in Figure 1.14. It localizes electrons by allowing their exchange through a quantum-conductor island (black circle) only via tunneling [67]. The circuit contains a tunneling junction with capacitance C_{TJ}, which is coupled to a quantum-conductor island, that itself is connected to a nontunneling (conventional) capacitor C_G. A voltage V_G is applied to

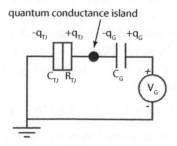

Fig. 1.14   Circuit known as the single-electron box and used for observing Coulomb blockade.

C_G with the tunneling junction grounded, as shown in Figure 1.14. To observe the single-electron effects, the total capacitance of the quantum-conductor island, $C_{TJ} + C_G$, must have a charging energy much greater than the thermal energy ($q_e^2/(C_{TJ}+C_G) \gg k_B T$). Let q_G and q_{TJ} be charges on the positive plates of the capacitor and the tunneling junction, respectively. Under stationary operation, the number of electrons localized in the quantum island is found by minimizing the free energy. The free energy E_{SEB} is not just the sum of electrostatic energies stored in each capacitor but must also contain a term representing the energy supplied automatically by the voltage sources when electrons tunnel in and out to the quantum-conductor island:

$$E_{SEB} = \frac{q_{TJ}^2}{2C_{TJ}} + \frac{q_G^2}{2C_G} - q_G V_G. \tag{1.56}$$

Noting that charge conservation on the conductive island demands $q_{TJ} = q_G - q_e N$, we can write E_{SEB} in the form

$$E_{SEB} = \frac{q_e^2}{2(C_{TJ} + C_G)} \left(N - \frac{C_G V_G}{q_e} \right)^2 - \frac{q_G^2}{2C_G}. \tag{1.57}$$

If the localized number of electrons is fixed, E_{SEB} is a quadratic function of V_G, as shown in Figure 1.15(a) for several values of N. As the quantum-conductor island is

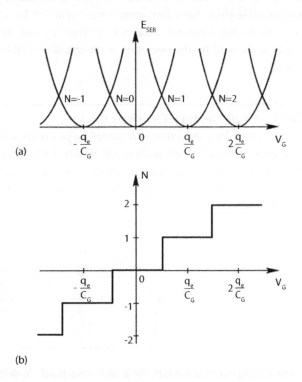

(a)

(b)

Fig. 1.15 (a) Free energy E_{SEB} as a function of gate voltage V_G for several values of N in the quantum-conductor island; (b) Number of localized electrons (N) versus V_G.

charge neutral when $V_G = 0$, no additional electrons can be added or removed without adding energy to the system equal to the charging energy (owing to Coulomb blockade). As the voltage V_G is increased, the energy of the configuration $N = 0$ increases, while that of $N = 1$ decreases, as seen in Figure 1.15(a). When $V_G = q_e/(2C_G)$, the energies of the two configurations are the same, and if V_G is increased further, the configuration $N = 1$ becomes the lowest energy configuration of the system. To attain this energy-favored state, one electron will tunnel through the capacitor C_{TJ}, bringing the number of electrons to one. Thus, as the voltage is swept, quantum states with more and more localized electrons become energetically favored, and electrons tunnel through the tunneling junction to the quantum-conductor island as needed to attain that state. The staircase in Figure 1.15(b) represents the number of electrons in the island as a function of gate voltage V_G.

Even though the single-electron box could deterministically change the number of electrons in the quantum-conductance island by changing the gate voltage, it is of limited use for applications. This is because no simple mechanism exists to count the number of electrons inside the island. However, the knowledge and insight gained by the operating principle of the single-electron box can be extended to produce the most basic single-electron device – the single-electron transistor (SET). The operation of SET is discussed in Aside 1.9. One important application of SETs is for sensing a small amount of charge [67]. In 2001, an SET-based electrometer exhibited a record sensitivity 3.2×10^{-6} electrons/$\sqrt{\text{Hz}}$ [68]. Such low values were realized by setting the bias voltages of the SET close to the Coulomb blockade voltage so that the current through the SET is strongly influenced by the potential of the quantum-conductor island as well as the gate potentials.

Aside 1.9   Single Electron Transistor (SET)

An SET is built by connecting two tunneling junctions and one conventional capacitor to a quantum-conductor island as shown in Figure 1.16. Since such a device has two current leads and one voltage lead, the circuit resembles that of a field-effect transistor, but there are significant operational differences. The first demonstration of the SET device was carried out in 1987 [69].

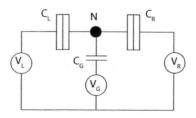

Fig. 1.16 Schematic of a single-electron transistor. The circuit is similar to a single-electron box but another tunneling junction is used to connect to the quantum-conductor island (black circle).

To enable a net current through the device, the biasing scheme is designed so that charge transfer is possible through both tunneling junctions. The free energy of SET is calculated similar to that for a single-electron box and has the form

$$E_{SET}(N) = \frac{q_e^2}{2(C_L + C_R + C_G)} \left(N - \frac{Q_I}{q_e}\right)^2 \tag{1.58}$$

where the induced charge $Q_I = C_G V_G + C_L V_L + C_R V_R$ has contributions from all voltage sources biasing the SET and N is the number of localized electrons in the quantum-conductor island. Tunneling of electrons into this island from the tunneling junction with capacitance C_L is possible provided $q_e V_L > [E_{SET}(N + 1) - E_{SET}(N)]$. Once there are $N + 1$ charges, it possible to remove one from the tunneling junction with capacitance C_R by setting $q_e V_R < [E_{SET}(N+1) - E_{SET}(N)]$. If both these conditions are met, a current will flow through the device. This can be achieved by symmetrically biasing the device such that $V_L = -V_R$.

1.6 Organization and Overview of the Material

In this introductory chapter we have tried to give a flavor of new phenomena that can occur in nanoscale quantum devices. After discussing in Section 1.1 new features that may appear at the nanoscale, we introduced in Section 1.2 several important length scales for nanosize objects that become important in later chapters. We discussed in Section 1.3 how quantum confinement leads to quantization of the electron's energies and how it affects the density of states in nanostructures. We emphasized in Section 1.4 the importance of the wave function's phase leading to quantum interference. We discussed in Section 1.5 charge transport mechanisms in naonscale devices and introduced the concept of quantum tunneling. The aim was to review the basic but important concepts that will enhance the reader's understanding and help her in understanding the material appearing in later chapters. The following is a brief summary of the topics covered in subsequent chapters.

We review in Chapter 2 the main classical and quantum-mechanical concepts needed for understanding the later material. We discuss in two sections the Lagrangian and Hamiltonian formalisms and introduce the important concepts such as a quantum state, eigenvalues and eigenstates, and the density operator. The time evolution of a quantum system is discussed in the Schrödinger, Heisenberg, and interaction pictures together with perturbation theory. Section 2.3 is devoted to the quantization of an electromagnetic field, whereas the concept of fermions and bosons is discussed in Section 2.4.

Chapter 3 is devoted to the linear response of a system. We begin by introducing the concept of impulse response in Section 3.1 and then focus on the three ensembles used in classical and quantum statistical mechanics for systems in thermal equilibrium. The Kubo formula governing the linear response of a system is discussed in Section 3.2 where we also introduce the concept of generalized susceptibility. The fluctuation-dissipation theorem is derived in Section 3.3 after considering the dynamic correlation function and applied in the context of Johnson–Nyquist noise. The focus of Section 3.4 is on the derivation of the dielectric function, and surface plasmons are discussed in Section 3.6.

Chapter 4 considers how dissipation and decoherence can occur in a quantum device through its interaction with its external environment. As an example, we discuss in Section 4.1 spontaneous emission in a two-level atomic system using the Jaynes–Cummings Hamiltonian and present details of theory developed by Weisskopf and Wigner. The Maser-equation approach is discussed in Section 4.2. The focus of Section 4.3 is on the derivation of the Lindblad equation and its applications to a damped harmonic oscillator and a damped two-level atom. Section 4.4 is devoted to the derivation of Redfield equation.

Current flow inside a quantum device is considered in Chapter 5. We discuss the main features of quantum transport in Section 5.1 and then use the Landauer–Büttiker method in Section 5.2 to drive an expression for the current using an approach based on the concept of a scattering matrix. Section 5.3 employs the nonequilibrium Green's function method to calculate the current flowing through a quantum device.

Chapter 6 focuses on quantum tunneling. After discussing the physics behind tunneling in Section 6.1, we present Gamow's theory of tunneling. The well-known Wentzel–Kramers–Brillouin (WKB) method is used in Section 6.2 to develop a quantum theory of tunneling. The time dependence of tunneling is analyzed in Section 6.3 using the transfer Hamiltonian method. The phenomena of sequential and resonant tunneling are discussed in Sections 6.4 and 6.5, respectively.

The focus of Chapter 7 is on quantum noise. We introduce the basic concepts in Section 7.1 where we also discuss the main sources of noise leading to thermal noise, shot noise, amplifier noise, and Brownian motion of a particle. The classical concept of spectral density is extended to the quantum domain in Section 7.2, where we also calculate the quantum spectral density when a harmonic oscillator and a two-level atom are coupled to a noise source. We discuss in Section 7.3 the role of quantum Langevin equations by considering the quantum theory of a laser. The Langevin formalism is also used in Section 7.4 to calculate the noise spectra associated with the intensity and phase noise of lasers.

Quantum-Mechanical Framework

> Quantum mechanics is not a theory about reality, it is a prescription for making the best possible predictions about the future if we have certain information about the past.
>
> G. 't Hooft, *Journ. Stat. Phys.* **53**, 323 (1988)

2.1 Review of Classical Mechanics

A mathematical description of a nanoscale device is typically based on the *equations of motion* describing how different parts making up the device change with time [70]. Such a description depends heavily on our understanding of the laws governing the device and on the approximations adopted in formulating the equations of motion. For example, a single device could have different equations of motion, depending on whether we use Newtonian mechanics, quantum mechanics, or relativistic mechanics to describe it [71]. So, one may ask whether a single set of equations of motion can be used if one agrees in advance on the physics describing the device. The answer is clearly no, because the equations of motion change with the type of coordinates used for locating various parts of a system relative to an agreed reference point (the origin of the coordinate axes). The only requirement for such coordinates is that they should be sufficiently unique to identify each and every part of a device.

2.1.1 Generalized Coordinates

The coordinates used in classical mechanics are referred to as *generalized coordinates* [72, 73]. As different choices of generalized coordinates result in different equations of motion for the same device, the equations of motion for any device are not unique. The situation is further complicated because specific properties of the components, referred to as the *constitutive relations*, are required to thread the physical laws governing the constraints restricting motion of the device [74]. As this process of deriving the equations of motion is somewhat ad hoc, there is no simple way to predict whether one has obtained a sufficient number of independent equations that can be solved to find time variations of generalized coordinates of the device. The Lagrangian and Hamiltonian methods avoid most of these issues and provide a structured approach to deriving the equations of motion described by a set of generalized coordinates [75]. The versatility of the Lagrangian method stems from the fact that, regardless of the physical phenomena of concern (related to electromagnetics, quantum mechanics, or Newtonian mechanics), the associated generic technique can be readily applied to analyze and learn about any system [76, 77].

Science and engineering utilize in practice few standard coordinate systems to represent points in two- or three-dimensional (3D) space [72, 73]. For example, coordinates based on rectilinear orthogonal axes are referred to as the Cartesian coordinates. Similarly, coordinates based on angles from a baseline are referred to as polar coordinates. It is conventional that coordinates of a specific point appear in a set with a predetermined order [78]. For example, the coordinates of a point in a 3D Cartesian system are written as (x, y, z), and the polar coordinates in a plane are denoted as (r, θ). However, for the generalized coordinates of a system, there is no prescription for the number of coordinates or their order in a set; the only restriction is that they must be scalars [79]. This gives us the freedom to be creative in inventing coordinates that are not only intuitive but also convenient to use in theory and computational work.

Among many possible generalized coordinates system available for representing a quantum device, a complete and independent set is preferred [80]. A generalized coordinate system is deemed *complete* if it can locate all parts of a device in all geometrically admissible configurations at all times. The *geometrically admissible* configurations are the set of configurations that satisfy the geometric constraints, but not necessarily the underlying physical principles valid for a particular configuration. A generalized coordinate system is deemed *independent* if, when all but one of them are fixed in value, there remains a continuous range of values for the unfixed coordinate. If both these conditions are simultaneously satisfied by a generalized coordinate system, it is referred to as a *complete and independent* generalized coordinate system.

When generalized coordinates are used to formulate the equations of motion for a device, we need to consider *admissible variations*, which represent hypothetical infinitesimal changes subject to geometric constraints of the device (but may not necessarily satisfy other physical principles applicable to the situation). The origin of this concept can be traced back to *virtual displacements* in classical mechanics, where infinitesimal changes of the system coordinates can occur while time is held constant [81]. It is called virtual rather than real since no actual displacement can take place instantaneously (as time is assumed frozen during such displacements). However, the importance of this concept lies in that it provides a procedural technique for studying the behavior of complex interacting systems described by generalized coordinate systems. As we shall see soon, such infinitesimal admissible variations are essentially test quantities that reveal interactions among forces (or generalized forces) internal to a system. As time is fixed when virtual displacements happen, unlike normal (real) displacements, they are defined with respect to a parameter η enumerating paths of the motion varied in a manner consistent with the geometric constraints. The symbol δ is traditionally used as the operator, which acts on generalized coordinates to generate variations as [82]

$$\delta \equiv \left. \frac{\partial}{\partial \eta} \right|_{\eta \to 0}. \tag{2.1}$$

Owing to this definition, the δ operator follows the same rules as the familiar differential operator. The size of the set of independent admissible variations is called the *degrees of freedom* of the system. It is a property of the system and thus has the same value no matter which set of generalized coordinates we use (there may be an infinite number of ways of doing so). A system is said to be *holonomic* if the degrees of freedom equals the number of generalized coordinates [83]; otherwise, the system is *nonholonomic*. Aside 2.1 illustrates an application of this concept to an N-particle system, subjected to several geometric constraints, using a 3D Cartesian coordinate system. In particular, it shows that admissible virtual displacements preserve the geometric constraints.

Aside 2.1 Admissible Virtual Displacements with Holonomic Constraints

Consider a system of N particles with the coordinates (x_i, y_i, z_i) for the ith particle ($i = 1 - N$). Suppose the motion of these particles is restricted by K holonomic constraints, denoted by the functions $\phi_j(x_1, y_1, z_1, \ldots, x_N, y_N, z_N) = 0$ for $j = 1, 2, \ldots, K$. It is important to note that these constraints are geometric in nature, and no explicit time dependence appears in any of them. The holonomic property assures that the constraints can always be written using the generalized coordinates. As admissible variations $(\delta x_i, \delta y_i, \delta z_i)$ are infinitesimal values, we can use differential calculus to find the relation [84]:

$$\frac{\partial \phi_j}{\partial x_1} \delta x_1 + \ldots + \frac{\partial \phi_j}{\partial x_N} \delta x_N = 0; \qquad j = 1, 2, \ldots K. \tag{2.2}$$

As all displacements in this equation are admissible variations, the resulting configuration should still adhere to the same geometric constraints. We can check this by expanding the constraint for each j in a Taylor series to the first order as

$$\phi_j(x_1 + \delta x_1, \ldots, x_N + \delta x_N) = \phi_j(x_1, \ldots, x_N) + \frac{\partial \phi_j}{\partial x_1} \delta x_1 + \ldots + \frac{\partial \phi_j}{\partial x_N} \delta x_N. \tag{2.3}$$

However, all terms on the right side vanish owing to the constraint itself and the relation in Eq. (2.2). The same argument holds in the y and z directions. Thus, admissible virtual displacements satisfy all constraints. This analysis cannot be used for nonholonomic constraints because they cannot be explicitly expressed as an equation involving only the generalized coordinates [83].

The concept of *virtual work* is fundamental in the Lagrangian and Hamiltonian methods because it is directly associated with the energy of a system [81]. It is formally defined as the total work done by all nonconservative forces acting on the system as a result of an admissible virtual displacement of the entire configuration. As this definition only involves nonconservative forces (e.g., friction), conservative forces (such as electrical and gravitational forces) must not be considered in virtual-work calculations [85].

2.1.2 Lagrangian Formalism

As current literature on nanoscale devices makes use of topology terminology, we introduce it on a need basis in this book. Aside 2.2 will help the reader in understanding the following material.

Consider a quantum device with N generalized coordinates denoted by $q_1, q_2 \ldots, q_N$. As each generalized coordinate is a scalar, the entire set represents a configuration vector, $\mathbf{q} = [q_1, \ldots, q_N]^T$, in the manifold $\mathbb{R}^N$. We denote the time derivative of this configuration vector (referred to as generalized velocities) by $\dot{\mathbf{q}} = [\dot{q}_1, \ldots, \dot{q}_N]^T$. This vector lies in the tangent space $T_{\mathbf{q}}\mathbb{R}^N$ and is said to be diffeomorphic to $\mathbb{R}^N$. Moreover, the combination $(\mathbf{q}, \dot{\mathbf{q}})$ lies in the tangent bundle $T\mathbb{R}^N$. The *Lagrangian* $\mathcal{L}$ is a real-valued function of this tangent bundle and is defined as [86]

$$\mathcal{L}(\mathbf{q}, \dot{\mathbf{q}}) = T(\mathbf{q}, \dot{\mathbf{q}}) - U(\mathbf{q}), \tag{2.4}$$

where $T(\mathbf{q}, \dot{\mathbf{q}})$ is the kinetic energy function and $U(\mathbf{q})$ is the potential energy function. The identification of the two terms as kinetic and potential energies does not always hold, but the generic properties derived from a Lagrangian mostly stay the same.

As the definition of the Lagrangian relies on various forms of energy associated with a system, it is often much easier to find it using elementary means. Most importantly, even though generalized coordinates are used to label each energy function, being energies, the values of T and U do not depend on the actual coordinates used. Thus, such an approach is essentially a coordinate-independent description of the system, allowing one to obtain equations of motion using nonconventional conditions (e.g., noninertial frames in strong gravitational fields).

Aside 2.2   Tangent Vectors, Tangent Spaces, and Tangent Bundles [86, 87]

A manifold is an abstract mathematical space that locally appears to be an Euclidean space, but globally may have a complicated structure. The most familiar manifold we know is the surface of the earth; it is locally flat but spherical globally. A manifold can be constructed by joining separate Euclidean spaces together, just like the surface of the earth can be depicted by joining maps of local regions together, and accounting for the resulting distortions resulting from the spherical nature of the global surface. Taking this analogy further, an atlas describes how a manifold is constructed by joining together simpler pieces that are represented by their own corresponding charts (also known as the local coordinate systems). If the charts are in the Euclidean space (denoted as $\mathbb{R}^N$ in N dimensions), the resulting manifold is called a *topological manifold*, and it behaves like an Euclidean space (i.e., the two are *homeomorphic*).

If the manifold M has additional properties such as differentiability, many familiar operations in calculus could be carried on it. For example, it is possible to define a real vector space called the *tangent space* and denoted as $T_q M$, at any point $q \in M$. The tangent space contains tangents to all possible curves passing through that point. If one collects all these tangent spaces for the manifold M (denoted as $\bigcup_{q \in M} T_q M$), the resulting structure is called the tangent bundle of the differentiable manifold M (denoted as TM). The *Lagrangian* that we discuss in this chapter is a natural energy function defined on the tangent bundle.

The action integral $\mathcal{S}(\mathbf{q})$ is a functional of the generalized coordinates $\mathbf{q}$ and velocities $\dot{\mathbf{q}}$ and is defined with the aid of the Lagrangian as [86]

$$\mathcal{S}(\mathbf{q}) = \int_{t_0}^{t_1} \mathcal{L}(\mathbf{q}, \dot{\mathbf{q}})\, dt, \tag{2.5}$$

where the integration is over a fixed time interval ranging from t_0 to t_1. As the generalized coordinates define a path in a manifold, we call $\mathbf{q}(t)$ a trajectory of the system. The action integral is defined for any trajectory taken by the system over the time duration covered by the integral. The actual trajectory of the system is selected by the *principle of stationary action* (also referred to as *Hamilton's action principle*). It states that the physical trajectory of the system must yield a stationary value of the action integral. In other words, infinitesimal variations in the integration path $\mathbf{q}(t)$ do not produce a corresponding infinitesimal change in the action integral $\mathcal{S}(\mathbf{q})$ when such variations occur around the physical trajectory, provided the initial and final configurations are held fixed.

The Lagrangian formalism is applicable to many branches of physics and can be used for systems involving either particles or waves [88]. In this approach, the full dynamics of a physical system is contained in its Lagrangian, and equations of motion are derived from it by invoking the principle of stationary action. As discussed before, the formalism uses virtual displacements in generalized coordinates but succeeds in generating the equations of motion applicable to real (measurable) quantities described by the system. However, as an extension of this process, it is possible to look at real displacements (as opposed to virtual displacements) on the action integral to derive a new set of equations. These are referred to in literature as the Jacobi equations.

Suppose $\mathbf{Q}(t)$ is the physical trajectory of a system. Consider a new trajectory $\mathbf{q}(\mathbf{t})$ resulting from an infinitesimal variation, $\delta\mathbf{q}(t)$, that is, $\mathbf{q}(t) = \mathbf{Q}(t) + \delta\mathbf{q}(t)$. As the initial and final configurations are fixed, $\delta\mathbf{q}(t_0) = \delta\mathbf{q}(t_1) = 0$. The change $\delta\mathcal{S}$ in the value of the action integral to the first order in $\delta\mathbf{q}$ is given by

$$\delta\mathcal{S}(\mathbf{Q}) = \int_{t_0}^{t_1} \left(\frac{\partial\mathcal{L}}{\partial\mathbf{q}} \delta\mathbf{q}(t) + \frac{\partial\mathcal{L}}{\partial\dot{\mathbf{q}}} \delta\dot{\mathbf{q}}(t) \right) dt, \tag{2.6}$$

where the functional derivatives are evaluated on the trajectory $\mathbf{Q}(t)$. If we integrate the second term in this integral using the "integration by parts" method and use the boundary conditions that $\delta\mathbf{q}(t_0) = \delta\mathbf{q}(t_1) = 0$, we obtain

$$\delta\mathcal{S}(\mathbf{Q}) = \int_{t_0}^{t_1} \left[\frac{\partial\mathcal{L}}{\partial\mathbf{q}} - \frac{d}{dt}\left(\frac{\partial\mathcal{L}}{\partial\dot{\mathbf{q}}} \right) \right] \delta\mathbf{q}(t)\, dt. \tag{2.7}$$

We apply the stationary action principle to this equation and demand that $\delta\mathcal{S}(\mathbf{Q}) = 0$ for any arbitrary infinitesimal variation $\delta\mathbf{q}(t)$ in the trajectory from its physical trajectory. This could only happen if the quantity inside the brackets vanishes. Setting it equal to zero, we obtain the well-known *Euler–Lagrange equations* given by

$$\frac{\partial\mathcal{L}}{\partial\mathbf{q}} = \frac{d}{dt}\left(\frac{\partial\mathcal{L}}{\partial\dot{\mathbf{q}}} \right). \tag{2.8}$$

Even though we derived these equations using the Lagrangian in Eq. (2.4), there are systems for which Euler–Lagrange equations are satisfied but they do not agree with the definition in Eq. (2.4) (see Aside 2.3). In field theory, it is customary to use the Euler–Lagrange equations as the definition of the Lagrangian, which could even be an explicit function of time. However, it is important to realize that Euler–Lagrange equations are necessary, but not sufficient, for realizing stationary action.

Aside 2.3   A Point Charge Subjected to Electric and Magnetic Fields [89]

Consider a point charge, with mass m and charge q, moving in an area where both the electric ($\mathbf{E}$) and magnetic ($\mathbf{B}$) fields are present. The Lorentz force experienced by the particle is given by $q(\mathbf{E}+\dot{\mathbf{r}}\times\mathbf{B})$, where $\dot{\mathbf{r}}$ is the velocity of the charge at point $\mathbf{r}$. The equation of the motion of the point charge is given by Newton's second law: $m\ddot{\mathbf{r}} = q(\mathbf{E} + \dot{\mathbf{r}} \times \mathbf{B})$. However, as we saw in Section 1.4, the scalar and vector potentials $V(\mathbf{r}, t)$ and $\mathbf{A}(\mathbf{r}, t)$ often play an important role. They are related to the $\mathbf{E}$ and $\mathbf{B}$ fields as indicated in Eq. (1.36).

As there is no simple strategy for finding the Lagrangian of a given system, we often need to guess its form and test it by applying the Euler–Lagrange equation to recover the known equation of motion. A suitable form for the charge particle is found to be

$$\mathcal{L}(\mathbf{r}, \dot{\mathbf{r}}, t) = \frac{1}{2}m\dot{\mathbf{r}}^2 - qV(\mathbf{r}, t) + q\dot{\mathbf{r}} \cdot \mathbf{A}(\mathbf{r}, t). \tag{2.9}$$

The first two terms are in the form of Eq. (2.4) but the third term depends on the vector potential. We apply the Euler–Lagrange equation given in Eq. (2.8) with $\mathbf{r}$ acting as $\mathbf{q}$. Using $\frac{\partial \mathcal{L}}{\partial \dot{\mathbf{r}}} = m\dot{\mathbf{r}} + q\mathbf{A}(\mathbf{r}, t)$, we obtain

$$\frac{d}{dt}\left(\frac{\partial \mathcal{L}}{\partial \dot{\mathbf{r}}}\right) = m\ddot{\mathbf{r}} + q\frac{\partial \mathbf{A}}{\partial t} + q(\dot{\mathbf{r}} \cdot \nabla)\mathbf{A}, \tag{2.10}$$

where we used the relation $\frac{d\mathbf{A}}{dt} = \frac{\partial \mathbf{A}}{\partial t} + (\dot{\mathbf{r}} \cdot \nabla)\mathbf{A}$. The $\mathbf{r}$ derivative is found to be

$$\frac{\partial \mathcal{L}}{\partial \mathbf{r}} = -q\nabla V + q\nabla(\dot{\mathbf{r}} \cdot \mathbf{A}) = -q\nabla V + q(\dot{\mathbf{r}} \cdot \nabla)\mathbf{A} + q\dot{\mathbf{r}} \times (\nabla \times \mathbf{A}), \tag{2.11}$$

where we used a well-known vector identity. If we equate these two derivatives and use the relations in Eq. (1.36), the equation of motion is found to be $m\ddot{\mathbf{r}} = q(\mathbf{E} + \dot{\mathbf{r}} \times \mathbf{B})$. This confirms that the chosen Lagrangian is appropriate for the charge particle.

Even though this Lagrangian can reproduce the equation of motion, it cannot be written as the difference of the kinetic energy ($\frac{1}{2}m\dot{\mathbf{r}}^2$) and some potential energy. The reason is that $U = qV(\mathbf{r}, t) - q\dot{\mathbf{r}} \cdot \mathbf{A}(\mathbf{r}, t)$ contains a velocity-dependent term. As a result, the vector $-\nabla U$ does not represent the force acting on the particle.

The Lagrangian is not even unique for any system. Consider two Lagrangians, $\mathcal{L}(\mathbf{q}, \dot{\mathbf{q}}, t)$ and $\mathcal{L}(\mathbf{q}, \dot{\mathbf{q}}, t) + \frac{d}{dt}f(\mathbf{q}, t)$, where $f(\mathbf{q}, t)$ is an arbitrary function that depends on $\mathbf{q}$ and t but not on $\dot{\mathbf{q}}$. It is easy to deduce from the form of the Euler–Lagrange equations (2.8) that both lead to the same equations of motion. Moreover, any scaling of the Lagrangian by a constant also leads to the same equations of motion.

2.1.3 The Hamiltonian

The concept of momentum is very useful for quantum mechanics. Consider what happens when the upper limit t_1 of the action integral in Eq. (2.5) is replaced with t, (i.e., $S(\mathbf{q}, t) = \int_{t_0}^{t} \mathcal{L}(\mathbf{q}, \dot{\mathbf{q}}) \, dt$). Now consider the variation of the action integral when path $\mathbf{q}(t)$ is changed by $\delta\mathbf{q}$. It is easy to show that Eq. (2.7) must be replaced with

$$\delta S(\mathbf{q}, t) = \int_{t_0}^{t} \left[\frac{\partial \mathcal{L}}{\partial \mathbf{q}} - \frac{d}{dt}\left(\frac{\partial \mathcal{L}}{\partial \dot{\mathbf{q}}} \right) \right] \delta\mathbf{q} \, dt + \frac{\partial \mathcal{L}}{\partial \dot{\mathbf{q}}} \cdot \delta\mathbf{q}, \qquad (2.12)$$

The integral vanishes if the Euler–Lagrange equation (2.8) is satisfied, and we obtain the simple result $\delta S = \mathbf{p} \cdot \delta\mathbf{q}$, where the *generalized momentum* (also referred to as the *canonical momentum* [86]) is defined as $\mathbf{p} = \frac{\partial \mathcal{L}}{\partial \dot{\mathbf{q}}}$.

Aside 2.4   Legendre Transformation [80, 86]

The Legendre transformation can be viewed as the conversion of one scalar function into another. The transformation is reversible under quite general conditions such that the resulting two functions are the Legendre transform of each other. It is common to call them the dual of one another.

Consider a function $\mathcal{A}(q_1, q_2, \ldots, q_N)$ of N variables. We define a new set of variables as $p_j = \frac{\partial \mathcal{A}}{\partial q_i}$ for $i = 1$ to N. The Legendre transformation of $\mathcal{A}$ is defined as

$$\mathcal{B}(p_1, p_2, \ldots, p_N) = \sum_{i=1}^{N} p_i q_i - \mathcal{A}(q_1, q_2, \ldots, q_N). \qquad (2.13)$$

Noting that $\frac{\partial \mathcal{B}}{\partial p_i} = q_i$ for $i = 1$ to N, the original function $\mathcal{A}$ can be viewed as being the Legendre transformation of $\mathcal{B}$, confirming that each is the dual of the other.

One may argue that $\mathcal{B}$ may be a function of both sets of variables p_i and q_i for $i = 1$ to N, and it is not justified to assume that $\mathcal{B}$ depends only on p variables. One way to resolve this issue is to calculate the total differential of $\mathcal{B}$:

$$d\mathcal{B} = \sum_{i=1}^{N} \left(p_i dq_i + q_i dp_i - \frac{\partial \mathcal{A}}{\partial q_i} dq_i \right) \rightarrow d\mathcal{B} = \sum_{i=1}^{N} \left[q_i dp_i + \left(p_i - \frac{\partial \mathcal{A}}{\partial q_i} \right) dq_i \right]. \quad (2.14)$$

From the definition of p_i, it follows that the coefficient of dq_is is zero for all $i = 1$ to N. Therefore, $d\mathcal{B}$ changes only with variables $p_1, p_2, \ldots, p_N$, confirming our original assertion.

Hamilton used the canonical momenta to define a new functional, now known as the *Hamiltonian* and denoted as $H(\mathbf{q}, \mathbf{p}, t)$. Aside 2.4 shows how the Legendre transformation of a Lagrangian can be used for this purpose. As discussed there, the Hamiltonian can be considered a dual of the Lagrangian. We assume that the relations $\mathbf{p} = \frac{\partial \mathcal{L}}{\partial \dot{\mathbf{q}}}$ are invertible in the sense that $\dot{\mathbf{q}}$ can be uniquely expressed in terms of $\mathbf{p}$ and $\mathbf{q}$. This implies that the mapping $(\mathbf{q}, \dot{\mathbf{q}}) \rightarrow (\mathbf{q}, \mathbf{p})$ exists, a property of the Lagrangian referred to as hyperregularity. Any Lagrangian failing to have this property is called degenerate, and the Lagrangian of most physical systems satisfies it. The corresponding Hamiltonian is defined as [90]:

$$H(\mathbf{q}, \mathbf{p}, t) = \mathbf{p} \cdot \dot{\mathbf{q}} - \mathcal{L}(\mathbf{q}, \dot{\mathbf{q}}, t). \tag{2.15}$$

It is important to realize that two different Lagrangians can produce the same Hamiltonian. As shown in Aside 2.5, we can use the preceding definition to find derivatives of the Hamiltonian with respect to $\mathbf{p}$, $\mathbf{q}$, and t. These derivatives lead to the *Hamilton's equations* in the form

$$\dot{\mathbf{q}} = \frac{\partial H}{\partial \mathbf{p}} \qquad \dot{\mathbf{p}} = -\frac{\partial H}{\partial \mathbf{q}}. \tag{2.16}$$

Hamilton's equations and the Euler–Lagrange equations are equivalent because it is possible to derive the latter from Hamilton's equations and to show that $\frac{\partial \mathcal{L}}{\partial \dot{\mathbf{q}}} = \mathbf{p}$ and $\frac{\partial \mathcal{L}}{\partial \mathbf{q}} = \dot{\mathbf{p}}$.

Aside 2.5   Partial Derivatives of the Hamiltonian

Definition of the Lagrangian assumes that the variables $\mathbf{q}$, $\dot{\mathbf{q}}$, and t are independent. Similarly, the variables $\mathbf{q}$, $\mathbf{p}$, and t used in the Hamiltonian are assumed to be independent. This implies that partial derivatives such as $\frac{\partial q}{\partial t}$ are zero. We can use this feature to calculate the partial derivatives of the Hamiltonian $H(\mathbf{q}, \mathbf{p}, t)$. Consider $\frac{\partial H}{\partial \mathbf{q}}$:

$$\frac{\partial H}{\partial \mathbf{q}} = \mathbf{p} \cdot \frac{\partial \dot{\mathbf{q}}}{\partial \mathbf{q}} - \left(\frac{\partial}{\partial \mathbf{q}} + \frac{\partial \dot{\mathbf{q}}}{\partial \mathbf{q}} \frac{\partial}{\partial \dot{\mathbf{q}}} + \frac{\partial t}{\partial \mathbf{q}} \frac{\partial}{\partial t} \right) \mathcal{L}(\mathbf{q}, \dot{\mathbf{q}}, t) = -\frac{\partial \mathcal{L}}{\partial \mathbf{q}}. \tag{2.17}$$

Application of the Euler–Lagrange equation then yields the simple result

$$\frac{\partial H}{\partial \mathbf{q}} = -\frac{d}{dt} \left(\frac{\partial \mathcal{L}}{\partial \dot{\mathbf{q}}} \right) = -\dot{\mathbf{p}}. \tag{2.18}$$

Similarly,

$$\frac{\partial H}{\partial \mathbf{p}} = \dot{\mathbf{q}} + \mathbf{p} \cdot \frac{\partial \dot{\mathbf{q}}}{\partial \mathbf{p}} - \frac{\partial \dot{\mathbf{q}}}{\partial \mathbf{p}} \frac{\partial \mathcal{L}}{\partial \dot{\mathbf{q}}} = \dot{\mathbf{q}}, \tag{2.19}$$

where we used the definition $\mathbf{p} = \frac{\partial \mathcal{L}}{\partial \dot{\mathbf{q}}}$. Finally,

$$\frac{dH}{dt} = \left(\dot{\mathbf{q}} \frac{\partial H}{\partial \mathbf{q}} + \dot{\mathbf{p}} \frac{\partial H}{\partial \mathbf{p}} + \frac{\partial H}{\partial t} \right) = \frac{\partial H}{\partial t}, \tag{2.20}$$

where we used Eqs. (2.18) and (2.19) to cancel the first two terms.

The chief advantage of the Hamilton's equations approach is that it enables us to find the Hamiltonian with a knowledge of the kinetic and potential energies (see Aside 2.6). As there is no systematic way of guessing the Lagrangian, this feature is quite useful in practice. Compared to the Lagrangian formalism, the Hamiltonian formalism also provides a simpler computational path. The reason is that the Hamiltonian equations (2.16) constitute a set of $2N$ first-order, ordinary differential equations for a system of N particles. In contrast, the Euler–Lagrange equations (2.8) constitute a set of N second-order, ordinary differential equations. We shall see later that Hamiltonian formalism also provides an easy path to set up quantum-mechanical equations for a given system.

Aside 2.6   Hamiltonian for a System of N Interacting Particles

We use the definition in Eq. (2.4) to write the Lagrangian in the matrix form

$$\mathcal{L}(\mathbf{q}, \dot{\mathbf{q}}) = \frac{1}{2}\dot{\mathbf{q}}^T M(\mathbf{q})\dot{\mathbf{q}} - U(\mathbf{q}), \tag{2.21}$$

where the first term represents the kinetic energy of all moving particles and the second term is their potential energy. The matrix $M(\mathbf{q})$ contains masses of all particles and is invertible for each value of $\mathbf{q}$ [91].

To find the Hamiltonian, we first find the momenta of all particles using

$$\mathbf{p} = \frac{\partial}{\partial \dot{\mathbf{q}}}\mathcal{L}(\mathbf{q}, \dot{\mathbf{q}}) = M(\mathbf{q})\dot{\mathbf{q}}. \tag{2.22}$$

Multiplying this equation with M^{-1}, we obtain $\dot{\mathbf{q}} = M^{-1}\mathbf{p}$, where M^{-1} is the inverse of matrix M. The Hamiltonian is easily found using the definition in Eq. (2.15) and is given by

$$H(\mathbf{q}, \mathbf{p}) = \frac{1}{2}\mathbf{p}^T (M^{-1})^T \mathbf{p} + U(\mathbf{q}). \tag{2.23}$$

As expected, the Hamiltonian is the sum of the kinetic and potential energies of all interacting particles and represents the total energy of the system. Because the Hamiltonian has no explicit time dependence, the total energy of the system is conserved on the phase-space trajectory for which Hamilton's action principle is satisfied.

The Hamiltonian can be used to express the time evolution of any physical quantity associated with the system. Consider a generic function $f(\mathbf{q}, \mathbf{p}, t)$ that depends on both $\mathbf{q}$ and $\mathbf{p}$ in addition to time t. The rate of change of $f(\mathbf{q}, \mathbf{p}, t)$ is found to be

$$\frac{df}{dt} = \left(\frac{\partial g}{\partial t} + \dot{\mathbf{q}}\frac{\partial f}{\partial \mathbf{q}} + \dot{\mathbf{p}}\frac{\partial f}{\partial \mathbf{p}} \right). \tag{2.24}$$

Substitution of the Hamiltonian equations (2.16) results in

$$\frac{df}{dt} = \frac{\partial f}{\partial t} + \left(\frac{\partial f}{\partial \mathbf{q}}\frac{\partial H}{\partial \mathbf{p}} - \frac{\partial f}{\partial \mathbf{p}}\frac{\partial H}{\partial \mathbf{q}} \right) = \frac{\partial f}{\partial t} + \{f, H\}, \tag{2.25}$$

where we introduced the *Poisson bracket* $\{f, H\}$ that plays a central role in both classical and quantum mechanics [73]. The functions f and H may depend on all variables in the set

$\{\mathbf{q}, \mathbf{p}\}$ or on a subset of these. A few important Poisson brackets that have found extensive use in quantum mechanics and can be derived easily are:

$$\{q_i, q_j\} = 0, \quad \{p_i, p_j\} = 0, \quad \{q_i, p_j\} = \delta_{ij},$$
$$\{f, q_j\} = -\frac{\partial f}{\partial p_j}, \qquad \{f, p_j\} = \frac{\partial f}{\partial q_j}. \tag{2.26}$$

If the function of interest does not depend on time explicitly ($\partial f / \partial t = 0$), the Poisson bracket provides a compact way to write the equation of motion as

$$\frac{df}{dt} = \{f, H\}. \tag{2.27}$$

An insightful application of this result is to write the Hamilton's equation of motion as $\dot{\mathbf{q}} = \{\mathbf{q}, H\}$ and $\dot{\mathbf{p}} = \{\mathbf{p}, H\}$.

Poisson brackets provide us with a simple way to find the conserved quantities for a system. Conserved quantities retain their values as the system evolves with time. As both the partial and total time derivatives of f vanish when f is conserved, we find from Eq. (2.25) that this is possible only when the Poisson bracket $\{f, H\}$ itself is zero. Another remarkable feature is that Poisson brackets can be used to create new conserved quantities by combining two known conserved quantities. This is sometimes referred to as the Poisson theorem. Suppose f_1 and f_2 are two conserved quantities in a dynamical system (i.e., $\{f_1, H\} = \{f_2, H\} = 0$). If we use the Jacobi identity (see Aside 2.7), we have the relation

$$\{H, \{f_1, f_2\}\} + \{f_1, \{f_2, H\}\} + \{f_2, \{H, f_1\}\} = 0. \tag{2.28}$$

It follows immediately that $\{H, \{f_1, f_2\}\} = 0$ (i.e., the new quantity $\{f_1, f_2\}$ is also a conserved quantity). However, caution should be exercised because this process does not always provide new conserved quantities for a finite-dimensional system. It is known that, for a mechanical system in N dimensions, the total number of conserved quantities is limited to $2N - 1$. If we know all of these conserved quantities, the operation $\{f_1, f_2\}$ on any two of them will generate trivial constants, or functions that are simple extensions of the original functions f_1 and f_2.

Aside 2.7   Key Properties of Poisson Brackets [92]

The definition of the Poisson bracket can be used to find many properties of such brackets. A trivial property is that $\{f, g\}$ is equal to zero if the functions f and g do not depend on the phase-space variables used for defining the Poisson bracket. Other useful properties are:

antisymmetry: $\{f, g\} = -\{g, f\}$

linearity: $\{f + g, h\} = \{f, h\} + \{g, h\}$

partial derivative: $\frac{\partial}{\partial t}\{f, g\} = \{\frac{\partial f}{\partial t}, g\} + \{f, \frac{\partial g}{\partial t}\}$

Leibniz's relation: $\{fg, h\} = f\{g, h\} + g\{f, h\}$

Jacobi identity: $\{f, \{g, h\}\} + \{g, \{h, f\}\} + \{h, \{f, g\}\} = 0$.

The total time derivative of the action is related to the Hamiltonian as

$$\frac{dS}{dt} = \mathcal{L} = -H(\mathbf{q}, \mathbf{p}, t) + \mathbf{p} \cdot \dot{\mathbf{q}}. \tag{2.29}$$

This equation gives us the differential of the action in the form

$$dS = -H(\mathbf{q}, \mathbf{p}, t)dt + \mathbf{p} \cdot d\mathbf{q} \equiv \frac{\partial S}{\partial t}dt + \frac{\partial S}{\partial \mathbf{q}} \cdot d\mathbf{q}. \tag{2.30}$$

The preceding identity provides us with the relations $H = -\frac{\partial S}{\partial t}$ and $\mathbf{p} = \frac{\partial S}{\partial \mathbf{q}}$. We can use them to show that the system dynamics is governed by the following first-order partial differential equation:

$$\frac{\partial S}{\partial t} + H\left(\mathbf{q}, \frac{\partial S}{\partial \mathbf{q}}, t\right) = 0, \tag{2.31}$$

This equation is known as the *Hamilton–Jacobi equation* [93] and is equivalent to the Hamilton's equations of motion in Eq. (2.16). For an N-dimensional system, Hamiltonian equations form a set of $2N$ first-order, ordinary differential equations, whereas the Hamilton–Jacobi equation is a single partial-differential equation in $N + 1$ dimensions.

It should be clear by now that the action S is the central quantity governing a system's motion in the phase space $(\mathbf{q}, \mathbf{p})$. The conserved quantities (or invariants) are pivotal to this motion because they shape the trajectory as it evolves. It is then legitimate to ask whether a direct connection exists between the action and the invariants of a system. The answer is affirmative, as an invariant exists that is directly related to the action. It is the action on a closed contour in phase space. Using Eq. (2.30), we can write the following contour integral over a closed contour γ in the phase space:

$$\oint_\gamma dS = \oint_\gamma [\mathbf{p} \cdot d\mathbf{q} - H(\mathbf{q}, \mathbf{p}, t)] \, dt. \tag{2.32}$$

It can be shown that this integral is invariant for any choice of the contour γ; it is known as the *Poincarè invariant*. It exists for all dynamical systems and is a universal invariant [94]. However, the practical value of the Poincarè invariant is limited because its calculation requires knowledge of the phase-space trajectory, which can be found only by solving the Hamilton equations or the Hamilton–Jacobi equation.

2.2 Fundamentals of Quantum Mechanics

The theoretical framework of quantum mechanics is built upon a small number of *postulates* based on the concept of a linear, unitary, vector space known as the Hilbert space [95, 96, 97]. In this section we use these postulates to introduce the physical states of a quantum system.

2.2.1 Concept of a State Vector

QM Postulate 1   *A quantum system is completely described by its state vector, which is a unit vector in the Hilbert space.*

However, this postulate does not specify properties of the Hilbert space (e.g., its dimension) or the state vector that represents the physical system. As we will see later, different branches of physics (such as QED) have emerged to address this issue [76, 98]). It is common to denote a state vector in the Hilbert space as $|\Psi\rangle$ in the bra-ket notation of Dirac [99]. We can represent this state in any coordinate system. In a coordinate system based on the generalized coordinates $\mathbf{q}$, the state vector $|\Psi\rangle$ specifies a function, called the *wave function* and written as $\Psi(\mathbf{q}) = \langle\mathbf{q}|\Psi\rangle$. In other words, the wave function is the projection of the state vector in a specific coordinate basis. For this reason, the terms state vector, quantum state, and wave function are often used interchangeably. In general, the wave function is a complex quantity at any point in the phase space and it is a continuous function of both time and $\mathbf{q}$ that is also differentiable.

As seen in Section 1.4, the states $|\Psi\rangle$ and $e^{i\theta}|\Psi\rangle$ represent the same physical system for any real value of θ. This is why sometimes physical states are identified as rays in the associated Hilbert space. Also, the superposition principle holds (i.e., any linear combination of state vectors is also a state vector). This is a consequence of the linear nature of the associated Hilbert space. The phenomenon of quantum interference discussed in Section 1.4 is a direct consequence of the superposition principle. However, care must be taken not to introduce any inadvertent errors. As an example, consider the state vector $|\Psi\rangle = \alpha_1|\Psi_1\rangle + \alpha_2|\Psi_2\rangle$ that is a linear combination of the state vectors $|\Psi_1\rangle$ and $|\Psi_2\rangle$. Here, α_1 and α_2 are two complex numbers. If the phases of these two states change so that they become $e^{i\theta_1}|\Psi_1\rangle$ and $e^{i\theta_2}|\Psi_2\rangle$, clearly each one of them represents the same physical system. However, the superposed state, $|\Psi'\rangle = \alpha_1 e^{i\theta_1}|\Psi_1\rangle + \alpha_2 e^{i\theta_2}|\Psi_2\rangle$, is different from the original state $|\Psi\rangle$. It is important to realize that even though global phase factors on state vectors can be ignored, relative phases of superposed states must be accounted for when representing quantum states.

The quantum states of two or more parts of a system can be combined to represent a *composite state* of the whole system. Suppose $\mathcal{H}_1$ and $\mathcal{H}_2$ are two independent Hilbert spaces of dimensions N and M, respectively. The composite system is denoted as $\mathcal{H}_1 \otimes \mathcal{H}_2$ and is referred to as the *tensor product* of its two independent parts with the dimension $N \times M$. Analogously, any state vector in the composite system is represented as $|\Psi_1\rangle \otimes |\Psi_2\rangle$ where $|\Psi_1\rangle \in \mathcal{H}_1$ and $|\Psi_2\rangle \in \mathcal{H}_2$. The compact notation, $|\Psi_1\rangle|\Psi_2\rangle$, is also used for $|\Psi_1\rangle \otimes |\Psi_2\rangle$. Aside 2.8 provides more details about the tensor products. A local operator $\mathcal{O}_1$ acting on $\mathcal{H}_1$ corresponds to $\mathcal{O}_1 \otimes I_2$ in the composite system. Similarly, $\mathcal{O}_2$ acting on $\mathcal{H}_2$ becomes $I_1 \otimes \mathcal{O}_2$, where I_1 and I_2 are identity matrices of dimensions N and M, respectively. Thus, the composite system provides a unified way to carry out operations specific to each Hilbert space.

A nice application of the tensor product naturally emerges when we consider a collection of quantum particles. If the quantum state of the jth particle belongs to the Hilbert space $\mathcal{H}_j$, the composite space of N particles can be represented by the tensor product $\mathcal{H}_1 \otimes$

$\mathcal{H}_2 \ldots \otimes \mathcal{H}_N$. Here we have used the property that a tensor product does not depend on the order, and no bracketing of terms is needed. This notation can be further simplified if all particles have the same Hilbert space. The tensor product is then written as $\mathcal{H}^{\otimes N}$.

The preceding representation breaks down when particles are indistinguishable, and one has to consider the permutation symmetry. This can be clearly seen by looking at a two-particle system in which two permutations of the particles' states, $|\Psi_1\rangle \otimes |\Psi_2\rangle$ and $|\Psi_2\rangle \otimes |\Psi_1\rangle$, describe the same two-particle configuration. Sometimes this is attributed to the situation that one is not able to track individual particles without disturbing the state of the system. If a permutation of the particles is irrelevant for the combined system, its state can only change by a numerical factor such that $|\Psi'\rangle = P |\Psi\rangle$. Since a second permutation must return the combined system to its original state, we require $P^2 = 1$ or $P = \pm 1$. The choice of $P = 1$ or $P = -1$ leads to two classes of particles: *bosons* are particles for which $P = 1$ and they obey the *Bose–Einstein statistics*. In contrast, $P = -1$ for *fermions* obeying the *Fermi–Dirac statistics* [100]. One can also say that such quantum states must follow either Bose–Einstein or Fermi–Dirac statistics. This is commonly referred to as the *Bose–Fermi alternative* or *symmetrization postulate*.

The combined state of two indistinguishable bosons or fermions can be obtained by symmetrizing or antisymmetrizing the tensor product $|\Psi_1\rangle \otimes |\Psi_2\rangle$ such that

$$|\Psi\rangle_{S,A} = \frac{1}{\sqrt{2}} \Big[|\Psi_1\rangle \otimes |\Psi_2\rangle \pm |\Psi_2\rangle \otimes |\Psi_1\rangle \Big], \tag{2.33}$$

where the factor of $1/\sqrt{2}$ ensures that the combined state is a unit vector in its Hilbert space (i.e., it is properly normalized). It is easy to see that $|\Psi\rangle_S$ does not change sign when we interchange 1 and 2, while $|\Psi\rangle_A$ does change its sign. We can generalize our notation $\mathcal{H}^{\otimes N}$ for a system of N identical particles by adding the subscripts "S" and "A". Thus, $(\mathcal{H}^{\otimes N})_S$ and $(\mathcal{H}^{\otimes N})_A$ denote the N-particle states that have been symmetrized or antisymmetrized appropriately. Other alternatives are theoretically possible and are known as *parastatistics* [101].

Even if one is ready to accept the Bose–Fermi alternative, it is not possible to make the correct choice of the statistics merely based on mathematical consistency. Rather, we note that this freedom of choice only exists in nonrelativistic quantum mechanics. In relativistic quantum mechanics, the Bose–Einstein statistics applies to particles with integer spin, and the Fermi–Dirac statistics applies to particles with half-integer spin. The opposite choice leads to violations of the axiomatic principles [102].

Aside 2.8   Properties of Tensor Products [103, 104]

A tensor product can be used to create a new Hilbert space out of two or more Hilbert spaces. Let $\mathcal{H}_\Psi$ and $\mathcal{H}_\Phi$ be two Hilbert spaces with the states $|\Psi_m\rangle$ and $|\Phi_m\rangle$. The tensor product $\mathcal{H}_\Psi \otimes \mathcal{H}_\Phi$ has elements of the form $\sum_m |\Psi_m\rangle \otimes |\Phi_m\rangle$. The tensor product $|\Psi\rangle \otimes |\Phi\rangle$ obeys the following properties:

- $c(|\Psi\rangle \otimes |\Phi\rangle) = (c |\Psi\rangle) \otimes |\Phi\rangle = |\Psi\rangle \otimes (c |\Phi\rangle)$, $c \in \mathbf{C}$
- $(|\Psi_1\rangle + |\Psi_2\rangle) \otimes |\Phi\rangle = |\Psi_1\rangle \otimes |\Phi\rangle + |\Psi_2\rangle \otimes |\Phi\rangle$
- $|\Psi\rangle \otimes (|\Phi_1\rangle + |\Phi_2\rangle) = |\Psi\rangle \otimes |\Phi_1\rangle + |\Psi\rangle \otimes |\Phi_2\rangle$

- Inner product: $\langle\Psi_1|\otimes\langle\Phi_1||\Psi_2\rangle\otimes|\Phi_2\rangle = \langle\Psi_1|\Psi_2\rangle\langle\Phi_1|\Phi_2\rangle$
- Tensor product of operators: $\mathcal{O}_\Psi\otimes\mathcal{O}_\Phi(|\Psi\rangle\otimes|\Phi\rangle) = \mathcal{O}_\Psi|\Psi\rangle\otimes\mathcal{O}_\Phi|\Phi\rangle$.

Using these properties, it is possible to combine orthonormal state vectors in each space to construct an orthonormal basis for the tensor-product space. In such a construction, sometimes the notation $|\Psi\otimes\Phi\rangle$ is used in place of $|\Psi\rangle\otimes|\Phi\rangle$ for simplicity.

It turns out that the partition of a quantum system into two or more subsystems is not always possible. This leads to the concept of *quantum entanglement* (or *quantum correlation*) [105, 106]. Suppose we have a quantum state $|\Psi\rangle$ for a quantum system described by the Hilbert space $\mathcal{H}$ composed of two Hilbert spaces $\mathcal{H}_1$ and $\mathcal{H}_2$ ($\mathcal{H}$ is also referred to as a bipartite Hilbert space). If $|\Psi\rangle$ *cannot* be written as $|\Psi\rangle = |\Psi\rangle_1\otimes|\Psi\rangle_2$, then $\mathcal{H}$ is called *entangled* with respect to $\mathcal{H}_1$ and $\mathcal{H}_2$. It is natural to think that this concept is related to classical correlations, and entanglement also occurs for classical particles. However, there are some fundamental differences. Classical correlations can often be traced back to some conservation law. For example, a particle can decay into two particles that fly apart in opposite directions with speeds that comply with the momentum conservation. Thus, knowing the momentum of one particle is sufficient to predict the momentum of the other. The momentum of each particle exists regardless of a measurement performed on the other, and it can be predicted by applying the conservation of total momentum. However, such a technique cannot be used for quantum entangled states, as illustrated in Aside 2.9.

Aside 2.9   Entanglement through Bell States [107, 108]

Consider two spin-$\frac{1}{2}$ particles. We represent the quantum state of each particle in a two-dimensional Hilbert space using the traditional "up" and "down" spin orientations, $|\uparrow\rangle_j$ and $|\downarrow\rangle_j$, where $j = 1$ and 2 for the two particles. The four *Bell states* for the composite system of two particles are denoted as

$$|\Phi^\pm\rangle = \frac{1}{\sqrt{2}}(|\uparrow\rangle_1|\uparrow\rangle_2\pm|\downarrow\rangle_1|\downarrow\rangle_2) \text{ and } |\Psi^\pm\rangle = \frac{1}{\sqrt{2}}(|\uparrow\rangle_1|\downarrow\rangle_2\pm|\downarrow\rangle_1|\uparrow\rangle_2). \quad (2.34)$$

Clearly none of them can be written as a product of single-particle spin states in the form $|\xi_1\rangle|\xi_2\rangle$. Both $|\Phi^\pm\rangle$ and $|\Psi^\pm\rangle$ are defined as a superposition of the two single-particle product states with a well-defined phase relation between them. For example, $|\Psi^\pm\rangle$ shows that if the spin of the first particle points up, the spin of the second one points down (and vice versa). More generally, states of the form $|\psi\rangle = a|\uparrow\rangle_1|\uparrow\rangle_2 + b|\downarrow\rangle_1|\downarrow\rangle_2$ are entangled for any two complex numbers a and b such that $|a|^2 + |b|^2 = 1$.

To further clarify the entangled state, consider a pair of spin-$\frac{1}{2}$ particles described by the Bell state $|\Psi^+\rangle$ as defined above. If we measure the state of one particle and find it in a specific state, say $|\uparrow\rangle_1$, we can immediately infer that the second particle, upon measurement, would always be found in the state $|\downarrow\rangle_2$. This is referred to as quantum entanglement and it appears similar to classical correlations. The crux lies in the fact that, unlike the classical case, a quantum state can only provide a probabilistic outcome with no definite answer. A measurement must be made for a definite answer.

It appears from the preceding discussion that information has been exchanged instantaneously between the two particles, as soon as the measurement is performed, regardless of the distance between them. This observation is the basis of the *Einstein–Podolsky–Rosen* (EPR) paradox [109]. According to this paradox, exploiting entanglement, an observer can make measurements on system A to make precise statements about system B that may be located very far away. It tries to answer the question whether a quantum-mechanical description of physical reality can be considered complete. We consider a theory to be complete if every element of physical reality can be mapped uniquely to it. That means we should observe exactly what is happening in the physical world, and the process of measuring should not distort it (i.e., measurement outcomes must be independent of the measurement process). In 1964, Bell investigated the EPR conclusion that the quantum description of physical reality is not complete by using it as a working hypothesis and quantified the EPR idea of a deterministic world [108, 110]. In a classical world, we would expect that measurement outcomes are independent of the measurement process, and the results obtained at one location are independent of any actions performed at distances where information cannot be exchanged even at the speed of light. Recent experiments show that quantum mechanics does properly predict the results of experiments that violate the EPR criteria of reality and locality [111, 112].

2.2.2 Eigenvalues and Eigenstates

QM Postulate 2 *Every measurable physical quantity is associated with a Hermitian operator and is observable only through a measurement of this operator's eigenvalues. When a certain eigenvalue is observed, the act of measurement changes (or collapses) the quantum state to the corresponding eigenstate of this operator.*

This postulate assigns a Hermitian operator to every measurement on a quantum system. Suppose the eigenstates of the operator $\hat{O}$ are given by the set $|\phi_j\rangle$ with the eigenvalues o_j such that $\hat{O}|\phi_j\rangle = o_j|\phi_j\rangle$), where j labels different states. For Hermitian operators ($\hat{O} = \hat{O}^\dagger$), the eigenvalues are always real, and j either takes discrete values or falls in a range of continuous values.

Discrete eigenspectra: When a quantum system is in the state $|\Psi\rangle$ (normalized such that $\langle\Psi|\Psi\rangle = 1$), the probability $\Pr(o_j)$ of measuring one of the nondegenerate eigenvalues o_j of the operator is given by

$$\Pr(o_j) = |\langle\phi_j|\Psi\rangle|^2 = |o_j|^2. \tag{2.35}$$

Here $o_j = \langle\phi_j|\Psi\rangle$ is the component of $|\Psi\rangle$ when projected onto the eigenstate vector $|\phi_j\rangle$. When eigenvalues are distinct and nondegenerate, the corresponding eigenstates form an orthonormal set. The situation changes if a specific eigenvalue o_j is k-fold degenerate. In this case, as there are k eigenstates giving the same eigenvalue o_j, we need to account for all of them when its probability is calculated. It turns out that one can always find linear combinations of the k eigenvectors that form an orthonormal set. If these vectors are given by $|\phi_j^{(i)}\rangle$, where $i = 1, 2, \ldots, k$, the probability of measuring o_j becomes

$$\Pr(o_j) = \sum_{i=1}^{k} |\langle \phi_j^{(i)}|\Psi\rangle|^2. \tag{2.36}$$

As many different orthonormal sets can exist for a given degenerate eigenvalue, one may wonder whether different choices would lead to different probabilities. However, it is possible to show that the probability is independent of the choice of the orthonormal set used to calculate it, as it should be on the physical grounds. Thus, if an operator is already in an eigenstate, then a measurement of that operator yields with certainty the corresponding eigenvalue. However, there is no prescription as to which eigenvalue of an observable operator would be observed with certainty. Instead, one can only specify the probability of observing that eigenvalue.

Continuous eigenspectra: If the eigenvalues o_c of the operator $\hat{O}$ are continuous with the eigenfunctions $|o_c\rangle$, we define the probability $d\Pr(o_c)$ that the measurement yields a value between o_c and $o_c + do_c$ as

$$d\Pr(o_c) = |\langle o_c|\Psi\rangle|^2 \, do_c. \tag{2.37}$$

It is also possible that an operator has some eigenvalues discrete and some eigenvalues continuous. This case can be handled by noting that both the discrete and continuous eigenstates can be recast in such a way that they form a basis for representing the operator. Such cases will be considered when we look at device models in later chapters.

QM Postulate 2 forms the basis for measuring a quantum system [113]. Quantum measurements on any system are performed by a collection of operators: $\mathcal{O}_m$ with $m = 1$ to M. These operators act on the Hilbert space associated with a specific system. As all measurements constitute a complete set of observable values, the underlying operators satisfy the completeness relation $\sum_{m \in M} \mathcal{O}_m^\dagger \mathcal{O}_m = I$. The situation before a measurement is called the *preparation*. If the system is prepared in the state $|\Psi\rangle$, then the probability of measuring the value for $\hat{o}_m$ is given by $\Pr(m) = \langle \Psi|\mathcal{O}_m^\dagger \mathcal{O}_m|\Psi\rangle$. Immediately after the measurement, the quantum state will change to a *postmeasurement* state $|\Psi_{pm}\rangle = \eta \mathcal{O}_m|\Psi\rangle$, where η is a complex number. The normalization condition, $\langle \Psi_{pm}|\Psi_{pm}\rangle = 1$, requires $\langle \Psi|\mathcal{O}_m^\dagger \mathcal{O}_m|\Psi\rangle |\eta|^2 = 1$. As we have seen before, global phase factors can be safely ignored. Thus, η can be specified using its magnitude. Thus, the prepared state $|\Psi\rangle$ collapses to a postmeasurement state given by

$$|\Psi_{pm}\rangle = |\eta|\mathcal{O}_m|\Psi\rangle = \frac{\mathcal{O}_m|\Psi\rangle}{\langle \Psi|\mathcal{O}_m^\dagger \mathcal{O}_m|\Psi\rangle^{1/2}}. \tag{2.38}$$

Two special types of measurements are widely used [114]: *projective measurements* (PM) and *positive operator-valued measurements* (POVM). The projective measurements rely on the eigenvalues of the operator to construct projective measurement operators. Without loss of generality, consider an operator $\mathcal{O}$ with nondegenerate discrete eigenstate $|o_m\rangle$, corresponding to the eigenvalue o_m. As the set of these eigenstates forms a complete orthonormal basis for the operator, we can represent the observable operator using its eigenstates and eigenvalues as $\mathcal{O} = \sum_m o_m |o_m\rangle \langle o_m|$. The projection operator for the mth eigenvalue is given by $|o_m\rangle \langle o_m|$. This means the observable operator is represented using projection operators, which form the basis for this measurement method. There are some nice properties of this representation. For example, we can calculate the average observable value of the operator $\mathcal{O}$ with the knowledge that only one of these eigenstates is observable. Consider the state $|\Psi_{pm}\rangle$. After noting $\mathcal{O}_m = |o_m\rangle \langle o_m|$, we can immediately find that the collapsed state is exactly an eigenstate such that $|\Psi_{pm}\rangle = |o_m\rangle$. Therefore the average observed value of the operator $\langle \mathcal{O} \rangle$ is the sum $\sum_m o_m \Pr(o_m)$, which can be written as

$$\langle \mathcal{O} \rangle = \sum_m o_m \langle \Psi | o_m \rangle \langle o_m | \Psi \rangle = \langle \Psi | \mathcal{O} | \Psi \rangle. \tag{2.39}$$

This way of calculating the average of an operator can be readily extended to any positive integer power of the operator by raising the corresponding eigenvalues to the same power. Using a Taylor's series to expand any function of the operator to polynomial powers, this recipe can also be used to calculate the average value of any functional of the operator.

The preceding procedure for the projection operators can be extended to describe the POVM scheme as well. Here, rather than specific projective operators, we consider a finite set of observable operators $\mathcal{S}_m$ defined using $\langle \Psi | \mathcal{S}_m | \Psi \rangle \geq 0$. If these operators also form a complete set, $\sum_m \mathcal{S}_m = I$. The condition, $\langle \Psi | \mathcal{S}_m | \Psi \rangle \geq 0$, for each operator ensures that their eigenvalues are always real and positive. Therefore, it is possible to define the operators $M_m = \sqrt{\mathcal{S}_m}$, which satisfy all the conditions of the original set of operators $\mathcal{S}_m$. Owing to the completeness property of this set, we obtain the analogous relation: $\sum_m M_m^\dagger M_m = I$. As this set of newly defined operators has properties similar to those of projective operators, the same measurement procedure can be adopted to this case (see Aside 2.10).

Aside 2.10   Quantum Measurement of a Qubit [115]

Classical information bits are represented by logic gates and take values 0 or 1 for their off and on states. We can designate these two possibilities as the $|0\rangle$ and $|1\rangle$ states, respectively. A quantum bit (or a qubit) is a generalization of this concept. The state of a qubit is a superposition state, $\alpha |0\rangle + \beta |1\rangle$, where α and β are two complex numbers such that $|\alpha|^2 + |\beta|^2 = 1$. They are called the qubit amplitudes, and the associated orthonormal states $|0\rangle$ and $|1\rangle$ are called the computational basis.

PM: Consider two measurements governed by the operators $\mathcal{O}_0 = |0\rangle \langle 0|$ and $\mathcal{O}_1 = |1\rangle \langle 1|$. They form a complete set because $\mathcal{O}_0^\dagger \mathcal{O}_0 + \mathcal{O}_1^\dagger \mathcal{O}_1 = I$. Suppose the qubit is currently in the state $|\Psi_1\rangle = \frac{1}{\sqrt{2}}(|0\rangle + |1\rangle)$. Then the probability of getting the state $|0\rangle$ is given by $\Pr(0) = \langle \Psi | \mathcal{O}_0^\dagger \mathcal{O}_0 | \Psi \rangle = \frac{1}{2}$. After the measurement, the prepared state, $|\Psi\rangle$, will collapse to the state $|0\rangle$. Once the qubit is in this state, if we perform the measurement again, the qubit remains in the same state. This property of projective measurements is known as repeatability. Essentially, repeated projected measurements do not change the state. The same conclusion holds for the state 1 as well.

POVM: Consider a POVM containing three measurement operators:

$$M_1 = \frac{\sqrt{2}}{1+\sqrt{2}} |1\rangle \langle 1|, \quad M_2 = \frac{1}{2+\sqrt{2}} (|0\rangle - |1\rangle)(\langle 0| - \langle 1|), \quad M_3 = I - M_2 - M_3.$$

(2.40)

These three operators are positive and satisfy the completeness relation $\sum_{m=1}^{3} M_m^{\dagger} M_m = I$. Thus, they are eligible as POVM operators.

Assume that a qubit is prepared such that either $|\Psi\rangle = |0\rangle$ or $|\Psi\rangle = |1\rangle$. We want to determine the exact state of the qubit by carrying out a measurement. If we calculate all the probabilities, we find that only $\langle 0|M_1|0\rangle = \langle 1|M_2|0\rangle = 0$, giving us the definitive information that determines the state of the system. The operator M_3 fails to provide this information, as it gives nonzero probabilities for both measurements.

Aside 2.11   Uncertainty principle [46, 96]

Uncertainty and errors are intrinsic to physical measurements. In his book *Philosophæ Naturalis Principia Mathematica* [116], even Isaac Newton found it necessary to take into account measurement errors. In the context of modern physics, Richard Feynman considered in his "lectures on physics" the implications of the uncertainty principle for electrons through the Young's double-slit experiment by demonstrating the impossibility of determining the path of an electron without disturbing the interference pattern.

A popular version of the uncertainty principle states: one cannot simultaneously know both the position and momentum of a moving particle. This is often interpreted as implying that it is impossible to simultaneously measure the position and momentum of a particle with an arbitrarily high precision. This statement does not mean that a particle cannot possess a definite position as well as a definite momentum, but only that the uncertainty principle prevents their simultaneous measurements, even if we work with a perfect measuring apparatus. Therefore, one must consider the standard deviation arising from an ensemble of similar measurements.

The most generic version of the uncertainty principle is based on properties of observable operators. Consider two observable operators A and B. The Cauchy–Schwarz inequality [117] implies that

$$|\langle \Psi|AB|\Psi\rangle|^2 \leq \langle \Psi|A^2|\Psi\rangle \langle \Psi|B^2|\Psi\rangle,$$

(2.41)

where $|\Psi\rangle$ is a quantum state of the system under consideration. If we define the *commutator* for these operators as $[A, B] = AB - BA$, and the anticommutator as $[A, B]_+ = AB + BA$, we can write the identity

$$|\langle \Psi|AB|\Psi\rangle|^2 = \frac{1}{4}|\langle \Psi|[A, B]|\Psi\rangle|^2 + \frac{1}{4}|\langle \Psi|[A, B]_+|\Psi\rangle|^2.$$

(2.42)

If we combine Eqs (2.41) and (2.42), we obtain

$$\langle \Psi|A^2|\Psi\rangle \langle \Psi|B^2|\Psi\rangle \geq \frac{1}{4}|\langle \Psi|[A, B]|\Psi\rangle|^2.$$

(2.43)

This is the uncertainty principle in its most general form. It can be recast as a condition on the standard deviations of the measurements on the operators A and B by replacing them with the operators $A - \langle A \rangle$ and $B - \langle B \rangle$ in the Eq. (2.43).

2.2.3 Density Operator

An alternative description of quantum states makes use of the *density operator*, or its representation in the form of a *density matrix* [118]. The utility of this operator comes from its property that it can handle the *pure* states, as well as *mixed* states for which the state of a system is known only in a probabilistic sense among several possible states.

For a pure state $|\Psi\rangle$ with $\langle \Psi | \Psi \rangle = 1$, the density operator ρ is defined as

$$\rho = |\Psi\rangle\langle\Psi| \quad \text{(for a pure state).} \tag{2.44}$$

One property of this density operator is that $\mathrm{Tr}(\rho) = 1$ when the pure state is normalized properly. In an orthonormal basis based on the states $\{|n\rangle\}$, the density operator becomes a density matrix with the elements $\rho_{mn} = \langle m|\rho|n\rangle$. The trace operation $\mathrm{Tr}(\rho)$ is then equal to $\sum_n \rho_{nn}$, or the sum of all diagonal elements. It is easy to show that the density matrix of a pure state is *Hermitian* ($\rho^\dagger = \rho$), *idempotent* ($\rho^2 = \rho$), and *positive definite* (i.e., $\langle\phi|\rho|\phi\rangle \geq 0$ for any state $|\phi\rangle$).

The preceding definition of a density operator concept can be extended to a common scenario where a quantum system could be in a number of pure states, but we only know the probability of finding the system in each state. If the system can be found in the pure state $|\Psi_m\rangle$ with the probability p_m, where m identifies one among all possible quantum states, the density operator of the system is defined as

$$\rho = \sum_m p_m |\Psi_m\rangle\langle\Psi_m| \quad \text{(for a mixed state).} \tag{2.45}$$

A good example is provided by the Stern–Gerlach device [119], where we do not know the exact state of each spin $\frac{1}{2}$ particle, but the physics of the problem tells us that the spin may point either "up" or "down" with equal probability. The density operator of a mixed state is also Hermitian, positive definite, and satisfies $\mathrm{Tr}(\rho) = 1$. However, it fails to be idempotent because $\rho^2 \neq \rho$ and $\mathrm{Tr}(\rho^2) < 1$. Aside 2.12 provides examples of the density operators in the pure and mixed states.

Aside 2.12   Density Operators of Pure and Mixed States [118]

Consider a two-level quantum system with pure states $|0\rangle$ and $|1\rangle$. Consider a new pure state, $|\Psi\rangle$, that is a superposition of these two states (i.e., $|\Psi\rangle = a|0\rangle + b|1\rangle$). The normalization condition, $\langle\Psi|\Psi\rangle = 1$, demands that the complex numbers a and b satisfy the relation $|a|^2 + |b|^2 = 1$. The density operator for this pure state case is given by

$$\rho = |\Psi\rangle\langle\Psi| = |a|^2 |0\rangle\langle0| + ab^* |0\rangle\langle1| + a^*b |1\rangle\langle0| + |b|^2 |1\rangle\langle1|. \tag{2.46}$$

It is common to introduce a column vector V containing two elements $|0\rangle$ and $|1\rangle$, and write the preceding equation as $\rho = V\bar{\rho}V^\dagger$, where the density matrix $\bar{\rho}$ is given by

$$\bar{\rho} = \begin{bmatrix} |a|^2 & ab^* \\ a^*b & |b|^2 \end{bmatrix}. \tag{2.47}$$

However, if we create a mixed state using the pure states $|0\rangle$ and $|1\rangle$, its density matrix takes the form

$$\rho = |a|^2 |0\rangle \langle 0| + |b|^2 |1\rangle \langle 1| \;\rightarrow\; \bar{\rho} = \begin{bmatrix} |a|^2 & 0 \\ 0 & |b|^2 \end{bmatrix}. \tag{2.48}$$

The interference terms, or the off-diagonal terms, are absent in this density matrix.

As we have seen in Aside 2.12, the density matrix of a quantum system clearly conveys the difference between the pure and mixed states. In a mixed state, probabilities are added very much like in a classical system, while in a pure state probability amplitudes are superimposed, resulting in the nondiagonal elements of a density matrix. The transition from a pure state to a mixed state can be observed through the loss of nondiagonal elements in a density matrix. Also, it is important to realize that a mixed state is different from a pure-state superposition of the form $|\Psi\rangle = \sum_m p_m |\Psi_m\rangle$, because all states making up this superposition are simultaneously present. The probabilities appearing in the coefficients are very much the deterministic coefficients making up this superposition. The density matrix for this state has the form

$$\rho = |\Psi\rangle \langle \Psi| = \sum_m p_m |\Psi_m\rangle \langle \Psi_m| + \sum_{m \neq n} \sqrt{p_m p_n} |\Psi_m\rangle \langle \Psi_n|. \tag{2.49}$$

The presence of interference terms in the preceding equation indicates that this is a pure state (and not a mixed state).

It is possible to quantify the purity of a quantum state by introducing the concept of *von Neumann entropy* $S(\rho)$, defined as [120]

$$S(\rho) = -\,\mathrm{Tr}(\rho \log_2 \rho) = -\sum_m \lambda_m \log_2 \lambda_m, \tag{2.50}$$

where λ_m are the eigenvalues of ρ. If a eigenvalue vanishes, we drop that eigenvalue from the sum, just as is done in classical information theory. The von Neumann entropy is zero for pure states because the associated projection operator has $\lambda_1 = 1$ as the only nonzero eigenvalue. However, for a maximally mixed state that corresponds to complete ignorance about the N mutually exclusive pure states, $\lambda_m = 1/N$ in Eq. (2.50), resulting in $S(\rho) = \log_2 N$. This is the maximum value that $S(\rho)$ can take. Any other scenario would have a value for $S(\rho)$ between 0 and $\log_2 N$. The von Neumann entropy is essentially a quantum analog of the well-known *Shannon entropy* used in communication theory.

Another method sometimes used for quantifying the purity of a state is to calculate the purity function $\mathcal{P}(\rho)$, defined as [121]

$$\mathcal{P}(\rho) = \mathrm{Tr}(\rho^2) = \sum_m \lambda_m^2, \tag{2.51}$$

where λ_m are the eigenvalues of ρ. This function has a value between 0 and 1 for any density operator. It is exactly equal to 1 for a pure state and provides a convenient way to detect the purity of a state without much calculations.

The density operator provides an easy way to calculate the average value of an operator, regardless of the pure or mixed characters of the states involved. Unlike the pure-state case where one can use the operation, $\langle \mathcal{O} \rangle = \langle \Psi | \mathcal{O} | \Psi \rangle$, to find the average of an operator $\mathcal{O}$, no single $|\Psi\rangle$ exists for a mixed state. However, it is possible to generalize the averaging concept by adopting ideas from the classical probability theory. The modified procedure consists of simply weighting the expectation values $\langle \Psi_m | \mathcal{O} | \Psi_m \rangle$ for each of the pure states $|\Psi_m\rangle$ contained in the mixed state by their respective classical probabilities p_m and sum the results over the entire ensemble to obtain

$$\langle \mathcal{O} \rangle = \sum_m p_m \langle \Psi_m | \mathcal{O} | \Psi_m \rangle = \mathrm{Tr}(\rho \mathcal{O}). \tag{2.52}$$

The result $\langle \mathcal{O} \rangle = Tr(\rho \mathcal{O})$ also holds for pure states. Thus, the density operator provides us with a unified way to calculate the average value of an operator.

The utility of the density operator becomes apparent especially when one needs the description of individual subsystems of a composite quantum device. The density operator of the composite device "AB" made by combining the subsystems "A" and "B" is provided by the tensor product $\rho_{AB} = \rho_A \otimes \rho_B$. We can recover the individual components, ρ_A and ρ_B, from ρ_{AB} by using $\rho_A = \mathrm{Tr}_B(\rho_{AB})$, where $\mathrm{Tr}_B(\cdots)$ is a partial trace operator over the subsystem B. When taking the partial trace, the probability amplitudes of B vanish given that $\mathrm{Tr}(\rho) = 1$ is true for any density operator. This procedure is called *tracing out* over a subsystem. An interesting application of this "tracing out" operation is *purification* in which the partial trace operation is used on a mixed state to generate a pure state. Suppose ρ_A is a density operator in the Hilbert space $\mathcal{H}_A$ with N dimensions. Then, it is possible to find a Hilbert space $\mathcal{H}_B$ and a pure state $|\Psi\rangle \in \mathcal{H}_A \otimes \mathcal{H}_B$ such that the partial trace over $|\Psi\rangle$ results in ρ_A (i.e., $\mathrm{Tr}_B(|\Psi\rangle \langle \Psi|) = \rho_A$).

In the case of multiple subsystems, if each one of them has the density matrix ρ_n ($n = 1$ to N), we can write the density matrix of the composite system as $\rho_C = \rho_1 \otimes \rho_2 \ldots \otimes \rho_N$. If all subsystems are identical but independent from each other, we can write the composite density matrix as $\rho_C = \rho^{\otimes N}$.

2.2.4 Canonical Quantization

We ask how quantum mechanics can be used for a physical device requiring quantum description. This seems a difficult task because quantum-mechanics results are not always intuitive or close to our day-to-day experience. The recipe comes from a procedure called *canonical quantization*, which provides a mathematical mapping from a classical system to the corresponding quantum system [122, 123]. However, "quantization" is not a well-posed problem, and there is a plethora of techniques developed over time. Here, we sketch one of the simplest recipes.

The process begins by defining a Hilbert space $\mathcal{H}$ associated with the system. Quantum states of the system are represented by state vectors, and observables by Hermitian operators, in this Hilbert space. The energy of the system is then expressed as a Hamiltonian, which is found using classical arguments with the aid of generalized coordinates and momenta ($\mathbf{q}$ and $\mathbf{p}$). The next step maps all $\mathbf{q}$ and $\mathbf{p}$ to operators as (a hat over these variables denotes their operator character)

$$\mathbf{q} \to \hat{\mathbf{q}}; \qquad \mathbf{p} \to \hat{\mathbf{p}} = -i\hbar\frac{\partial}{\partial \mathbf{q}}. \tag{2.53}$$

We also convert classical Poisson brackets to operator commutators (Lie brackets of quantum observables) as follows (more accurate versions of these relations are called Weyl relations) [122, 124]:

$$[\hat{q}_m, \hat{p}_n] = i\hbar\delta_{mn}, \quad [\hat{q}_m, \hat{q}_n] = 0, \quad [\hat{p}_m, \hat{p}_n] = 0. \tag{2.54}$$

There are many subtle ambiguities in this conversion process. For example, classically identical terms such as $p_m q_n^3$ and $q_n^3 p_m$ are not the same in their operator versions because these two operators are not Hermitian. One way of overcoming this problem is by replacing each one of them with the symmetric form $\frac{1}{2}(\hat{p}_m \hat{q}_n^3 + \hat{q}_n^3 \hat{p}_m)$. Many other ambiguous steps need to be tackled using arguments based on the axioms of quantum mechanics. The most crucial and nontrivial step is the introduction of the numerically small but nonzero parameter $\hbar = h/(2\pi)$, where the Planck constant h is a hallmark of all quantum systems [125].

Once the Hamiltonian has been constructed using the preceding recipe, it becomes possible to invoke the machinery of quantum mechanics for fully describing a quantum device. Interestingly, the results obtained using classical mechanics nearly hold in most cases, except for small $\hbar$-dependent corrections. This is the basis for the *correspondence principle*, which states that quantum mechanics approaches the classical description if $\hbar$-dependent contributions are negligibly small. Aside 2.13 shows an application of this principle using a classical pendulum as an example.

One may ask why the Hamiltonian is preferred over a Lagrangian in quantum mechanics. In the Lagrangian formalism, the operators $\mathbf{q}$ and $\dot{\mathbf{q}}$ do not commute and the Hamiltonian defined by the Legendre transformation is not uniquely determined. Moreover, it is considerably harder to set up equations of motion using $\frac{\partial}{\partial \mathbf{q}}$ and $\frac{\partial}{\partial \dot{\mathbf{q}}}$ when $\mathbf{q}$ and $\dot{\mathbf{q}}$ are operators. These difficulties are remedied by adopting the Hamiltonian, which provides us a simple way to set up the equation of motion for any function χ of $\mathbf{p}$ and $\mathbf{q}$ variables as $i\hbar\dot{\chi} = [\chi, H]$. In this formalism, conserved quantities are found by setting $\dot{\chi} = 0$, or by using $[\chi, H] = 0$.

Aside 2.13   Canonical Quantization Example

Consider a pendulum made of mass m, hung from a rigid rod of length l. The pendulum traces a circle as it rotates in the vertical plane. If we use polar coordinates, a single generalized coordinate θ, representing the angle from the vertical, can describe the pendulum motion. The first step is to calculate the kinetic and potential energies classically when the pendulum is located at an angle θ. Noting that the position $x = l\theta$, the kinetic energy is found to be $T = \frac{1}{2}ml^2\dot{\theta}^2$. The potential energy at this position is $V = mgl(1 - \cos\theta)$. The Lagrangian for this pendulum is given by

$$\mathcal{L}(\theta, \dot{\theta}) = T - V = \frac{1}{2}ml^2\dot{\theta}^2 - mgl(1 - \cos\theta). \tag{2.55}$$

The canonical momentum associated with the angle θ is given by $p_\theta = \frac{\partial \mathcal{L}}{\partial \dot{\theta}} = ml^2\dot{\theta}$. Using this result, the Hamiltonian H can be written as

$$H(p_\theta, \theta) = p_\theta\dot{\theta} - \mathcal{L}(\theta, \dot{\theta}) = \frac{p_\theta^2}{2ml^2} + mgl(1 - \cos\theta). \tag{2.56}$$

For a quantum description of this pendulum we map θ and p_θ to corresponding operators, $\hat{\theta}$ and $\hat{p}_\theta$, and impose the commutation relation $[\hat{\theta}, \hat{p}_\theta] = i\hbar$. If we assume that the angle θ remains relatively small in practice and use the series expansion for $\cos\theta$, the Hamiltonian can be written as $\hat{H} = \hat{H}_0 + \delta\hat{H}$, where

$$\hat{H}_0 = \frac{\hat{p}_\theta^2}{2ml^2} + \frac{mgl}{2}\hat{\theta}^2, \tag{2.57}$$

and $\delta\hat{H}$ contains the higher-order terms. By dividing the Hamiltonian into a harmonic part $\hat{H}_0$ and a nonharmonic part $\delta\hat{H}$, we can use the known techniques to treat each part separately. In particular, the nonharmonic part can be treated using the perturbation techniques discussed later.

Inspired by the quantization of harmonic oscillators, we introduce the creation operator $a^\dagger$ and the annihilation operator a using the relations

$$\hat{\theta} = \sqrt{\frac{\hbar}{2b}}(a^\dagger + a) \text{ and } \hat{p}_\theta = i\sqrt{\frac{\hbar b}{2}}(a^\dagger - a). \tag{2.58}$$

It is easy to verify that $[a, a^\dagger] = 1$ provides the correct commutator relation $[\hat{\theta}, \hat{p}_\theta] = i\hbar$. The parameter b is found using $\hat{H}_0$. Substitution of the preceding variables in $\hat{H}_0$ shows that, if we choose $b = mg\sqrt{gl}$, $\hat{H}_0$ can be written in the form

$$\hat{H}_0 = \hbar\omega\left(a^\dagger a + \frac{1}{2}\right), \tag{2.59}$$

where we used the classical expression $\omega = \sqrt{g/l}$ for the angular frequency of a harmonic oscillator. It is common to introduce $\hat{n} = a^\dagger a$ as the number operator and write $\hat{H}_0$ as $\hat{H}_0 = \hbar\omega(\hat{n} + \frac{1}{2})$, enabling us to use the full machinery of quantum mechanics.

2.2.5 Time Evolution of a State Vector

QM Postulate 3   *The evolution of the state vector of a closed quantum system from its initial value $|\Psi(t_0)\rangle$ at time $t = t_0$ is governed by a unitary transformation $U(t, t_0)$ such that the state vector at time t is $|\Psi(t)\rangle = U(t, t_0)|\Psi(t_0)\rangle$.*

The unitary operator $U(t, t_0)$ governs the entire dynamics of a quantum system, subject to the restriction that the system is closed (i.e., it is not interacting with any other system).

No quantum system is really closed if it is interacting with its surroundings. However, it is possible to isolate a system to the extent that, to a very good approximation, it can be viewed as a closed system. It is also possible to consider it a part of a larger closed system that is undergoing unitary evolution.

There are several representations (sometimes called *pictures*) of the unitary operators in quantum mechanics [46, 95, 96]. It is possible to establish connections among them by exploiting properties of unitary transformations. Three commonly used representations are known as the Schrödinger picture, the Heisenberg picture, and the interaction picture. The Schrödinger picture is mainly used when the Hamiltonian describing a quantum system does not depend on time. The other two pictures have features that make them attractive for time-dependent Hamiltonians. The Heisenberg picture is more closely related to classical mechanics, where the dynamical variables (which become operators in quantum mechanics) evolve in time. The interaction picture is the preferred choice when quantum particles are interacting with an electromagnetic field, and the Hamiltonian is explicitly time dependent.

The Schrödinger Picture

The Schrödinger picture is preferred when dealing with a Hamiltonian with no explicit time dependence ($\partial H / \partial t = 0$). In this picture, quantum states evolve with time but the operators do not change with time. The evolution of the quantum state $|\Psi(t)\rangle$ is governed by the Schrödinger equation

$$i\hbar \frac{d\,|\Psi(t)\rangle}{dt} = H\,|\Psi(t)\rangle . \tag{2.60}$$

This equation has the solution $|\Psi(t)\rangle = \exp[-\frac{i}{\hbar}H(t-t_0)]\,|\Psi(t_0)\rangle$. It follows that the time evolution is governed by the unitary operator $U(t,t_0) = \exp[-\frac{i}{\hbar}H(t-t_0)]$. It is easy to see that $U(t,t_0)$ satisfies

$$i\hbar \frac{\partial U(t,t_0)}{\partial t} = HU(t,t_0), \tag{2.61}$$

with the initial condition $U(t_0,t_0) = 1$ (i.e., the time-evolution operator itself satisfies the Schrödinger equation). The preceding equation remains valid even for time-dependent Hamiltonians, but it is not easy to solve it in such cases.

The Heisenberg Picture

In the Heisenberg picture, operators evolve in time but the quantum state remains unchanged. However, even though the operators become time dependent, they continue to satisfy the commutation relation such as $[\mathbf{q}(t),\mathbf{p}(t)] = i\hbar$ at all times. The time evolution of any Heisenberg operator $\mathcal{O}(t)$ is governed by

$$i\hbar \frac{d}{dt}\mathcal{O}(t) = \frac{\partial}{\partial t}\mathcal{O}(t) + [\mathcal{O}(t),H], \tag{2.62}$$

where the Hamiltonian $H(\mathbf{q}, \mathbf{p})$ is a function of both $\mathbf{q}$ and $\mathbf{p}$ that vary with time. The Heisenberg picture can be used for all Hamiltonians irrespective of whether or not H depends on time explicitly.

If the Schrödinger and the Heisenberg pictures are used to describe the time evolution of the same quantum device, they should make the same predictions for any measurement made on this device. This requires that the matrix element of any operator between any two quantum states should be identical in the two pictures. Mathematically,

$$\langle \Psi_a(t)|\mathcal{O}(t_o)|\Psi_b(t)\rangle = \langle \Psi_a(t_0)|\mathcal{O}(t)|\Psi_b(t_0)\rangle. \tag{2.63}$$

Using the relation $|\Psi(t)\rangle = U(t, t_0)|\Psi(t_0)\rangle$ valid for any quantum state, the left side of this equation can be written as

$$\langle \Psi_a(t)|\mathcal{O}(t_o)|\Psi_b(t)\rangle = \langle \Psi_a(t_0)|U^\dagger(t, t_0)\mathcal{O}(t_0)U(t, t_0)|\Psi_b(t_0)\rangle, \tag{2.64}$$

where $U(t, t_0) = \exp[-\frac{i}{\hbar}H(t - t_0)]$. A comparison of the two preceding equations leads to the important relation, $\mathcal{O}(t) = U^\dagger(t, t_0)\mathcal{O}(t_0)U(t, t_0)$, that connects the Schrödinger and the Heisenberg pictures.

The Interaction Picture

The interaction picture handles Hamiltonians for which time dependence can be separated out of the main Hamiltonian as a perturbation, resulting in the form

$$H(t) = H_0 + V(t), \tag{2.65}$$

where H_0 is the time-independent part with known eigenstates $|\Psi_0\rangle$ obtained by solving $H|\Psi_0\rangle = E_0|\Psi_0\rangle$. The second part $V(t)$ depends on time but it is relatively small and acts as a perturbation. Often this term represents interaction of a quantum device with an external field.

In the interaction picture, both the quantum state $|\Psi(t)\rangle$ and the operators $\mathcal{O}(t)$ are allowed to change with time as

$$|\widetilde{\Psi}(t)\rangle = \exp\left(\frac{i}{\hbar}H_0 t\right)|\Psi(t)\rangle, \qquad \widetilde{\mathcal{O}}(t) = \exp\left(\frac{i}{\hbar}H_0 t\right)\mathcal{O}(t)\exp\left(-\frac{i}{\hbar}H_0 t\right). \tag{2.66}$$

Substituting the form of $|\Psi(t)\rangle$ in the Schrödinger equation, we obtain

$$i\hbar\frac{d|\widetilde{\Psi}(t)\rangle}{dt} = \widetilde{V}(t)|\widetilde{\Psi}(t)\rangle, \qquad \widetilde{V}(t) = \exp\left(\frac{i}{\hbar}H_0 t\right)V(t)\exp\left(-\frac{i}{\hbar}H_0 t\right). \tag{2.67}$$

This equation is identical to the original Schrödinger equation with $\widetilde{V}$ acting as an effective Hamiltonian. Its solution can be written as

$$|\widetilde{\Psi}(t)\rangle = \widetilde{U}(t, t_0)|\widetilde{\Psi}(t_0)\rangle, \tag{2.68}$$

where $\widetilde{U}(t, t_0)$ is a unitary operator with $\widetilde{U}(t_0, t_0) = 1$. It is found by solving its evolution equation,

$$i\hbar\frac{d\widetilde{U}(t, t_0)}{dt} = \widetilde{V}(t)\widetilde{U}(t, t_0). \tag{2.69}$$

It is not easy to solve Eq. (2.69) because the operator $\tilde{V}$ varies with time. However, the concept of *time ordering* can be used to write its solution as

$$\tilde{U}(t, t_0) = \mathcal{T} \left\{ \exp\left(-\frac{i}{\hbar} \int_{t_0}^{t} d\tau \tilde{V}(\mathcal{T}) \right) \right\}, \tag{2.70}$$

where $\mathcal{T}$ is the time-ordering operator, whose properties are discussed in Aside 2.14. Its presence ensures that the operators are ordered from right to left such that their time arguments go from earlier times to later times. In many applications making use of the interaction picture, the perturbation $\tilde{V}(t)$ is weak compared to H_0. To the first order, $\tilde{U}(t, t_0)$ can then be approximated as

$$\tilde{U}(t, t_0) \approx 1 - \frac{i}{\hbar} \int_{t_0}^{t} d\tau \tilde{V}(\mathcal{T}). \tag{2.71}$$

This simplified form of the time-evolution operator forms the basis for the *Kubo formula* in the linear response theory discussed in Section 3.2.

Aside 2.14   Time Ordering, Normal Ordering, and Wick's Theorem [126]

Time ordering: For a set of time-dependent operators, $\mathcal{A} = \{\mathcal{A}_n \mid n = 1, 2, \ldots, N\}$, the time-ordering operator $\mathcal{T}$ is defined as

$$\mathcal{T} \{\mathcal{A}_1 \mathcal{A}_2 \ldots \mathcal{A}_N\} = \zeta_{AB} \mathcal{B}_1 \mathcal{B}_2 \ldots \mathcal{B}_N, \tag{2.72}$$

where the set $\mathcal{B} = \{\mathcal{B}_n \mid n = 1, 2, \ldots, N\}$ has the same elements as the set $\mathcal{A}$, but they are arranged such that operators with an earlier time appear to the right of the operators with later times. The parameter ζ_{AB} takes values ± 1 and depends not only on the number of permutations but also on whether the particle involved is a fermion or boson. It is defined as

$$\zeta_{AB} = \begin{cases} -1 & \text{for fermions if the number of permutations is odd} \\ +1 & \text{otherwise.} \end{cases} \tag{2.73}$$

Time ordering obeys the distributive property:

$$\mathcal{T} \left\{ \mathcal{A}_1^{(1)} \mathcal{A}_2^{(1)} \ldots \mathcal{A}_N^{(1)} + \mathcal{A}_1^{(2)} \mathcal{A}_2^{(2)} \ldots \mathcal{A}_N^{(2)} \right\} = \mathcal{T} \left\{ \mathcal{A}_1^{(1)} \mathcal{A}_2^{(1)} \ldots \mathcal{A}_N^{(1)} \right\}$$
$$+ \mathcal{T} \left\{ \mathcal{A}_1^{(2)} \mathcal{A}_2^{(2)} \ldots \mathcal{A}_N^{(2)} \right\}. \tag{2.74}$$

Normal ordering: Consider a set of operators, $\mathcal{A} = \{\mathcal{A}_n \mid n = 1, 2, \ldots, N\}$, where each element is an annihilation or creation operator. The normal ordering operator is defined as

$$\mathcal{N} \{\mathcal{A}_1 \mathcal{A}_2 \ldots \mathcal{A}_N\} \equiv \, : \mathcal{A}_1 \mathcal{A}_2 \ldots \mathcal{A}_N := \zeta_{AC} \mathcal{C}_1 \mathcal{C}_2 \ldots \mathcal{C}_N, \tag{2.75}$$

where the set $\mathcal{C} = \{\mathcal{C}_n \mid n = 1, 2, \ldots, N\}$ has the same elements as the set $\mathcal{A}$ but the indices are permuted such that all annihilation operators appear to the right of all creation operators. The parameter ζ_{AC} is defined identical to ζ_{AB}. Normal ordering also obeys the distributive property:

$$\mathcal{N}\left\{\mathcal{A}_1^{(1)}\mathcal{A}_2^{(1)}\dots\mathcal{A}_N^{(1)} + \mathcal{A}_1^{(2)}\mathcal{A}_2^{(2)}\dots\mathcal{A}_N^{(2)}\right\} = \mathcal{N}\left\{\mathcal{A}_1^{(1)}\mathcal{A}_2^{(1)}\dots\mathcal{A}_N^{(1)}\right\}$$
$$+\mathcal{N}\left\{\mathcal{A}_1^{(2)}\mathcal{A}_2^{(2)}\dots\mathcal{A}_N^{(2)}\right\}. \tag{2.76}$$

Wick's theorem: Wick's theorem [126, 127] provides a way to translate from time to normal ordering. Time ordering only takes into account the time stamp of the operators. In contrast, normal ordering only takes into account the creation or annihilation character of the operators, regardless of the time stamp attached to those operators. Wick's theorem states

$$\mathcal{T}\{\mathcal{A}_1\mathcal{A}_2\dots\mathcal{A}_N\} = \mathcal{N}\left\{\mathcal{A}_1\mathcal{A}_2\dots\mathcal{A}_N\right\} + \mathcal{N}\left\{_1\mathcal{A}_1\mathcal{A}_2\mathcal{A}_3\dots\mathcal{A}_N\right\} + \dots, \tag{2.77}$$

where all possible contractions are to be made. The contraction between two operators A and B is defined as $\overline{AB} = \langle 0|\mathcal{T}\{AB\}|0\rangle$, where $|0\rangle$ denotes the vacuum state.

2.2.6 Perturbation Theory

There is a formal way to make approximations in quantum mechanics known as perturbation theory, and it is widely used in modeling quantum devices and for estimating their performance. The technique is somewhat different depending on whether the Hamiltonian depends explicitly on time or not.

We consider here the time-independent perturbation theory [96]. It expresses the quantum states of a perturbed Hamiltonian, $H_0 + V$, using the eigenstates and energies of its nonperturbed part H_0 that are found by solving $H_0|\Psi_0\rangle = E|\Psi_0\rangle$. It is possible to find the perturbed quantum state $|\Psi\rangle$ of the full Hamiltonian with the same energy E using

$$|\Psi\rangle = |\Psi_0\rangle + K(E)V|\Psi\rangle, \tag{2.78}$$

where the operator $K(E) = (E - H_0)^{-1}$ is a type of Green's function known as the Lippmann–Schwinger kernel [128]. Noting that $\det(E - H_0) = 0$ (because E is an eigenvalue of H_0), the kernel K is singular. This singularity is eliminated by replacing E with $E + i\epsilon$, where ϵ is an infinitesimally small real number such that

$$K(E) = \lim_{\epsilon \to 0} \frac{1}{E - H_0 + i\epsilon}. \tag{2.79}$$

As our goal is to express $|\Psi\rangle$ in terms of $|\Psi_0\rangle$, we need to find a way to eliminate $|\Psi\rangle$ from the right side of Eq. (2.78). This is done by defining the transfer matrix operator T such that $V|\Psi\rangle = T|\Psi_0\rangle$. Using it in Eq. (2.78) and simplifying we obtain a self-consistent equation $T = V + VK(E + i\epsilon)T$, which can be solved iteratively to get

$$T = V + VK(E + i\epsilon)V + VK(E + i\epsilon)VK(E + i\epsilon)V + \dots. \tag{2.80}$$

When the eigenfunctions of H_0 form a complete and orthonormal set, it is possible to find matrix elements of the operator T, which essentially represent transition probabilities between any two eigenstates. Suppose $\{|\Psi_{0j}\rangle\}$ is the complete set of orthonormal eigenfunctions of the unperturbed Hamiltonian H_0 such that $H_0\Psi_{0j} = E_j\Psi_{0j}$. The matrix elements $|\Psi_{0m}\rangle$ and $|\Psi_{0n}\rangle$ can be calculated from Eq. (2.80) as

$$\langle\Psi_{0m}|T|\Psi_{0n}\rangle = \langle\Psi_{0m}|V|\Psi_{0n}\rangle + \sum_j \langle\Psi_{0m}|V|\Psi_{0j}\rangle\, K(E_j + i\epsilon)\, \langle\Psi_{0j}|V|\Psi_{0n}\rangle + \ldots . \quad (2.81)$$

This equation is known as the *Born series for T-matrix* [129]. The first three terms in this series are referred to as the first, second, and third Born approximations [96, 130]. The structure of each term provides insight into the underlying physical process. A nonzero V_{jm} creates an intermediate state $|\Psi_{0j}\rangle$, which lasts until it connects to another state $|\Psi_{0k}\rangle$ by the Lippmann–Schwinger kernel, and so on. The whole process starts at the initial state and continues until the kernel K hits the final state. The *Feynman diagrams* are traditionally used to represent such a process [131].

Equation (2.81) can be used to calculate the transition rate Γ_{mn} between any two stationary states using the generalized Fermi's golden rule given by [132]

$$\Gamma_{mn} = \frac{2\pi}{\hbar}|\langle\Psi_{0m}|T|\Psi_{0n}\rangle|^2\delta(E_m - E_n). \quad (2.82)$$

Here we have used Fermi's rule in the context of the T matrix. The original version, referred to as *Fermi's golden rule*, differs from it by the substitution $T \to V$.

2.3 Quantization of Electromagnetic Fields

The design and analysis of quantum devices inevitably require consideration of their interaction with one or more external fields. Electromagnetic fields are the most widely manipulated fields in such devices. In this section, we consider how electromagnetic interactions are handled for nanoscale quantum devices. As all physical laws do not depend on the motion of observers in different inertial frames, *Lorentz invariance* plays a critical role, requiring that Maxwell's equations be invariant under Lorentz transformations [94, 98]; see Aside 2.15. We start with the standard Maxwell's equations and recast them in the covariant form. This approach provides us a way to present two different formulations found in literature. Familiarity with both formulations is necessary for modeling nanoscale quantum devices.

2.3.1 Maxwell's Equations and Gauge Invariance

In the Système Internationale (SI) units adopted in this book, Maxwell's equations can be written as [133]

$$\nabla \cdot \mathbf{B} = 0, \qquad\qquad \nabla \times \mathbf{E} + \frac{\partial\mathbf{B}}{\partial t} = 0, \qquad\qquad (2.83)$$

$$\nabla \cdot \mathbf{E} = \rho/\epsilon_0, \qquad \nabla \times \mathbf{B} - \frac{1}{c^2}\frac{\partial \mathbf{E}}{\partial t} = \mu_0 \mathbf{J},$$

where ϵ_0 is the vacuum permittivity, μ_0 is the vacuum permeability, and c is the speed of light ($c = 1/\sqrt{\mu_0\epsilon_0}$). Here $\mathbf{E}$ is the electric field and $\mathbf{B}$ is the magnetic flux density. It is easy to show that the charge density ρ and the current density $\mathbf{J}$ satisfy the continuity equation to ensure charge conservation:

$$\nabla \cdot \mathbf{J} + \frac{\partial \rho}{\partial t} = 0. \tag{2.84}$$

The fields $\mathbf{E}$ and $\mathbf{B}$ interact with a quantum device by exerting a force on every charged particle. This is called the Lorentz force and is given by $\mathbf{F} = q(\mathbf{E} + \mathbf{v} \times \mathbf{B})$, where $\mathbf{v}$ is the velocity of the charged particle.

Equations (2.83) are not covariant, as they do not retain their form after a Lorentz transformation [134, 135]. Aside 2.15 describes the Lorentz transformation and the tensor notation used hereafter. Noting that moving charges imply current flow, it is evident that a charge at rest in one inertial frame will appear as a current in another inertial frame. The quantity that satisfies this property is the contravariant current density: $J^\alpha = (c\rho, \mathbf{J})$. It combines charge and current densities into a single 4-vector (a vector with four components) that ensures charge conservation under Lorentz transformations (see Aside 2.15).

Aside 2.15   Tensor Notation and Lorentz Transformation [76, 77, 98, 136]

We first discuss the concept of a 4 vector. The position vector in 3D needs three numbers, x, y, z, to specify it uniquely. To unify space and time, a vector in four dimensions is introduced with the components $x^\mu = (ct, x, y, z)$ to represent a space-time point. In the following discussion, the Greek indices ($\mu, \nu, \ldots$) take four possible values (0, 1, 2, 3, whereas the Latin indices ($i, j, k, \ldots$) take three possible values (1, 2, 3). The use of a superscript to denote a vector may be confusing to readers not versed in the theory of relativity. In fact, we need to use two kinds of vectors. A vector with a superscript is called a contravariant vector; if a subscript is used, it is called a covariant vector. For example, the gradient of a scalar quantity S is a covariant vector denoted as $\delta_\mu S = \frac{\partial S}{\partial x^\mu}$.

Tensors constitute a generalization of the vector concept and require multiple indices to represent them; the rank of a tensor equals the number of indices required to represent it. It is useful to employ Einstein's summation convention on repeated indices whenever no ambiguity is likely to occur [137]. A useful tensor is the Levi–Civita tensor (also known as the alternating tensor). It is defined as

$$\epsilon^{ij\ldots l} = \epsilon_{ij\ldots l} = \begin{cases} 1, & \text{if even permutations put indices in the order 1,2,}\ldots \\ -1, & \text{if odd permutations put indices in the order 1,2,}\ldots \\ 0, & \text{if two or more indices are the same.} \end{cases} \tag{2.85}$$

This tensor satisfies the following useful identities:

$$\epsilon_{ijk}\epsilon_{imn} = \delta_{jm}\delta_{kn} - \delta_{jn}\delta_{km}, \qquad \epsilon^{ijk}\epsilon_{imn} = \delta^j{}_m\delta^k{}_n - \delta^j{}_n\delta^k{}_m. \tag{2.86}$$

The space-time metric can be used to convert between the covariant and contravariant vectors. This metric is a second-rank tensor defined such that $g_{\mu\nu} = 0$ when $\mu \neq \nu$. It can be represented in the form of a diagonal matrix with the elements $g_{00} = +1, g_{11} = g_{22} = g_{33} = -1$. The signature of the space-time metric used here is $(+, -, -, -)$, which corresponds to the hyperbolic representation of the Minkowski space [138]. The distance ds between two neighboring space-time points satisfies

$$ds^2 = dx_\mu dx^\mu = g_{\mu\nu} dx^\mu dx^\nu = c^2 dt^2 - dx^2 - dy^2 - dz^2. \qquad (2.87)$$

We also use the following notation for derivatives: $\partial_\mu = \frac{\partial}{\partial x^\mu}$ and $\partial^\mu = \frac{\partial}{\partial x_\mu}$. Using this notation, we can introduce the 4-velocity u^α with components $(c, \mathbf{u})$, where $\mathbf{u}$ is the 3D velocity. The 4-momentum p^α can also be introduced using $(E/c, \mathbf{p}) = m_0 u^\alpha$, where $\mathbf{p}$ is the 3D momentum, E is the total energy, and m_0 is the rest mass of the particle. The differential operators can be generalized to define the 4-gradient as

$$\partial^\nu = \frac{\partial}{\partial x_\nu} = \left(\frac{1}{c} \frac{\partial}{\partial t}, -\nabla \right), \qquad (2.88)$$

where ∇ is the usual 3D gradient vector defined as $\nabla = \left(\frac{\partial}{\partial x}, \frac{\partial}{\partial y}, \frac{\partial}{\partial z} \right)$. The Laplacian operator $\nabla^2 = \nabla \cdot \nabla$ becomes the d'Alembertian operator in four dimensions and is defined as

$$\Box = \partial^\mu \partial_\mu = g^{\mu\nu} \partial_\nu \partial_\mu = \frac{1}{c^2} \frac{\partial^2}{\partial t^2} - \nabla^2. \qquad (2.89)$$

We can now discuss the Lorentz transformation. It takes the simple form:

$$x'^\mu = \Lambda^\mu{}_\nu x^\nu, \qquad \Lambda^\mu{}_\nu = \frac{\partial x'^\mu}{\partial x^\nu}. \qquad (2.90)$$

The inverse transformation can be written as

$$x^\mu = \bar{\Lambda}^\mu{}_\nu x'^\nu, \qquad \bar{\Lambda}^\mu{}_\nu = \frac{\partial x^\mu}{\partial x'^\nu}. \qquad (2.91)$$

Here $\Lambda^\mu{}_\nu$ and $\bar{\Lambda}^\mu{}_\nu$ are two tensors whose matrix elements obey the relation $\Lambda^\mu{}_\nu \bar{\Lambda}^\eta{}_\nu = \delta^\eta{}_\nu$, which implies $\bar{\Lambda}\Lambda = I$ or $\bar{\Lambda} = \Lambda^{-1}$. Owing to these relations, a contravariant 4-vector transforms as $A'^\mu = \Lambda^\mu{}_\nu A^\nu$ and $A^\mu = \bar{\Lambda}^\mu{}_\nu A'^\nu$. In contrast, the covariant 4-vector transforms as $A'_\mu = \bar{\Lambda}^\nu{}_\mu A_\nu$ and $A_\mu = \Lambda^\nu{}_\mu A'_\nu$.

The next step requires the introduction of vector and scalar potentials. Justification for such an approach is that some physical phenomena stem from these potentials. An example is provided by the Aharonov–Bohm effect discussed in Section 1.4. The relationship between the potentials and fields is governed by the relations:

$$\mathbf{B} = \nabla \times \mathbf{A}, \qquad \mathbf{E} = -\nabla\phi - \frac{\partial \mathbf{A}}{\partial t}. \qquad (2.92)$$

An equivalent way to satisfy these relations is to introduce the *gauge-field* 4-vector as $A^\mu = (\phi/c, \mathbf{A})$ and define the *field-strength tensor* $F^{\mu\nu}$ as

$$F^{\mu\nu} = \partial^\mu A^\nu - \partial^\nu A^\mu. \tag{2.93}$$

The preceding definition automatically satisfies the two relations between the fields and potentials in Eq. (2.92). This can be seen by expanding $F^{\mu\nu}$ and noting that its elements contain the components of $\mathbf{E}$ and $\mathbf{B}$ as follows:

$$F^{\mu\nu} = \begin{pmatrix} 0 & -E_1/c & -E_2/c & -E_3/c \\ E_1/c & 0 & -B_3 & B_2 \\ E_2/c & B_3 & 0 & -B_1 \\ E_3/c & -B_2 & B_1 & 0 \end{pmatrix}. \tag{2.94}$$

The associated covariant tensor $F_{\mu\nu}$ is found using the relation $F_{\mu\nu} = g_{\mu\alpha} g_{\nu\beta} F^{\alpha\beta}$. It has the form

$$F_{\mu\nu} = \begin{pmatrix} 0 & E_1/c & E_2/c & E_3/c \\ -E_1/c & 0 & -B_3 & B_2 \\ -E_2/c & B_3 & 0 & -B_1 \\ -E_3/c & -B_2 & B_1 & 0 \end{pmatrix}. \tag{2.95}$$

The main difference between $F^{\mu\nu}$ and $F_{\mu\nu}$ is a change in the signs of the components of the electric field. Owing to the simple structure of these tensors, we can extract the electrical and magnetic fields by using

$$E_i = cF_{0i} = -cF^{0i}; \qquad B_i = -\frac{1}{2}\epsilon_{ijk} F^{jk}. \tag{2.96}$$

Using these definitions, we can combine two Maxwell's equations, $\nabla \cdot \mathbf{E} = \rho/\epsilon_0$ and $\nabla \times \mathbf{B} - \frac{1}{c^2}\frac{\partial \mathbf{E}}{\partial t} = \mu_0 \mathbf{J}$, into one covariant equation

$$\partial_\mu F^{\mu\nu} = \mu_0 J^\nu, \tag{2.97}$$

where the 4-current is defined as $J^\nu = (c\rho, \mathbf{J})$. The other two Maxwell's equations in Eq. (2.83) can also be incorporated in the tensor equation

$$\partial_\sigma F_{\mu\nu} + \partial_\mu F_{\nu\sigma} + \partial_\nu F_{\sigma\mu} = 0. \tag{2.98}$$

The apparent complexity of this equation can be reduced by introducing a dual tensor, $G_{\mu\nu} = \frac{1}{2}\epsilon_{\mu\nu\sigma\tau} F^{\sigma\tau}$. It enables us to write Eq. (2.98) in the compact form

$$\partial_\mu G^{\mu\nu} = 0. \tag{2.99}$$

A direct evaluation of the terms of this equation confirms that it contains the remaining two Maxwell's equations: $\nabla \cdot \mathbf{B} = 0$, and $\nabla \times \mathbf{E} = -\partial \mathbf{B}/\partial t$.

The gauge-field 4-vector $A^\mu = (\phi/c, \mathbf{A})$ has some interesting properties. It is easy to see that $\mathbf{B}$ remains unchanged if we add to $\mathbf{A}$ the gradient of a scalar function. However, we also need to modify the scalar potential to keep the electric field unchanged. Such changes in the vector and scalar potentials are referred to as a *gauge transformation* [139]. For a given scalar function χ, the gauge transformation is written as

$$\phi \to \phi - \frac{\partial \chi}{\partial t}; \qquad \mathbf{A} \to \mathbf{A} + \nabla\chi. \tag{2.100}$$

These two relations can be combined in the following compact form:

$$A^\mu \to A^\mu - \partial^\mu \chi. \tag{2.101}$$

It is easy to check that such a gauge transformation does not change the field strength tensor:

$$F^{\mu\nu} \to \partial^\mu(A^\nu - \partial^\nu\chi) - \partial^\nu(A^\mu - \partial^\mu\chi) \to F^{\mu\nu}. \tag{2.102}$$

This gauge freedom is often useful for solving specific problems because one can impose a suitable *gauge condition* to simplify the underlying mathematics. The gauge condition fixes the gauge and eliminates the redundant degrees of freedom, while ensuring that the observable quantities derived at the end of calculation remain gauge invariant [139]. Several gauge conditions are discussed in Aside 2.16.

Aside 2.16   Common Gauge Transformations [140]

Great care should be exercised when using gauge transformations because their incorrect use can lead to situations where the resulting equations, though mathematically correct, describe unphysical scenarios [141]. However, if one is prepared to accept this limitation, the results are often valuable for constructing simplified descriptions. For example, the widely used Coulomb gauge [142] assumes that the scalar potential is felt instantaneously at any location (i.e., it violates special relativity). However, it is widely used in quantum mechanics because it enables the static and dynamic interactions to be separated. Similarly, the Kirchhoff gauge [143] assumes imaginary propagation speeds; velocity gauge allows superluminal propagation; and a zero propagation speed is used in the static Coulomb gauge [144].

Lorenz gauge is defined by the gauge condition [145]

$$\partial_\mu A^\mu = 0 \to \nabla \cdot \mathbf{A} + \frac{1}{c^2}\frac{\partial\phi}{\partial t} = 0. \tag{2.103}$$

This gauge is Lorentz covariant and is widely used for quantization of fields. When it is used for the inhomogeneous Maxwell's equations (with sources), it decouples the wave equations for the vector and scalar potentials such that $\partial^2\mathbf{A} = \mu_0\mathbf{J}$ and $\partial^2\phi = \rho/\epsilon_0$. In the 4-vector notation, we obtain a single equation $\partial_\mu\partial^\mu A^\nu = \mu_0 J^\nu$. The continuity condition also takes the form, $\partial^\mu J_\mu = 0$. In the absence of charges and currents, the resulting homogeneous equations are much easier to solve, increasing the value of this transformation. Caution should be exercised because the Lorenz gauge does not fully specify the scalar and vector potentials. Any function χ with the property $\partial^2\chi = 0$ applied to Eq. (2.101) does not alter the observed field values.

Coulomb gauge is defined by the gauge condition [142]

$$\nabla \cdot \mathbf{A} = 0. \tag{2.104}$$

Unlike the Lorenz gauge, the Coulomb gauge is not covariant. However, if the scalar potential does not depend on time, this gauge becomes equivalent to the Lorenz gauge. It is widely used in quantum-mechanical calculations, where the vector potential is quantized but the Coulomb interaction is treated classically. Under this gauge, the scalar potential is set by the *Poisson equation*, $\nabla^2 \phi = \rho/\epsilon_0$, implying that ϕ is experienced instantaneously at any location. This can be clearly seen by considering the *velocity gauge* where the scalar potential is assumed to propagate with an arbitrary speed v, rather than the speed of light [142]. With this change, ϕ is obtained by solving

$$\nabla^2 \phi - \frac{1}{v^2}\frac{\partial^2}{\partial t^2} = \frac{\rho}{\epsilon_0}, \qquad \nabla \cdot \mathbf{A} + \frac{c}{v^2}\frac{\partial \phi}{\partial t} = 0, \qquad (2.105)$$

where the velocity-gauge condition is also specified. It is easy to see that both equations reduce to those in the Coulomb gauge in the limit $v \rightarrow \infty$. As a result, the retardation effects do not appear in the scalar potential if the Coulomb gauge is used. However, the associated vector potential can still exhibit the retardation effects. A special case of Coulomb gauge is the *Weyl gauge* (also referred to as the *radiation gauge*) where the scalar potential is set to zero ($\phi = 0$).

In the classical electromagnetic theory, gauge invariance is realized through measurements of the electric and magnetic fields. Even though the scalar and vector potentials change, the associated fields stay the same in different gauges. However, when electromagnetic theory is coupled with quantum mechanics, it becomes apparent that the vector potential can have measurable effects in phenomena such as the Aharanov–Bohm effect [146]. Therefore, attention is required when fixing a gauge to ensure that the predicted phenomena are physically realizable. The introduction of the gauge-field 4-vector A^μ in Eq. (2.93) reduces the number of redundant degrees of freedom in the field components of Maxwell's equations and provides a complete set of dynamical variables for the underlying theory.

Even though we have used $\mathbf{E}$ and $\mathbf{B}$ as the main vector fields in Maxwell's equations, two other vectors, $\mathbf{D} = \epsilon_0 \mathbf{E} + \mathbf{P}$ and $\mathbf{H} = \mathbf{B}/\mu_0 - \mathbf{M}$, become important when electromagnetic interactions are studied in materials. The quantities $\mathbf{P}$ and $\mathbf{M}$ are called *material polarization* and *material magnetization*, respectively. We have already seen that both $\mathbf{E}$ and $\mathbf{B}$ can be recovered from the second-rank tensor $F_{\mu\nu}$. Similarly, we can incorporate pairs $\{\mathbf{D}, \mathbf{H}\}$ and $\{\mathbf{P}, \mathbf{M}\}$ into tensors $H_{\mu\nu}$ and $P_{\mu\nu}$, respectively. The resulting relation is $H_{\mu\nu} = F_{\mu\nu}/\mu_0 + P_{\mu\nu}$. It can be expanded to reveal the following matrix equation:

$$\begin{pmatrix} 0 & cD_x & cD_y & cD_z \\ -cD_x & 0 & -H_z & H_y \\ -cD_y & H_z & 0 & -H_x \\ -cD_z & -H_y & H_x & 0 \end{pmatrix} = \frac{F_{\mu\nu}}{\mu_0} + \begin{pmatrix} 0 & cP_x & cP_y & cP_z \\ -cP_x & 0 & M_z & -M_y \\ -cP_y & -M_z & 0 & M_x \\ -cP_z & M_y & -M_x & 0 \end{pmatrix}. \qquad (2.106)$$

It is easy to show by direct substitution that the four Maxwell's equations inside a material can be written in the following compact form:

$$\partial_\mu H^{\mu\nu} = J^\nu. \qquad (2.107)$$

2.3.2 Invariants and Lagrangian of the Electromagnetic Field

The tensor notation helps us to find functions that remain invariant during the interaction of matter with radiation. All invariants are zero-rank tensors. A widely used technique for finding invariants makes use of the operator identity $\partial_\mu \partial_\nu = \partial_\nu \partial_\mu$. As an example, if we deploy this identity in Eq. (2.107) after applying the operator ∂_ν and note that $H_{\mu\nu} = -H_{\nu\mu}$, we obtain an invariant, $\partial_\nu J^\nu = 0$, which is the continuity equation reflecting charge conservation. Three other important invariants are:

$$F^{\mu\nu} F_{\mu\nu} = 2(\mathbf{B}^2 - \mathbf{E}^2/c^2), \quad F^{\mu\nu} G_{\mu\nu} = -\frac{4}{c}(\mathbf{B} \cdot \mathbf{E}), \quad J_\mu A^\mu = \rho\phi - \mathbf{J} \cdot \mathbf{A}, \quad (2.108)$$

where the last invariant corresponds to the energy density of a charge interacting with an electromagnetic field. The validity of these invariants can be easily established by evaluating each directly.

One important Lorentz invariant of the electromagnetic field is its action S, which is defined using the Lagrangian $\mathcal{L}$ as [147]

$$S = \int \mathcal{L}(A, \partial^\nu A)\, d^4x. \quad (2.109)$$

The assumption that action is an integral over a Lagrangian that is a functional of the fields and their derivatives is called the *locality hypothesis* [148]. Also, it is important to recognize that some of these quantities are not physically observable (the definition of S is purely mathematical). Note also the four-dimensional integral resulting from the 4-vector x^μ. The 4-vector A depends on x^μ but the Lagrangian does not depend on x^μ explicitly. As before, action should be stationary (i.e., $\delta S = 0$). We use this condition to derive the Euler-Lagrange equations in Aside 2.17.

Aside 2.17   Euler–Lagrange Equation for Covariant Functions

We start with the action defined in Eq. (2.109). Its variation can be written as

$$\delta S = \int \left(\frac{\partial \mathcal{L}}{\partial A_\mu} \delta A_\mu + \frac{\partial \mathcal{L}}{\partial(\partial^\nu A_\mu)} \delta(\partial^\nu A_\mu) \right) d^4x. \quad (2.110)$$

We rearrange this expression by noting that $\delta(\partial^\nu A_\mu) = \partial^\nu(\delta A_\mu)$:

$$\delta S = \int \left(\frac{\partial \mathcal{L}}{\partial A_\mu} \delta A_\mu + \partial^\nu \left(\frac{\partial \mathcal{L}}{\partial(\partial^\nu A_\mu)} \delta A_\mu \right) - \partial^\nu \left(\frac{\partial \mathcal{L}}{\partial(\partial^\nu A_\mu)} \right) \delta A_\mu \right) d^4x. \quad (2.111)$$

The second term inside the integral is a 4-divergence. We use the divergence theorem (a generalization of the Gauss theorem) to show that it is zero because δA_μ vanishes on the boundary of the integral. As a result, we obtain

$$\delta S = \int \left(\frac{\partial \mathcal{L}}{\partial A_\mu} - \partial^\nu \frac{\partial \mathcal{L}}{\partial(\partial^\nu A_\mu)} \right) \delta A_\mu\, d^4x. \quad (2.112)$$

As the action principle demands $\delta\mathcal{S} = 0$ for all variations of A^μ, we immediately obtain the Euler–Lagrange equation

$$\frac{\partial\mathcal{L}}{\partial A_\mu} = \partial^\nu\frac{\partial\mathcal{L}}{\partial(\partial^\nu A_\mu)}. \tag{2.113}$$

The preceding derivation highlights that, if we add a divergence term to a known Lagrangian, the equations of motion resulting from that Lagrangian do not change. Recall the earlier discussion that many different Lagrangians can lead to the same equations of motion. For example, noting that the interaction of a charge with an electromagnetic field has the form $J^\mu A_\mu$, the Lagrangian for this case can be easily obtained.

Before we can apply the Euler–Lagrange equations, we need to find a Lagrangian appropriate for Maxwell's equations. This Lagrangian should be based on the 4-vector A^μ. Using the gauge invariance, Lorentz invariance, and the requirement that the equations of motion must be linear partial differential equations, the most general form of the Lagrangian can be written as

$$\mathcal{L}(A, \partial^\nu A) = \chi_1 A^\mu A_\mu + \chi_2 \partial_\mu A^\nu \partial^\mu A_\nu + \chi_3 \partial_\mu A^\nu \partial_\nu A^\mu + \chi_4 (\partial_\mu A^\mu)^2, \tag{2.114}$$

where χ_1, χ_2, χ_3, and χ_4 are constants that need to be fixed by ensuring that this Lagrangian produces Maxwell's equation correctly. As shown in Aside 2.18, Maxwell's equations are obtained if we choose $\chi_1 = 0$ and $\chi_2 + \chi_3 + \chi_4 = 0$. Using them, the Lagrangian has the form

$$\mathcal{L}(A, \partial_\mu A) = \chi_3(\partial_\mu A^\nu \partial_\nu A^\mu - \partial_\mu A^\nu \partial^\mu A_\nu) + \chi_4[(\partial_\mu A^\mu)^2 - \partial_\mu A^\nu \partial^\mu A_\nu]. \tag{2.115}$$

If we now use the definition of $F^{\mu\nu}$ in Eq. (2.93) and demand that all fields vanish at infinity, we can write it in the simple form

$$\mathcal{L}(A, \partial^\nu A) = -\frac{1}{4\mu_0}F_{\mu\nu}F^{\mu\nu}, \tag{2.116}$$

where the prefactor $-\frac{1}{4\mu_0}$ was chosen to yield the correct equations of motion.

As we saw earlier, $F^{\mu\nu}F_{\mu\nu}$ is an invariant and has the value $F^{\mu\nu}F_{\mu\nu} = 2(\mathbf{B}^2/c^2 - \mathbf{E}^2)$. As shown in Aside 2.18, the Euler–Lagrange equations of this Lagrangian are the Maxwell's equations. If the 4-currents are also added to this (see the discussion at the end of Aside 2.17), the Lagrangian that would generate Maxwell's equations interacting with charges becomes

$$\mathcal{L}(A, \partial^\nu A) = -\frac{1}{4\mu_0}F_{\mu\nu}F^{\mu\nu} - J_\alpha A^\alpha. \tag{2.117}$$

Aside 2.18   Lagrangian for the Electromagnetic Field [149]
Using the Lagrangian given in Eq. (2.114), we can easily calculate the following two derivatives:

$$\tfrac{\partial\mathcal{L}}{\partial A_\mu} = 2\chi_1 A^\mu \tag{2.118}$$

$$\frac{\partial \mathcal{L}}{\partial(\partial^\nu A_\mu)} = 2\chi_2 \partial_\nu A^\mu + 2\chi_3 \partial^\mu A_\nu + 2\chi_4 g^\mu{}_\nu (\partial_\sigma A^\sigma). \tag{2.119}$$

Substituting these derivatives in the Euler–Lagrange equation (2.113), we obtain

$$2\chi_1 A^\mu = \partial_\nu \left(2\chi_2 \partial_\nu A^\mu + 2\chi_3 \partial^\mu A_\nu + 2\chi_4 g^\mu{}_\nu (\partial_\sigma A^\sigma)\right), \tag{2.120}$$

which can be simplified to obtain

$$\chi_1 A^\mu = \chi_2 (\partial^\mu \partial_\mu) A^\mu + (\chi_3 + \chi_4) \partial^\mu (\partial_\sigma A^\sigma). \tag{2.121}$$

This equation can be put in the form $\partial^\mu(\partial_\sigma A^\sigma) - (\partial^\mu \partial_\mu) A^\mu = 0$ if we choose $\chi_1 = 0$ and $\chi_3 + \chi_4 = -\chi_2$.

To establish the equivalence between the preceding expression and Eq. (2.116), we need to show that they differ from each other by a divergence factor. As we saw in Aside (2.17), a divergence factor does not alter the resulting equations of motion. Using the expression for $F^{\mu\nu}$ as given in Eq. (2.93), we obtain

$$F_{\mu\nu}F^{\mu\nu} = (\partial_\mu A_\nu - \partial_\nu A_\mu)(\partial^\mu A^\nu - \partial^\nu A^\mu) \rightarrow 2(\partial_\mu A_\nu \partial^\mu A^\nu - \partial_\nu A_\mu \partial^\mu A^\nu). \tag{2.122}$$

This version can be recast as a sum of two terms where the first term is equal to Eq. (2.114) within a constant and the second term is a divergence that does not alter the Lagrangian. The final resulting equation is

$$F_{\mu\nu}F^{\mu\nu} = 2\partial_\mu A_\nu \partial^\mu A^\nu - 2(\partial^\mu A_\mu)^2 - 2\partial_\nu (A_\mu \partial^\mu A^\nu - A^\nu \partial^\mu A_\mu). \tag{2.123}$$

The main reason for finding the Lagrangian in (2.116) is to use it for quantizing the electromagnetic field. As we have seen before, the process of quantization requires identification of the generalized coordinates and their momenta as well as finding the commutation relations among these variables. We can use the 4-potential A^ν for the generalized coordinates. As before, the generalized momenta should be related to a derivative of A^ν. The right choice is its derivative with respect to $\partial_0 A^\nu$, resulting in the definition $M^\nu = \frac{\partial \mathcal{L}}{\partial(\partial_0 A^\nu)}$ for the generalized momenta M^ν.

Noting that the Lagrangian, $\mathcal{L}(A, \partial_\mu A) = \frac{1}{2}\epsilon_0(\mathbf{E}^2 - c^2\mathbf{B}^2)$, does not contain any terms containing $\dot{\phi}$, the generalized momentum of A^0 is zero. Owing to this constraint, it is not possible to elevate all A^μ to operator status. This can be understood by noting that A^μ are not the observable variables in Maxwell's equations. We can address this issue by adopting a gauge condition to eliminate any redundancies. The natural choices are (i) the radiation gauge or (ii) the Lorenz gauge. It is a nontrivial task to prove that these gauge conditions do not alter any of the physically observable properties of the electromagnetic field. In practice, it suffices to demonstrate that different gauge choices yield results that differ from each other at most by a unitary transformation.

2.3.3 Quantization in the Radiation Gauge

In this subsection we adopt the radiation gauge to quantize the electromagnetic field [150]. The general strategy we adopt is similar to the process we followed earlier for the

quantum-mechanical formulation of a nanoscale device. In that case, the classical description employed the generalized coordinates and momenta (q^μ, p^μ), and we mapped them to
the corresponding Heisenberg operators $(\hat{q}^\mu, \hat{p}^\nu)$ and imposed the commutation relations:

$$[q^\mu, p^\nu] = i\hbar\delta^{\mu\nu}, \qquad [q^\mu, q^\nu] = 0, \qquad [p^\mu, p^\nu] = 0. \tag{2.124}$$

In the 4-vector notation, the generalized coordinates become $q^\mu = (t, x^1, x^2, x^2) \equiv (t, \mathbf{x})$.
The corresponding generalized momenta are $p^\mu = (p^0, \mathbf{p})$, where $p^0 = \frac{E}{c}$ and E is the
energy of the object being quantized (see Aside 2.15).

The same strategy can be used for quantizing the electromagnetic field. As the
Lagrangian is described using the 4-potential A^μ, we use the mapping: $q^\mu \rightarrow A^\mu$ and
$p^\mu \rightarrow M^\mu$, where $M^\mu = \frac{\partial \mathcal{L}}{\partial(\partial_0 A^\mu)}$. In close analogy, the commutation relations at any time t
are

$$[A^\mu(t, \mathbf{x}), M^\nu(t, \mathbf{y})] = i\hbar\delta^{\mu\nu}\delta^{(3)}(\mathbf{x} - \mathbf{y}), \quad [A^\mu, A^\nu] = 0, \quad [M^\mu, M^\nu] = 0, \tag{2.125}$$

where all operators are evaluated at the same time t. Recall that, following the notation in
Aside 2.15, $x^\mu = (t, \mathbf{x})$ and $y^\mu = (t, \mathbf{y})$.

The radiation gauge fixes the redundancy in A^μ by imposing the gauge conditions: $\phi = 0$
and $\nabla \cdot \mathbf{A} = 0$. As these conditions are clearly not Lorentz covariant, we must explicitly
verify at the end of the quantization process that the outcome of any experiment does not
depend on the inertial reference frame in which it was observed. In the radiation gauge,
$A^0 = 0$. The remaining three components satisfy the equation of motion $\partial^\mu \partial_\mu A^\nu = 0$ for
$\nu = 1, 2, 3$. This equation can be solved by taking the Fourier transform of A^ν with respect
to the spatial coordinates as

$$\widetilde{A}^\nu(\mathbf{k}) = \int A^\nu(\mathbf{x}) \exp(-i\mathbf{k} \cdot \mathbf{x}) \, d^3x, \tag{2.126}$$

where $\mathbf{k} = (k_1, k_2, k_3)$. It is easy to see that $\widetilde{A}^\nu(\mathbf{k})$ satisfies $(\partial_0{}^2 + k^2)\widetilde{A}^\nu(\mathbf{k}) = 0$. This is
an ordinary differential equation with the solutions of the form $\exp(\pm ikx^0)$ or $\exp(\pm i\omega t)$,
where $\omega = ck$ is the frequency of the electromagnetic wave and k is its propagation
constant. Choosing the negative sign and taking the inverse Fourier transform, we obtain

$$\mathbf{A}(t, \mathbf{x}) = \frac{1}{(2\pi)^3} \int \widetilde{\mathbf{A}}(\mathbf{k}) \exp(i\mathbf{k} \cdot \mathbf{x} - i\omega t) \, d^3k, \tag{2.127}$$

where $\widetilde{\mathbf{A}}(\mathbf{k}) = \widetilde{\mathbf{A}}^*(-\mathbf{k})$. It is clear that both $\mathbf{k}$ and $-\mathbf{k}$ provide independent solutions to this
equation. Therefore, their superposition represents the general solution of this equation.
Noting that the condition $\nabla \cdot \mathbf{A} = 0$ implies $\mathbf{k} \cdot \widetilde{\mathbf{A}}(\mathbf{k}) = 0$, two polarization unit vectors
$\mathbf{e}_\zeta(\mathbf{k})$ are introduced such that $\mathbf{e}_\zeta(\mathbf{k}) \cdot \mathbf{k} = 0$, where $\zeta = 1, 2$. Owing to the radiation gauge
condition $\mathbf{k} \cdot \widetilde{\mathbf{A}}(\mathbf{k}) = 0$, $\widetilde{\mathbf{A}}(\mathbf{k})$ must be a linear combination of $\mathbf{e}_1(\mathbf{k})$ and $\mathbf{e}_2(\mathbf{k})$:

$$\widetilde{\mathbf{A}}(\mathbf{k}) = \mathbf{e}_1(\mathbf{k})a_1(\mathbf{k}) + \mathbf{e}_2(\mathbf{k})a_2(\mathbf{k}), \tag{2.128}$$

where a_1 and a_2 are the complex amplitudes of this linear combination. Using these results,
the general solution for $\mathbf{A}(t, \mathbf{x})$ can be written as

$$\mathbf{A}(t, \mathbf{x}) = \frac{1}{(2\pi)^3} \sum_{\zeta=1,2} \int \mathbf{e}_\zeta(\mathbf{k}) \left[a_\zeta(\mathbf{k})e^{(i\mathbf{k} \cdot \mathbf{x} - i\omega t)} + a_\zeta^*(\mathbf{k})e^{(-i\mathbf{k} \cdot \mathbf{x} - i\omega t)} \right] d^3k. \tag{2.129}$$

To proceed further, we make $\mathbf{A}(t, \mathbf{x})$ a Hermitian operator. This can be achieved by making each amplitude a Heisenberg operator $\hat{a}_\zeta(\mathbf{k})$. The conjugate of the amplitude, $a_\zeta^*(\mathbf{k})$, is mapped to the adjoint Heisenberg operator $\hat{a}_\zeta^\dagger(\mathbf{k})$. They satisfy the following commutation relations:

$$\left[\hat{a}_\zeta(\mathbf{k}), \hat{a}_{\zeta'}^\dagger(\mathbf{k}')\right] = (2\pi)^3 \delta^{(3)}(\mathbf{k} - \mathbf{k}')\delta_{\zeta\zeta'}; \quad \left[\hat{a}_\zeta(\mathbf{k}), \hat{a}_{\zeta'}(\mathbf{k}')\right] = \left[\hat{a}_\zeta^\dagger(\mathbf{k}), \hat{a}_{\zeta'}^\dagger(\mathbf{k}')\right] = 0.$$
(2.130)

These commutation relations are central to the quantization of an electromagnetic field because they enable us to map its classical description to the quantum domain. To see whether we have succeeded in this mission, we need to find the generalized momenta M^μ and test for the commutator relationships given in Eq. (2.125).

If we pick A^0, the associated momentum $M^0 = 0$ for the Lagrangian given in (2.116). As $A^0 = 0$ in the radiation gauge, both A^0 and M^0 can be discarded as dynamic variables. This is a consequence of reducing the degrees of freedom of the vector potential by fixing it to the radiation gauge. For the three remaining components A^i ($i = 1, 2, 3$) we find $M^i = -E^i$, where E^i is the ith component of the electric field. After some tedious calculations, it can be shown that the commutator $[A^i(t, \mathbf{x}), E^j(t, \mathbf{y})] = -i\hbar\delta^{ij}\delta^{(3)}(\mathbf{x} - \mathbf{y})$, as expected for the quantization scheme used here.

We can now construct the state vectors relevant to our quantization scheme. We start by defining the vacuum state $|0\rangle$ in the Hilbert space (known as the Fock space) as $\hat{a}_\zeta(\mathbf{k})|0\rangle = 0$ for all $\mathbf{k}$ and $\zeta = 1, 2$. We can construct the next state $|\mathbf{k}\rangle$ by using the creation operator as $\hat{a}_\zeta^\dagger(\mathbf{k})|0\rangle$. The remaining states in the Fock space are generated by using the same procedure. The associated Hamiltonian can be obtained by normal ordering of the classical expression for the electromagnetic energy $\mathcal{E}_{em} = \frac{1}{2}\int(\mathbf{E}\cdot\mathbf{E} + \mathbf{B}\cdot\mathbf{B})\,d^3x$. The results are given by

$$H = \frac{\hbar\omega}{2}\sum_{\zeta=1,2}\int[a_\zeta^\dagger(\mathbf{k})a_\zeta(\mathbf{k}) + a_\zeta(\mathbf{k})a_\zeta^\dagger(\mathbf{k})]\frac{d^3k}{(2\pi)^3}.$$
(2.131)

This expression can be simplified using the commutator relations given in Eq. (2.130) to obtain

$$H = \int\sum_{\zeta=1,2}\hbar\omega a_\zeta^\dagger(\mathbf{k})a_\zeta(\mathbf{k})\frac{d^3k}{(2\pi)^3}.$$
(2.132)

The operator $a_\zeta^\dagger(\mathbf{k})a_\zeta(\mathbf{k})$ is the number operator for photons of momentum $\mathbf{p} = \hbar\mathbf{k}$. The single-particle states containing such photons are generated from the vacuum state through $\sqrt{2\hbar\omega}a_\zeta^\dagger(\mathbf{k})|0\rangle$ and are identified with photons of momentum $\mathbf{p}$ and energy $\hbar\omega$.

We discuss briefly what happens when the Lorenz gauge is used in place of the radiation gauge to quantize the electromagnetic field. Since this gauge is covariant, the scalar potential does not vanish and A^0 has a finite value. However, regardless of the value of A^0, the associated conjugate momentum vanishes for the Lagrangian given in (2.116), and it is not possible to define the commutator relationship for them. To alleviate this problem, we modify the Lagrangian (2.116) by adding a divergence term that does not alter the resulting Euler–Lagrange equation. The modified Lagrangian has the form

$$\mathcal{L}(A, \partial^\nu A) = -\frac{1}{4\mu_0} F_{\mu\nu} F^{\mu\nu} + \frac{\lambda}{2} \left(\partial_\mu A^\mu\right)^2. \tag{2.133}$$

Here the variable λ plays the role of a Lagrange multiplier. The simplest choice is to set $\lambda = 1$. However, if one uses this Lagrangian for quantizing an electromagnetic field, non-physical states are found in addition to the physically valid states. A technique known as Gupta–Bleuler quantization shows how to recover physical states using properties of the Fock space [150]. With this technique, one can get nonzero conjugate momenta for all four dynamic variables.

2.4 Second Quantization

In the preceding section we quantized an electromagnetic wave by using commutator relations that were similar to those used for describing the quantum behavior of a particle. One distinctive feature of this quantization procedure was that we used the creation and annihilation operators, often referred to as the ladder operators, that enable one to build the system states starting from the vacuum state. In this section we focus on the single-particle Fock states and build many-particle states by filling up each single-particle state with a certain number of identical particles. This formalism is known as *second quantization* [151, 152]. It plays a significant role in any study of many-particle systems that are central to modeling quantum devices. In particular, it is the method of choice for indistinguishable particles because the traditional wave function is not suitable for carrying out a complex symmetrization procedure needed to describe many-particle states. The second quantization method addresses this issue by counting the number of particles that occupy each state. Because this process does not refer to any labeling of particles, it contains no redundant information and provides a compact description of many-particle states.

2.4.1 Many-Particle States for Fermions and Bosons

We begin with single-particle (Fock) states $|n\rangle$, where n is any positive integer. The single-particle states of several particles are written as $|n\rangle_\zeta$, where ζ takes integer values $1, 2, \ldots$. They can be used to write the *occupation-number representation* of a many-particle state consisting of n_1 particles in the state $|n_1\rangle_1$, n_2 particles in the state $|n_2\rangle_2$, and so on. Here, we have adopted a notation where a subscript outside the ket identifies the particle, but a subscript within the ket identifies its state. It is common to write such a many-particle state as

$$|[n_\zeta]\rangle = |n_1, n_2, \ldots, n_\zeta, \ldots\rangle, \tag{2.134}$$

where $N = \sum_\zeta n_\zeta$ is the total number of particles. Owing to the Pauli exclusion principle, n_ζ can take only two values (0 or 1) for fermions but it can be any nonnegative integer for bosons:

$$n_\zeta = \begin{cases} 0, 1 & \text{for fermions,} \\ 0, 1, 2, 3, \ldots & \text{for bosons.} \end{cases} \tag{2.135}$$

Among this set of many-particle states, some states have special significance. The state with all occupation numbers equal to zero is called the vacuum state $|0\rangle = |0, 0, \ldots\rangle$. A state with one non-zero occupation number is denoted as $|n_\zeta\rangle = |0, 0, \ldots, n_\zeta, \ldots\rangle$. The symmetrized and antisymmetrized multiparticle states are defined using the S_+ and S_- operations as

$$S_\pm |n_1, n_2, \ldots, n_\zeta, \ldots\rangle = \frac{1}{\sqrt{N}} \sum_{P \in S_N} (\pm 1)^P P |n_1, n_2, \ldots, n_\zeta, \ldots\rangle, \tag{2.136}$$

where the summation is taken over all the $N!$ elements in the permutation group S_N and the permutation operator P has the property

$$(-1)^P = \begin{cases} 1 & \text{for even permutations} \\ -1 & \text{for odd permutations.} \end{cases} \tag{2.137}$$

Using these definitions, the many-particle states of the bosons and fermions can be written as

$$|n_1, n_2, \ldots, n_\zeta, \ldots\rangle_\pm = \frac{1}{\sqrt{n_1! n_2! \ldots n_\zeta! \ldots}} S_\pm |n_1, n_2, \ldots, n_\zeta, \ldots\rangle, \tag{2.138}$$

where $\pm$ correspond to bosons and fermions, respectively. Notice that the denominator in the preceding equation reduces to 1 for fermions in view that $n_j! = 1$ ($n_j = 0$ or 1 for all j).

In practice, dealing with the sum of permutations in Eq. (2.136) is cumbersome and inefficient. A better way is to look for an algebraic operation that inherently encodes the symmetry or antisymmetry property and frees us from this unwieldy representation. The technique offering this possibility uses operators that either lower the number of particles in a given state by one (annihilation operator) or increase it by one (creation operator). The creation and annihilation operators are different for bosons and fermions because their states have different symmetry properties, as described earlier.

We define the creation operator $\hat{a}_\zeta^\dagger$ and the annihilation operator $\hat{a}_\zeta$ for every ζ as

$$\hat{a}_\zeta^\dagger |n_1, n_2, \ldots, n_\zeta, \ldots\rangle_\pm = \sqrt{n_\zeta + 1}(\pm 1)^{s_\zeta} |n_1, n_2, \ldots, n_\zeta + 1, \ldots\rangle_\pm, \tag{2.139}$$

$$\hat{a}_\zeta |n_1, n_2, \ldots, n_\zeta, \ldots\rangle_\pm = \sqrt{n_\zeta}(\pm 1)^{s_\zeta} |n_1, n_2, \ldots, n_\zeta - 1, \ldots\rangle_\pm, \tag{2.140}$$

where $s_\zeta = \sum_{j=1}^{\zeta-1} n_j$. The quantity s_ζ is also known as the *Jordan–Wigner string* in the case of fermions. Its value not only demands a predefined ordering of the single-particle states but also requires knowing the fermion occupation numbers of all the preceding states. Owing to this dependency of the local state on the fermion group considered, the creation and annihilation operators are considered nonlocal for the fermions in some sense. Recall that the occupation number n_ζ is either 0 and 1 for fermions. It is possible to show that these operators satisfy the following commutation relations:

$$[\hat{a}_j^\dagger, \hat{a}_k^\dagger]_\pm = [\hat{a}_j, \hat{a}_k]_\pm = 0, \qquad [\hat{a}_j, \hat{a}_k^\dagger]_\pm = \delta_{j,k}, \tag{2.141}$$

where we have used the definition $[\alpha, \beta]_{\pm} = \alpha\beta \mp \beta\alpha$. Applying the creation operator repeatedly to the vacuum state, it is possible to generate every state vector for both bosons and fermions as follows:

$$|n_1, n_2, \ldots, n_\zeta, \ldots\rangle_+ = \prod_\zeta \frac{1}{\sqrt{n_\zeta!}} (\hat{a}_\zeta^\dagger)^{n_\zeta} |0\rangle, \tag{2.142}$$

$$|n_1, n_2, \ldots, n_\zeta, \ldots\rangle_- = \prod_\zeta (\hat{a}_\zeta^\dagger)^{n_\zeta} |0\rangle. \tag{2.143}$$

So far we have assumed that ζ takes integer values. It is possible to extend the preceding formulation to the case where ζ varies continuously. For this purpose, we replace the last commutation relation in Eq. (2.141) with $[\hat{a}_{\mathbf{p}_1}, \hat{a}_{\mathbf{p}_2}^\dagger]_{\pm} = \delta(\mathbf{p}_1 - \mathbf{p}_2)$, where $\mathbf{p}_1$ and $\mathbf{p}_2$ are two continuous variables representing each particle's momentum. The resulting creation and annihilation operators can be transferred to the physical space using the Fourier theory. We define new creation and annihilation operators, $\hat{A}_\mathbf{x}^\dagger$ and $\hat{A}_\mathbf{x}$, at a spatial point $\mathbf{x}$ using the relation:

$$\hat{A}_\mathbf{x} = \left(\frac{\hbar}{2\pi}\right)^{3/2} \int_\mathbf{p} \hat{a}_\mathbf{p} \exp\left(\frac{i}{\hbar}\mathbf{p} \cdot \mathbf{x}\right) d^3\mathbf{p}. \tag{2.144}$$

One can easily show that following commutator relations are satisfied by these operators:

$$[\hat{A}_{\mathbf{x}_1}^\dagger, \hat{A}_{\mathbf{x}_2}^\dagger]_{\pm} = [\hat{A}_{\mathbf{x}_1}, \hat{A}_{\mathbf{x}_2}]_{\pm} = 0, \qquad [\hat{A}_{\mathbf{x}_1}, \hat{A}_{\mathbf{x}_2}^\dagger]_{\pm} = \delta(\mathbf{x}_1 - \mathbf{x}_2). \tag{2.145}$$

Aside 2.19 shows an example where such operators are used for describing a quantum system. Their main advantage is that they can create or destroy a particle at a given location in the physical space (indicated by $\mathbf{x}$).

Aside 2.19   Position and Momentum Representations of a Particle's Quantum State

As we saw earlier in this chapter, a quantum state $|\zeta\rangle$ can be described in many different basis. If we use the position coordinates ($\mathbf{x}$) for the basis using $\Psi(\mathbf{x}) = \langle\mathbf{x}|\zeta\rangle$, we recover the wave function appearing in the Schrödinger equation. However, we can also use the momentum basis and form the function $\Phi(\mathbf{p}) = \langle\mathbf{p}|\zeta\rangle$. Since each representation forms a complete, orthornormal basis, we can expand $\Phi(\mathbf{p})$ in the $\Psi(\mathbf{x})$ basis as

$$\Phi(\mathbf{p}) = \langle\mathbf{p}|\zeta\rangle = \int \langle\mathbf{p}|\mathbf{x}\rangle \langle\mathbf{x}|\zeta\rangle \, d\mathbf{x} = \int \langle\mathbf{p}|\mathbf{x}\rangle \Psi(\mathbf{x}) \, d\mathbf{x}. \tag{2.146}$$

Thus, we need to find the quantity $\langle\mathbf{p}|\mathbf{x}\rangle$. We know from elementary quantum mechanics that the following Fourier transform exists:

$$\Phi(\mathbf{p}) = \int \langle\mathbf{x}|\zeta\rangle \exp\left(-\frac{i}{\hbar}\mathbf{p} \cdot \mathbf{x}\right) d\mathbf{x}. \tag{2.147}$$

This relation immediately implies that

$$\langle\mathbf{p}|\mathbf{x}\rangle = \exp\left(-\frac{i}{\hbar}\mathbf{p} \cdot \mathbf{x}\right). \tag{2.148}$$

In some cases $\mathbf{p}$ may take only discrete values. In this case the inverse Fourier relation becomes

$$\Psi(\mathbf{x}) = \sum_{\mathbf{p}} \Phi(\mathbf{p}) \exp\left(\frac{i}{\hbar} \mathbf{p} \cdot \mathbf{x}\right). \tag{2.149}$$

Consider a state $|\Psi\rangle$ resulting from the application of operator $\hat{A}_{\mathbf{x}}^{\dagger}$ on the vacuum state $|0\rangle$:

$$|\Psi\rangle = \hat{A}_{\mathbf{x}}^{\dagger} |0\rangle = \sum_{\mathbf{p}} \exp\left(-\frac{i}{\hbar} \mathbf{p} \cdot \mathbf{x}\right) \hat{a}_{\mathbf{p}}^{\dagger} |0\rangle. \tag{2.150}$$

As we discussed earlier, this operation should create a single particle at point, $\mathbf{x}$. It is instructive to check whether this is indeed the case. We invoke the occupation number operator (see Aside 2.20), which counts the number of particles in the state $|\Psi\rangle$. As the operator $\hat{a}_{\mathbf{p}_0}^{\dagger} \hat{a}_{\mathbf{p}_0}$ counts the number of particles with momentum $\mathbf{p}_0$, we apply it to the state $|\Psi\rangle$ to obtain

$$\sum_{\mathbf{p}_0} \hat{a}_{\mathbf{p}_0}^{\dagger} \hat{a}_{\mathbf{p}_0} |\Psi\rangle = \sum_{\mathbf{p}_0} \sum_{\mathbf{p}} \exp\left(-\frac{i}{\hbar} \mathbf{p} \cdot \mathbf{x}\right) \hat{a}_{\mathbf{p}_0}^{\dagger} \hat{a}_{\mathbf{p}_0} \hat{a}_{\mathbf{p}}^{\dagger} |0\rangle. \tag{2.151}$$

However, we know from the commutation relations that $\langle 0|\hat{a}_{\mathbf{p}_0} \hat{a}_{\mathbf{p}}^{\dagger}|0\rangle = \delta_{\mathbf{p}_0 \mathbf{p}}$. Use of this result in Eq. (2.151) results in

$$\sum_{\mathbf{p}_0} \hat{a}_{\mathbf{p}_0}^{\dagger} \hat{a}_{\mathbf{p}_0} |\Psi\rangle \rightarrow |\Psi\rangle. \tag{2.152}$$

This result shows that the state $|\Psi\rangle$ is an eigenstate of the occupation number operator with an eigenvalue of 1. As the eigenvalue of the occupation number operator represents the number of particles in the associated eigenstate, the state $|\Psi\rangle$ represents a single particle. Suppose this particle was created at some position $\mathbf{x}_0$. We can show this by considering the average $\langle \mathbf{x}_0 | \Psi \rangle$:

$$\langle \mathbf{x}_0 | \Psi \rangle = \sum_{\mathbf{p}} \exp\left(-\frac{i}{\hbar} \mathbf{p} \cdot \mathbf{x}\right) \langle \mathbf{x}_0 \hat{a}_{\mathbf{p}}^{\dagger} |0\rangle| = \sum_{\mathbf{p}} \exp\left[-\frac{i}{\hbar} \mathbf{p} \cdot (\mathbf{x} - \mathbf{x}_0)\right]. \tag{2.153}$$

This expression reduces to $\delta(\mathbf{x} - \mathbf{x}_0)$, confirming that the particle was created at the position $\mathbf{x}_0$, as expected.

2.4.2 Representation of Operators

It is important to consider how various operators are represented using the creation and annihilation operators associated with second-quantization formulation. Suppose $\hat{A}$ is an operator for N particles that only depends on their coordinates. It can be written as a sum of individual single-particle operators as

$$\hat{A} = \sum_{k=1}^{N} A_k. \tag{2.154}$$

We can represent $\hat{A}$ in a complete basis using the matrix elements $A_{\zeta\eta} = \langle\zeta|\hat{A}|\eta\rangle$ as

$$\hat{A} = \sum_{\zeta\eta} |\zeta\rangle\,\langle\zeta|\,\hat{A}\,|\eta\rangle\,\langle\eta| \rightarrow \sum_{\zeta\eta} A_{\zeta\eta}\,|\zeta\rangle\,\langle\eta| \rightarrow \sum_{\zeta\eta} A_{\zeta\eta} \sum_{k=1}^{N} |\zeta\rangle_k\,\langle\eta|_k. \qquad (2.155)$$

To represent the operator $\hat{A}$ using the annihilation and creation operators, we need to investigate its action on a general multiparticle state $|n_1, n_2, \ldots, n_\zeta, \ldots\rangle_\pm$. In the general case $\zeta \neq \eta$, we obtain

$$\sum_{k=1}^{N} |\zeta\rangle_k\,\langle\eta|_k\,|n_1, n_2, \ldots, n_\zeta, \ldots\rangle_\pm \rightarrow \sum_{k=1}^{N} \frac{|\zeta\rangle_k\,\langle\eta|_k}{\sqrt{n_1!\,n_2!\ldots n_\zeta!\ldots}} S_\pm\,|n_1, n_2, \ldots, n_\zeta, \ldots\rangle.$$

$$(2.156)$$

As shown in Aside 2.20, this equation reduces to the particle-number operator of that state. The same result holds for $\zeta = \eta$ as well. Thus the operator $\hat{A}$ has the following form in the second-quantization representation:

$$\hat{A} = \sum_{\zeta\eta} A_{\zeta\eta}\hat{a}_\zeta^{\dagger}\hat{a}_\eta, \qquad (2.157)$$

where $A_{\zeta\eta} = \langle\zeta|\hat{A}|\eta\rangle$ corresponds to single-particle states. Essentially, the operator $\hat{A}$ is a superposition over all processes that use $\hat{a}_\eta$ to annihilate a single particle in the state $|\eta\rangle$, scatter it through the matrix element $A_{\zeta\eta}$, and then use $\hat{a}_\zeta^{\dagger}$ to create that particle in the final state $|\zeta\rangle$. Thus, the whole process can be viewed as a form of scattering process that brings particles from the initial state to the final state via all possible single-particle scattering events.

Aside 2.20   Occupation Number Operator

The occupation (or particle) number operator $\hat{n}_\zeta$ for each state $|\zeta\rangle$ is defined as [153]

$$\hat{n}_\zeta = \hat{a}_\zeta^{\dagger}\hat{a}_\zeta \qquad (2.158)$$

such that

$$\hat{n}_\zeta\,|n_1, n_2, \ldots, n_\zeta, \ldots\rangle_\pm = n_\zeta\,|n_1, n_2, \ldots, n_\zeta, \ldots\rangle_\pm. \qquad (2.159)$$

The total occupation number (or particle number) for the state, $|n_1, n_2, \ldots, n_\zeta, \ldots\rangle_\pm$, can be defined as $\hat{N} = \sum_\zeta \hat{n}_\zeta$. It is straightforward to show that

$$\hat{N}_\zeta\,|n_1, n_2, \ldots, n_\zeta, \ldots\rangle_\pm = \sum_\zeta n_\zeta\,|n_1, n_2, \ldots, n_\zeta, \ldots\rangle_\pm \rightarrow N\,|n_1, n_2, \ldots, n_\zeta, \ldots\rangle_\pm.$$

$$(2.160)$$

The situation changes if we relax the single-particle association for the operator $\hat{A}$ and assume that it is associated with two interacting particles. Then the second-quantization representation of the operator is given by

$$\hat{A} = \sum_{\zeta_1 \zeta_2 \eta_1 \eta_2} A_{\zeta_1 \zeta_2 \eta_1 \eta_2} \hat{a}_{\zeta_1}^{\dagger} \hat{a}_{\zeta_2}^{\dagger} \hat{a}_{\eta_1} \hat{a}_{\eta_2}, \tag{2.161}$$

where $A_{\zeta_1 \zeta_2 \eta_1 \eta_2} = \langle \zeta_1 \zeta_2 | \hat{A} | \eta_1 \eta_2 \rangle$ for two-particle states. Very much like the single-particle operator, this representation embodies the idea that a two-particle operator represents all processes that annihilate two particles from their initial states, scatter them through the matrix element $A_{\zeta_1 \zeta_2 \eta_1 \eta_2}$, and create two particles in two new states.

Aside 2.21   Tight-Binding Models

For materials that are formed from closed-shell atoms or ions, the free-electron model is not adequate for describing the motion of electrons. Tight-binding models are simplified band models for electrons in solids interacting only with neighboring atoms [154]. Even though they look deceptively simple, tight-binding models can be used to calculate intricate properties of solids such as surface states or plasmonic response. Unlike the free-electron models, the tight-binding models assume that an electron remains mostly bound to its own atom, except for an occasional transfer to a neighboring atom. A particularly simple example is given by the Hamiltonian

$$\hat{H} = -\kappa \sum_{i,j} (\hat{a}_i^{\dagger} \hat{a}_j + \hat{a}_j^{\dagger} \hat{a}_i), \tag{2.162}$$

where $\hat{a}_i^{\dagger}$ creates an electron on site i, while $\hat{a}_j$ annihilates an electron on a neighboring site j. The product $\hat{a}_i^{\dagger} \hat{a}_j$ describes intuitively the hopping of an electron from the site j to the site i. A positive value of the hopping parameter ($\kappa > 0$) indicates that each hopping lowers the kinetic energy of the system.

The Hamiltonian in Eq. (2.162) can be diagonalized by expanding $\hat{a}_j$ into a Fourier series (owing to the periodic nature of the atoms in a solid), resulting in the operators $\hat{c}_{\mathbf{k}} = \sum_j e^{-i\mathbf{k}\cdot\mathbf{r}_j} \hat{a}_j$, where $\mathbf{k}$ is a vector in the reciprocal space. The Hamiltonian can then be written in the form $\hat{H} = \sum_{\mathbf{k}} \epsilon_{\mathbf{k}} \hat{c}_{\mathbf{k}}^{\dagger} \hat{c}_{\mathbf{k}}$. In a one-dimension lattice, one finds that $\epsilon_{\mathbf{k}} = -2\kappa \cos(kd)$, where d is the lattice spacing.

3 Linear Response Theory

The sciences do not try to explain, they hardly even try to interpret, they mainly make models. By a model is meant a mathematical construct which, with the addition of certain verbal interpretations, describes observed phenomena. The justification of such a mathematical construct is solely and precisely that it is expected to work - that is correctly to describe phenomena from a reasonably wide area. Furthermore, it must satisfy certain esthetic criteria - that is, in relation to how much it describes, it must be rather simple.

John von Neumann

3.1 Linear Response of a System

Linear systems have been studied in diverse disciplines including physics, mathematics, and engineering [155, 156, 157]. In electrical engineering, when electronic probing is used to monitor a system, its interaction with the probe is considered a small perturbation to the system; if it were not, we would not be probing the system, but the system would be modified by the probe [158]! Consequently, the results of probing can be expressed in terms of a linear response function that depends on the properties of the monitored system (but not on the probe). Owing to its fundamental importance, there is no doubt that the linear response of systems will continue to play an important role for as long as one can foresee.

Linear response theory describes mathematically changes induced in the properties of a system as a result of external probing. The primary aim is to model the system's response without considering details of the probe–system interaction. One surprising but highly useful result is that the linear response function of the system is predominantly determined by the eigenvalues and eigenfunctions of the unperturbed system [159]. As a consequence, it is possible to determine the eigenvalues (excitation energies) of a system from its frequency response to external probing. It should be stressed that many systems behave nonlinearly, and their response to an external probe is not always linear. However, by treating the external stimulus as "perturbative" (i.e., relatively small), linear response theory can be used for nonlinear systems as well. If this assumption appears to be too restrictive, one may resort to the Lagrangian formalism and solve the equations of motion as described in Section 2.1. However, owing to computational complexity of such an approach, we are likely to miss valuable physical insights into the system's behavior. Linear response theory makes a compromise by making reasonable predictions from general physical principles, albeit with some loss of accuracy [160].

3.1.1 General Formalism and Impulse Response

The response of a linear system to an input signal $\mathbf{v}(t)$ is governed by [161]

$$\mathbf{y}(t) = \int_{-\infty}^{\infty} \mathbf{g}(t, \tau)\mathbf{v}(\tau)\,d\tau, \tag{3.1}$$

where $\mathbf{y}(t)$ is the output signal. The function $g(t, \tau)$ is called the *impulse response* because it is the output one receives if the system is excited by an impulse. Its two arguments denote the time τ at which an impulse was applied and the time t at which the system's output is observed.

The infinite integration limits in Eq. (3.1) can be removed by considering the causality requirement. Suppose that the input was applied at $\tau = t_0$. The system is called *relaxed* at time t_0 if it has a null response before this instance. For a relaxed system, we can replace the lower limit of integration by t_0. If the system is causal, then the output cannot occur before the input is applied. This can only be assured if $\mathbf{g}(t, \tau) = 0$ when $\tau > t$. As a result, the upper limit of the integral in Eq. (3.1) can be replaced with t. Thus, the response of a relaxed, causal, linear system can be written as [155, 161]:

$$\mathbf{y}(t) = \int_{t_0}^{t} \mathbf{g}(t, \tau)\mathbf{v}(\tau)\,d\tau. \tag{3.2}$$

Many linear systems are also time invariant in the sense that their impulse response does not change if all times are shifted by a constant amount, say t_c. This is possible only if $\mathbf{g}(t + t_c, \tau + t_c) = \mathbf{g}(t, \tau)$. Choosing $t_c = -\tau$ gives us $\mathbf{g}(t, \tau) = g(t - \tau, 0)$. Replacing $\mathbf{g}(t - \tau, 0)$ with $\mathbf{g}(t - \tau)$ for a time-invariant system, its linear response takes the form [161]:

$$\mathbf{y}(t) = \int_{t_0}^{t} \mathbf{g}(t - \tau)\mathbf{v}(\tau)\,d\tau. \tag{3.3}$$

As t_0 is an arbitrary reference time, $t_0 = 0$ is widely used, especially in engineering. Without loss of generality, we adopt this convention here. To simplify the following discussion, we also assume that the input $v(t)$ and the output $y(t)$ are scalar functions of time. This assumption is not restrictive, and the conclusions reached here remain valid for the vector functions as well. With these simplifications, the response of a linear time-invariant (LTI) system is governed by

$$y(t) = \int_{0}^{t} g(t - \tau)v(\tau)\,d\tau \quad \rightarrow \quad y(t) = g(t) * v(t), \tag{3.4}$$

where the compact form of $y(t)$ makes use of the *convolution* operator $*$ that is employed commonly to describe the input-output relationship for linear systems.

It is well known that the convolution operation becomes a product if one takes the Fourier transform of all functions appearing in the convolution. The same property also

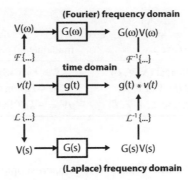

Fig. 3.1 Relationship for a linear time-invariant system between the time-domain functions and their Laplace transforms in the s–domain or Fourier transforms in the ω domain.

holds for the Laplace transforms. As the lower limit is not infinite in Eq. (3.4), it is often necessary to employ the Laplace transform of the input function defined as

$$\mathcal{L}\{v(t)\}(s) = V(s) = \int_0^t v(t)\exp(-st)\,dt, \tag{3.5}$$

where s is a complex number and $V(s)$ denotes the Laplace transform of $v(t)$ in the s domain. By using the convolution property of the Laplace transform, Eq. (3.4) in the s domain takes the form

$$\mathcal{L}[y(t)] = \mathcal{L}[g(t)]\mathcal{L}[v(t)] \quad \rightarrow \quad Y(s) = G(s)V(s). \tag{3.6}$$

In the case of Fourier transforms, s becomes purely imaginary ($s = i\omega$), and the corresponding relation is written as $Y(\omega) = G(\omega)V(\omega)$. Figure 3.1 shows both of these relations in a graphic form.

Laplace domain results can be converted to the time domain using the *inverse Laplace transform*, also known as the *Mellin inverse formula* [162]:

$$v(t) = \mathcal{L}^{-1}\{V(s)\}(t) = \lim_{\omega \to \infty} \frac{1}{2\pi i} \int_{\gamma - i\omega}^{\gamma + i\omega} V(s)\exp(st)\,ds, \tag{3.7}$$

where the integration is done along a vertical line located at $\Re(s) = \gamma$ in the complex plane such that γ is greater than the real part of all singularities of $V(s)$. This ensures that the contour path lies in the region of convergence of $V(s)$. If all singularities of $V(s)$ happen to lie in the left half-plane, γ can be set to zero, and the preceding integral becomes identical to the inverse Fourier transform. In practice, this integral is difficult to evaluate when $\gamma \neq 0$. Sometimes one can bypass the integral by decomposing $V(s)$ into a sum of known transforms and construct the inverse by inspection. However, this "inspection" method is applicable only in a few limited cases.

3.1.2 Equilibrium Ensembles in Classical Statistical Mechanics

As the linear system is assumed to be in thermal equilibrium before the external perturbation is applied, we discuss the three equilibrium ensembles used commonly in statistical

Table 3.1 Three ensembles used in thermodynamics [159].		
ensemble (fixed)	fundamental parameter	total differential
microcanonical (N, V, E)	$S = k_B \ln(S_{\text{states}})$	$dS = \frac{1}{T}dE + \frac{1}{T}pdV - \frac{1}{T}\mu dN$
canonical (N, V, T)	$F = -k_B T \ln(Z_{NVT})$	$dF = -sdT - pdV + \mu dN$
grand-canonical (μ, V, T)	$pV = -k_B T \ln(Z_{\mu VT})$	$d(pV) = SdT + pdV + Nd\mu$

mechanics. The microstates of any macroscopic system represents all possible states that it may take. An ensemble is an idealization consisting of virtual copies of all microstates of a macroscopic system, subject to the constraints imposed on the system [159]. Such an ensemble is used to calculate the probability distribution function, whose knowledge enables one to compute the average properties of the system. Ensembles considered in statistical thermodynamics compute these probabilities by applying the laws of classical or quantum mechanics that govern the system's evolution. There are three different ways one can formulate system's equilibrium; these are known as microcanonical, canonical, and grand-canonical ensembles. For a macroscopic system in equilibrium, all three ensembles give the same final result. As such, choice of the ensemble is dictated by the nature of the physical system under consideration and the properties one is interested in.

Microcanonical ensemble: A Microcanonical ensemble is based on the system's total energy, governed by the Hamiltonian $H(\mathbf{q}, \mathbf{p})$, in the range $[E, E + \Delta E]$ with a fixed number N of particles inside a fixed volume V. The system stays in equilibrium by not exchanging energy or particles with its environment. This ensemble assigns equal probability density, $\rho_{eq}(\mathbf{q}, \mathbf{p})$, to every microstate in the energy range $[E, E + \Delta E]$. All other microstates are given a probability of zero,

$$\rho_{eq}(\mathbf{q}, \mathbf{p}) = \begin{cases} 1/\Omega, & H(\mathbf{q}, \mathbf{p}) \in [E, E + \Delta E] \\ 0, & \text{otherwise,} \end{cases} \tag{3.8}$$

where we have defined Ω as the phase-space volume of the region in which $H(\mathbf{q} \text{ is in the range } \mathbf{p}) \in [E, E + \Delta E]$:

$$\Omega(E, E + \Delta E) = \int_{H(\mathbf{q}, \mathbf{p})} d\mathbf{q} d\mathbf{p}. \tag{3.9}$$

Using the density of states in the form $D_{os}(E) = \int \delta[E - H(\mathbf{q}, \mathbf{p})] \, d\mathbf{q} d\mathbf{p}$ (see Section 1.3.2), the probability density can be written as $\rho_{eq}(\mathbf{q}, \mathbf{p}) = \delta[E - H(\mathbf{q}, \mathbf{p})]/D_{os}(E)$.

The fundamental parameter that relates the microcanonical ensemble to thermodynamics is the *entropy* (see Table 3.1). Based on information theory, entropy is a measure of uncertainty (or lack of information). If a system has many equally probable outcomes, say N_{states}, then the entropy of the system is defined as $S = k_B \ln(N_{\text{states}})$ [163]. This definition can be readily ported to the microcanonical ensemble by counting the number of microscopic states in the energy range $[E, E + \Delta E]$. This number is given by $N_{\text{states}} = \Omega(E, E + \Delta E)/(N! \, h^{3N})$, where h is the Planck constant. Here we used the uncertainty principle (see Aside 2.11) of quantum mechanics stating that an infinitesimal volume

$d\mathbf{q}\,d\mathbf{p}$ can only be located with an uncertainty of h^3. We also used the fact that N particles can be arranged in $N!$ ways.

Canonical ensemble: A canonical ensemble is in equilibrium with its environment at a constant temperature T and has a fixed number of particles (N) inside a constant volume V [159]. This ensemble does not pose any constraints on the system's total energy E, which can be exchanged with the environment. The probability density for a system with the Hamiltonian $H(\mathbf{q}, \mathbf{p})$ is given by

$$\rho_{eq}(\mathbf{q}, \mathbf{p}) = \frac{\exp\left[-H(\mathbf{q}, \mathbf{p})/k_B T\right]}{\int \exp\left[-H(\mathbf{q}, \mathbf{p})/k_B T\right] d\mathbf{q}d\mathbf{p}}. \tag{3.10}$$

The partition function Z_{NVT} for this ensemble is defined as

$$Z_{NVT} = \int \frac{1}{N!\,h^{3N}} \exp\left(-\frac{H(\mathbf{q}, \mathbf{p})}{k_B T}\right) d\mathbf{q}d\mathbf{p}. \tag{3.11}$$

As seen in Table 3.1, this quantity relates to the Gibb's free energy through the relation $F = -k_B T \ln(Z_{NVT})$ and provides a way to link thermodynamic variables to the equilibrium distribution.

Grand-canonical ensemble: A grand-canonical ensemble can exchange both energy and particles with its environment. It has a constant volume V at a constant temperature T and a constant chemical potential μ [159]. The probability density for a system containing N particles with the Hamiltonian $H(\mathbf{q}, \mathbf{p})$ is given by

$$\rho_{eq}(\mathbf{q}, \mathbf{p}) = \frac{\exp\left(-[H(\mathbf{q}, \mathbf{p}) - \mu N]/k_B T\right)}{\sum_{N=0}^{\infty} \int \exp\left(-[H(\mathbf{q}, \mathbf{p}) - \mu N]/k_B T\right) d\mathbf{q}d\mathbf{p}}. \tag{3.12}$$

The partition function $Z_{\mu VT}$ for this ensemble can be written using the partition function Z_{NVT} of the canonical ensemble as

$$Z_{\mu VT} = \sum_{N=0}^{\infty} \exp\left(\mu N/k_B T\right) Z_{NVT}. \tag{3.13}$$

As seen in Table 3.1, this quantity relates to the system pressure p through the relation $pV = -k_B T \ln(Z_{\mu VT})$, and provides a way to link thermodynamic variables to the equilibrium distribution.

3.1.3 Equilibrium Ensembles in Quantum Statistical Mechanics

The linear response theory applicable to quantum systems considers near-equilibrium fluctuations, which are ensemble-averaged small changes induced by an external probe. The underlying assumption is that the macroscopic properties of a linear system are the results of ensemble-averaged microscopic properties. When a system is interacting with its surrounding environment, both its energy and the number of particles can fluctuate (because particles can either leave or enter from the environment). In either case, the system is considered to be in equilibrium with its environment if its temperature and chemical potential (for each kind of particle) remain constant. Enforcing reasonable constraints, it is possible to adapt the classical equilibrium distributions to a quantum system.

We have seen in Section 2.2 that one can associate with any physical quantity O an observable $\mathcal{O}$, which is a Hermitian operator in the Hilbert space of the quantum device considered. When we carry out measurements, the only possible values for the physical quantity O are the eigenvalues λ_O of this Hermitian operator $\mathcal{O}$. This implies that, unlike in classical physics, the value of a physical quantity is not uniquely determined for a particular microstate. Rather, the probabilistic nature of the events needs to be taken into account using the Born interpretation of the state vector, as discussed in Section 2.2. A state vector does not describe the properties of one single device. Rather, it describes the properties of a statistical ensemble of such devices prepared under the same conditions. Knowledge of the state vector $|\Psi\rangle$ enables us to determine the expectation value of the observable as $\langle O \rangle = \langle \Psi | \mathcal{O} | \Psi \rangle$.

To extend the classical equilibrium concept to quantum devices, we define a quantum macrostate as an ensemble of microstates. Then, if the density operator ρ of the ensemble is known, the expectation value of $\mathcal{O}$ is calculated using

$$\langle O \rangle = \mathrm{Tr}\,(\rho\mathcal{O}). \tag{3.14}$$

Because the trace of an operator is independent of the basis used, $\mathrm{Tr}\,(\rho\mathcal{O})$ does not depend on the basis of the eigenvectors representing the underlying states. Here, the density operator ρ corresponds to the probability distribution in classical statistical physics. Thus, it is important to realize that ρ performs two different roles simultaneously. First, it provides the quantum average on the state; second, it performs the statistical average on the state vectors of the environment.

When we consider the density operator of the equilibrium distribution, denoted by ρ_{eq}, the corresponding ensemble must be stationary (i.e., $\partial\rho_{eq}/\partial t = 0$). It follows from the *Liouville equation*,

$$\frac{\partial}{\partial t}\rho_{eq} = \frac{1}{i\hbar}[H_{eq}, \rho_{eq}], \tag{3.15}$$

that ρ_{eq} commutes with the Hamiltonian of the ensemble that itself does not depend on time because of the equilibrium nature of the ensemble. This feature can be used to calculate the expectation value of $\mathcal{O}$ in the Heisenberg picture by using

$$\langle \widetilde{\mathcal{O}}(t) \rangle_{eq} = \mathrm{Tr}\,[\rho_{eq}\widetilde{\mathcal{O}}(t)], \qquad \widetilde{\mathcal{O}}(t) = U^{\dagger}(t)\mathcal{O}U(t), \tag{3.16}$$

where $U(t) = \exp(-\frac{i}{\hbar}H_{eq}t)$. As the trace operation is invariant under a cyclic permutation, it follows that

$$\langle \widetilde{\mathcal{O}}(t) \rangle_{eq} = \mathrm{Tr}\left[\rho_{eq}U^{\dagger}(t)\mathcal{O}U(t)\right] = \mathrm{Tr}\,(\rho_{eq}\mathcal{O}) = \langle \mathcal{O} \rangle_{eq}. \tag{3.17}$$

This result is expected because, even though the operator is time-dependent, its ensemble average over an equilibrium distribution should be a constant.

To apply these considerations to the microcanonical ensemble, we assume that the eigenvalues of the Hamiltonian H of a quantum device are given by the set $\{E_n\}$ with the eigenfunctions $|n\rangle$ (i.e., $H|n\rangle = E_n|n\rangle$). When we want to identify the number N of particles in the system, we modify the notation by incorporating N in the eigenvalue equation as $H|n\rangle = {}_N E_n|n\rangle$ (i.e., ${}_N E_n$ is the nth energy level of a quantum device with N particles).

As the microcanonical ensemble has a constant number of particles inside a fixed volume V with a constant energy E, the matrix elements of its equilibrium density matrix can be calculated using

$$(\rho_{eq})_{n,m} = \langle n|\rho_{eq}|m\rangle = \langle n|\delta(E - H)|m\rangle = \delta(E - {}_N E_n)\delta_{n,m}. \tag{3.18}$$

Owing to this result, the associated partition function, $Z_{NVE} = \text{Tr}(\rho_{eq}) = \sum_n \delta(E - {}_N E_n)$, is equal to the density of states $D_{os}(E)$.

The canonical ensemble has a fixed number N of particles inside a constant volume V at a constant temperature T. Correspondingly, the matrix elements of the equilibrium density matrix are calculated as

$$(\rho_{eq})_{n,m} = \langle n|\exp[-H/(k_B T)]|m\rangle = \exp\left[-{}_N E_n/(k_B T)\right]\delta_{n,m}. \tag{3.19}$$

The partition function in this case is given by $Z_{NVT} = \text{Tr}(\rho_{eq}) = \sum_n \exp\left(-{}_N E_n/k_B T\right)$.

Finally, the grand-canonical ensemble has a fixed volume V at a constant temperature T, and a constant chemical potential μ. The matrix elements of the equilibrium density matrix for N particles are calculated as

$$({}_N \rho_{eq})_{n,m} = \langle n|\exp\left[-(H - \mu N)/k_B T\right]|m\rangle = \exp\left[-({}_N E_n - \mu N)/k_B T\right]\delta_{n,m}. \tag{3.20}$$

The associated partition function for this case is given by

$$Z_{\mu T} = \sum_N \text{Tr}\,({}_N \rho_{eq}) = \sum_{N,n} \exp\left[-({}_N E_n - \mu N)/k_B T\right]. \tag{3.21}$$

Aside 3.1 shows how these concepts can be used to derive the Bose–Einstein and Fermi–Dirac distributions.

Aside 3.1 Fermi–Dirac and Bose–Einstein Distributions

Indistinguishable particles play a central role in the operation of quantum devices [164]. As discussed in Section 2.4, such particles are called *bosons* when their spin is zero or an integer (in multiples of $\hbar$), and their statistics are described by the *Bose–Einstein* distribution function. Examples of bosons include photons, phonons, and Higgs particles. In contrast, if the particles have half-integer spins, they are called *fermions*, and their statistics are governed by the *Fermi–Dirac* distribution function. Examples of fermions include electrons, muons, and protons. Both types of distributions play important roles in practice. For example, the behavior of electrons in metals and semiconductors at a finite temperature depends on the Fermi–Dirac distribution. Similarly, the properties of lasers, spasers, and Bose–Einstein condensates are governed by the Bose–Einstein distribution [164, 165].

A crucial question that we need to answer is: which equilibrium distribution must we adopt for bosons and fermions? As all three equilibrium ensembles require that we know the exact number of states in each of the allowed energies, counting states becomes an essential consideration. The canonical and the microcanonical ensembles require us to sum the states over a fixed number of particles. In practice, counting of states is much more complicated for a system with a fixed number of particles than for a system with a fixed chemical potential. This consideration suggests the grand-canonical distribution as a possible choice. However, to apply the grand-canonical ensemble, we need to know the chemical potential.

Application of the grand-canonical ensemble becomes much easier in situations where the particles can be treated as being independent with negligible interactions. In this situation, the total energy of the system can be written as a sum over energies of the single-particle states. This is the case for a collection of *strongly degenerate* fermions or bosons. In the case of fermions, strongly degenerate means that the lowest single-particle states are occupied with a probability close to unity. In the case of bosons, strongly degenerate means that a significant fraction of the particles are in the lowest energy state.

Fermi–Dirac distribution: Consider a fermion with the eigenstate $|\mathbf{k}\rangle$ and energy $E_\mathbf{k}$. Owing to the *Pauli exclusion principle*, the occupation number $n_\mathbf{k}$ for this state can only have values 0 or 1. By adopting the grand-canonical ensemble as the equilibrium distribution, the average occupation number $\langle n_\mathbf{k} \rangle$ is found to be [164, 165]:

$$\langle n_\mathbf{k} \rangle = \frac{\mathrm{Tr}\,(\rho_{eq} n_\mathbf{k})}{\mathrm{Tr}\,(\rho_{eq})} = \frac{\sum_{n_\mathbf{k}=0}^{1} n_\mathbf{k} \exp\left[-(n_\mathbf{k} E_\mathbf{k} - \mu n_\mathbf{k})/k_B T\right]}{\sum_{n_\mathbf{k}=0}^{1} \exp\left[-(n_\mathbf{k} E_\mathbf{k} - \mu n_\mathbf{k})/k_B T\right]} = \frac{1}{\exp\left[(E_\mathbf{k} - \mu)/k_B T\right] + 1}.$$
(3.22)

This is known as the Fermi–Dirac distribution and represents the probability that energy level $E_\mathbf{k}$ is occupied by a fermion. Even though a distribution function describes the number of particles occupying a given state, the Fermi–Dirac distribution can be interpreted as a probability distribution because at most one fermion is allowed in each state. In practice, the chemical potential μ is replaced by the Fermi energy E_F.

Figure 3.2 shows the Fermi–Dirac distribution as a function of the energy ratio $(E_\mathbf{k} - \mu)/k_B T$. The classical *Maxwell–Boltzmann* distribution in the form $\exp\left[-(E_\mathbf{k} - \mu)/k_B T\right]$ is also shown for comparison. A useful observation is that for states with energy slightly larger than the Fermi level $E_F = \mu$ (by a few $k_B T$), we can approximate the Fermi–Dirac distribution with the classical distribution. This approximation has proved quite useful for describing the behavior of electrons and holes inside semiconductors and the distribution of optical phonons in solids at low temperatures. Given that $\langle n_\mathbf{k} \rangle$ is a function of energy $E_\mathbf{k}$, we can calculate the number of fermions in the energy range $[E_i, E_f]$ by using the density of states as

$$N_f = \int_{E_i}^{E_f} D_{os}(E_\mathbf{k}) \langle n_\mathbf{k} \rangle \, dE_\mathbf{k}.$$
(3.23)

Bose–Einstein distribution: Consider a boson with the eigenstate $|\mathbf{k}\rangle$ and energy $E_\mathbf{k}$. Unlike the fermions, the occupation number $n_\mathbf{k}$ of bosons can take any integer value from 0 to ∞. By adopting the grand-canonical ensemble as the equilibrium distribution, we can calculate the average occupation number as [164, 165]

$$\langle n_\mathbf{k} \rangle = \frac{\mathrm{Tr}\,(\rho_{eq} n_\mathbf{k})}{\mathrm{Tr}\,(\rho_{eq})} = \frac{\sum_{n_\mathbf{k}=0}^{\infty} n_\mathbf{k} \exp\left[-(n_\mathbf{k} E_\mathbf{k} - \mu n_\mathbf{k})/k_B T\right]}{\sum_{n_\mathbf{k}=0}^{\infty} \exp\left[-(n_\mathbf{k} E_\mathbf{k} - \mu n_\mathbf{k})/k_B T\right]} = \frac{1}{\exp\left[(E_\mathbf{k} - \mu)/k_B T\right] - 1}.$$
(3.24)

This function is called the Bose–Einstein distribution; it is also plotted in Figure 3.2 using $\mu = 0$. It was first derived by Bose in 1924 for phonons and generalized later by Einstein

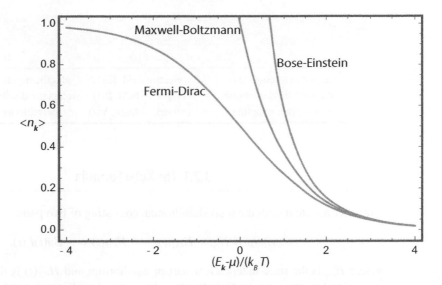

Fig. 3.2 Average population distribution of single-particle states for bosons (Bose–Einstein) and fermions (Fermi–Dirac) under the grand-canonical ensemble. Both distributions approach the classical distribution (Maxwell–Boltzmann) when $(F_k - \mu)/k_B T \gg 1$.

for atoms. Just like the Fermi–Dirac distribution, we can approximate the Bose–Einstein distribution with the Maxwell–Boltzmann distribution when $(E_k - \mu) \gg k_B T$. A celebrated application of the Bose–Einstein distribution is for *blackbody radiation*, which can be treated as an ideal gas of photons with a variable number of particles. The condition of energy minimum gives $\mu = 0$, and the energy of each photon is $h\nu$ for the radiation of frequency ν. The density of states for photons of frequency ν from Section 1.5 is given as $D_{os}(\nu) = 8\pi \nu^2/c^3$. Multiplying this density of states with the average energy of photons in each state ($\langle n_\mathbf{k} \rangle h\nu$) yields the Planck's blackbody radiation formula:

$$d\rho_\nu = \frac{8\pi h}{c^3} \frac{\nu^3 d\nu}{\exp\left(\frac{h\nu}{k_B T}\right) - 1}.$$

(3.25)

3.2 Linear Response Function

In this section we discuss the linear response theory as formulated by Kubo [166], Mazo [167], and others [168]. We consider a system in thermal equilibrium and calculate the system's response to an external stimulus applied at time $t = 0$ using the grand-canonical ensemble of Section 3.1.3. This choice enables us to access many results known in the disciplines of systems engineering and control engineering.

Table 3.2 Some examples of perturbations to the equilibrium Hamiltonian.

	$F(t)$	$\mathcal{B}$
electric field coupling	electric field, $\mathbf{E}(t)$	electric dipole moment, $\mathbf{d}$
magnetic field coupling	magnetic field, $\mathbf{B}(t)$	magnetic dipole moment, $\boldsymbol{\mu}$
electro-optic coupling	external voltage, $\mathbf{V}(t)$	permittivity tensor, ϵ

3.2.1 The Kubo Formula

Consider a system with the total Hamiltonian consisting of two parts,

$$H_{total} = H_{eq} + H_{ext}(t), \qquad H_{ext}(t) = -\mathcal{B}u(t)F(t), \tag{3.26}$$

where H_{eq} is the time-independent part in equilibrium and $H_{ext}(t)$ is the time-dependent part with the external field $F(t)$. The Heaviside step function $u(t)$ equals 0 for $t < 0$ and 1 for $t > 0$ [169]. It is introduced to ensure that the external influence is turned on at $t = 0$. Here $\mathcal{B}$ is a centered operator corresponding to an observable B such that $\langle \mathcal{B} \rangle_{eq} = 0$. This is not a real restriction because any operator can be converted to such a form by subtracting its ensemble average from it. Even though we have included only one external stimulus in the Hamiltonian, the following analysis can be readily extended to multiple stimuli because a linear system's responses to different perturbing stimuli add up independently in view of the superposition principle. Some examples of external perturbations are given in Table 3.2.

We calculate the ensemble averages of the quantities involved using the density operator ρ introduced in Section 2.2.3. Based on the discussion in Section 2.2.5, it satisfies the following equation in the interaction picture:

$$\frac{\partial \rho(t)}{\partial t} + \frac{i}{\hbar}[H_{eq}, \rho(t)] = \frac{i}{\hbar}u(t)F(t)[\mathcal{B}, \rho(t)], \tag{3.27}$$

The formal solution of this equation for $t \geq 0$ is given by the integral equation

$$\rho(t) = \rho_{eq}(t) + \frac{i}{\hbar}\int_0^t U(t-\tau)F(\tau)[\mathcal{B}, \rho(\tau)]U^{\dagger}(t-\tau)\,d\tau, \tag{3.28}$$

where $U(t) = \exp(-\frac{i}{\hbar}H_{eq}t)$ and $\rho_{eq}(t)$ is the solution of the homogeneous equation

$$\frac{\partial \rho_{eq}}{\partial t} + \frac{i}{\hbar}[H_{eq}, \rho_{eq}] = 0. \tag{3.29}$$

The integral equation can be used to obtain an approximate solution for $\rho(t)$ through an iterative procedure that begins by substituting $\rho(\tau) = \rho_{eq}(\tau)$ on the right side and then uses the new solution to obtain the next-order solution.

Suppose there is another operator $\mathcal{A}$ for the system of interest, and we want to know how its observed values change as a result of the perturbation applied to the system. As the system was in equilibrium before the perturbation, its average value for $t < 0$ is given by $\langle \mathcal{A} \rangle_{eq} = \text{Tr}[\mathcal{A}\rho_{eq}]$. After $t > 0$, its ensemble average is given by $\langle \mathcal{A}(t) \rangle = \text{Tr}[\mathcal{A}\rho(t)]$.

Clearly, the perturbation-induced change, $\Delta\mathcal{A}(t) = \mathcal{A}(t) - \langle\mathcal{A}\rangle_{eq}$, is a centered operator whose ensemble average is given by

$$\langle\Delta\mathcal{A}(t)\rangle = \text{Tr}\left[\frac{i}{\hbar}\int_0^t \mathcal{A}U(t-\tau)F(\tau)[\mathcal{B}, \rho_{eq}]U^\dagger(t-\tau)\,d\tau\right].\tag{3.30}$$

This expression can be simplified by noting that the trace operation is invariant under cyclic permutation (i.e., $\text{Tr}[ABC] = \text{Tr}[BCA]$). The result is given by

$$\langle\Delta\mathcal{A}(t)\rangle = \text{Tr}\left[\frac{1}{i\hbar}\int_0^t \rho_{eq}[\mathcal{B}, \tilde{\mathcal{A}}(t-\tau)]F(\tau)\,d\tau\right] = \frac{1}{i\hbar}\int_0^t \langle[\mathcal{B}, \tilde{\mathcal{A}}(t-\tau)]\rangle_{eq}F(\tau)\,d\tau,\tag{3.31}$$

where we adopted the notation used in Section 2.2.5 and defined $\tilde{\mathcal{A}}(t)$ in the interaction picture as $\tilde{\mathcal{A}}(t) = U^\dagger(t)\mathcal{A}U(t))$. We also moved the trace operation inside the integral and used the notation $\langle[\mathcal{B}, \tilde{\mathcal{A}}(t-\tau)]\rangle_{eq} = \text{Tr}\left[\rho_{eq}[\mathcal{B}, \tilde{\mathcal{A}}(t-\tau)]\right]$.

As Eq. (3.31) is in the form of a convolution, we write it as

$$\langle\Delta\mathcal{A}(t)\rangle = \int_0^t \chi_{AB}(t-\tau)F(\tau)\,d\tau,\tag{3.32}$$

where $\chi_{AB}(t)$ is the *retarded linear response function*, defined through the *Kubo formula*:

$$\chi_{AB}(t) = \frac{1}{i\hbar}u(t)\langle[\mathcal{B}, \tilde{\mathcal{A}}(t)]\rangle_{eq},\tag{3.33}$$

where $u(t)$ ensures causality of the response function. The Fourier transform of this function is known as the *generalized susceptibility* (or generalized admittance). As required for any causal response function, this susceptibility satisfies the *Kramers–Kronig relations* described in Aside 3.2.

Aside 3.2   Kramers–Kronig Relations

We write the Fourier transform of $\chi_{AB}(t)$ as

$$\mathcal{F}\{\chi_{AB}\}(\omega) = \int_{-\infty}^{\infty} u(t)\chi_{AB}(t)\exp(i\omega t)\,dt,\tag{3.34}$$

where $\mathcal{F}\{\ldots\}$ denotes the Fourier transform operation and $u(t)$ is the step function. Using the convolution theorem, we obtain

$$\mathcal{F}\{\chi_{AB}\}(\omega) = \mathcal{F}\{u\}(\omega) * \mathcal{F}\{\chi_{AB}\}(\omega), \qquad \mathcal{F}\{u\}(\omega) = \pi\delta(\omega) - \frac{1}{i\omega},\tag{3.35}$$

where we used a known result for the Fourier transform of the Heaviside step function $u(t)$. After a few algebraic manipulations, we obtain

$$\mathcal{F}\{\chi_{AB}\}(\omega) = \frac{1}{i\pi}\text{P.V.}\int_{-\infty}^{\infty} \frac{\mathcal{F}\{\chi_{AB}\}(\omega')}{\omega - \omega'}\,d\omega',\tag{3.36}$$

where P.V. denotes *Cauchy's principle value* of the integral. Separating the real and imaginary parts of $\mathcal{F}\{\chi_{AB}\}(\omega)$, we obtain the Kramers–Kronig relations [170]

$$\Re[\mathcal{F}\{\chi_{AB}\}(\omega)] = \frac{1}{\pi}\text{P.V.}\int_{-\infty}^{\infty} \frac{\Im[\mathcal{F}\{\chi_{AB}\}(\omega')]}{\omega' - \omega}\,d\omega'\tag{3.37}$$

$$\Im[\mathcal{F}\{\chi_{AB}\}(\omega)] = -\frac{1}{\pi}\text{P.V.} \int_{-\infty}^{\infty} \frac{\Re[\mathcal{F}\{\chi_{AB}\}(\omega')]}{\omega' - \omega} \, d\omega'. \qquad (3.38)$$

These two relations show us how the real and imaginary parts of a causal response function are related. They are of immense importance in modeling quantum devices. For example, if the absorption coefficient (related to the imaginary part of the permittivity) is known for a material over the entire frequency range, the refractive index (related to the real part of the permittivity) can be calculated from it (and vice versa).

In practice, it is useful to write the Kramers–Kronig relations using only positive frequencies. The Fourier transform of a real function $f(t)$ satisfies the relation $\mathcal{F}\{f(t)\}(-\omega) = \mathcal{F}^*\{f(t)\}(\omega)$. Using it, we obtain the modified Kramers–Kronig relations (for causal real functions) containing positive frequencies:

$$\Re[\mathcal{F}\{\chi_{AB}\}(\omega)] = \frac{2}{\pi}\text{P.V.} \int_{0}^{+\infty} \frac{\omega'\Im[\mathcal{F}\{\chi_{AB}\}(\omega')]}{\omega'^2 - \omega^2} \, d\omega' \qquad (3.39)$$

$$\Im[\mathcal{F}\{\chi_{AB}\}(\omega)] = -\frac{2}{\pi}\text{P.V.} \int_{0}^{+\infty} \frac{\omega\Re[\mathcal{F}\{\chi_{AB}\}(\omega')]}{\omega'^2 - \omega^2} \, d\omega'. \qquad (3.40)$$

Let us consider how the Kramers–Kronig relations can be applied to analyze passive dielectric materials. For such a material, the permittivity $\epsilon(\omega)$ represents its response to an electromagnetic field. However, $\epsilon(\omega)$ does not vanish for large frequencies but has a finite value as $\omega \to \infty$. This can be understood by noting that the material cannot respond fast enough to very large frequencies. However, one requirement in deriving the Kramers–Kronig relations was that the Fourier transform for the linear response function $\chi(t)$ must vanish at infinity. Therefore, we must use the difference $\epsilon(\omega) - \epsilon_\infty$ for formulating the Kramers–Kronig relations.

The preceding example shows that causality alone is sufficient to establish the Kramers–Kronig relations in a passive dielectric medium. This is not the case for an active dielectric medium or a magnetic medium because of the presence of instabilities [171]. The theory of complex analytic functions is used in such situations to derive the Kramers–Kronig relations. More specifically, if singularities exist in the upper-half complex plane, the contour is adjusted such that the integral is taken on a line above the singularities.

In practice, the integrals appearing in the Kramers–Kronig relations converge slowly. A way to improve the convergence is to use the subtractive Kramers–Kronig relations as described in Ref. [170]. The idea is to incorporate independent measurements of the real part of the permittivity at one or more reference wave numbers to minimize errors due to extrapolations of the data. This process can be used to derive multiply-subtractive Kramers–Kronig relations if the convergence still remains an issue.

3.2.2 Properties of Linear Response Function

As seen in Eq. (3.32), the response of an observable $\mathcal{A}$ for a linear system can be written as the convolution integral

$$\langle \Delta \mathcal{A}(t) \rangle = \int_0^t \chi_{AB}(t-\tau)F(\tau)\,d\tau = \chi_{AB}(t) * F(t). \tag{3.41}$$

We emphasize that both $\mathcal{A}$ and $\mathcal{B}$ are Hermitian operators, as discussed in Section 2.2. It is easy to verify that the commutator of two Hermitian operators is anti-Hermitian (i.e., $[A, B]^\dagger = -[A, B]$). As a result, the expectation value of the commutator in Eq. (3.33) is a purely imaginary number, making $\chi_{AB}(t)$ a real-valued function. It is important to note that our derivation holds even when the operators are not Hermitian. However, the retarded linear response function can take complex values in that situation.

It is common to call $\chi_{AB}(t)$ the "retarded" linear response function. The reason is that it describes the response of the observable $\mathcal{A}$ at time t to an impulse applied to the system at an earlier time $t - \tau$ through $\mathcal{B}$. This observation follows from the commutative property of the convolution operator in Eq. (3.41). Using $F(t) = \delta(t)$ for an impulse, we find

$$\langle \Delta \mathcal{A}(t) \rangle = \chi_{AB}(t) * \delta(t) = \delta(t) * \chi_{AB}(t) = \int_0^t \chi_{AB}(\tau)\delta(t-\tau)\,d\tau = \chi_{AB}(t). \tag{3.42}$$

This equation shows clearly the response of a linear system at time t when an impulse is applied at time $t - \tau$.

Another important quantity is the *linear response function* denoted by $\mathcal{K}_{AB}(t)$. It determines the system's response caused by the perturbation $H_{ext}(t) = -\mathcal{B}F(t)$ without assuming that $F(t)$ is applied to the system at time $t = 0$. It is defined similar to $\chi_{AB}(t)$ in Eq. (3.33) but without the step function $u(t)$:

$$\mathcal{K}_{AB}(t) = \frac{1}{i\hbar} \langle [\mathcal{B}, \tilde{\mathcal{A}}(t)] \rangle_{eq}. \tag{3.43}$$

This function applies when t varies in the range $(-\infty, \infty)$. It is easy to show that the two response functions are related to each other as

$$\mathcal{K}_{AB}(t) = \begin{cases} \chi_{AB}(t) & \text{if } t > 0 \\ -\chi_{BA}(-t) & \text{if } t < 0. \end{cases} \tag{3.44}$$

Here we used the cyclic permutation property of the trace operator for $t < 0$.

Frequently, the quantity of interest is not the observable $\mathcal{A}$ but its time derivative, $\dot{\mathcal{A}} = d\mathcal{A}/dt$. An example is when one is interested in the current flowing through a device but the charge is measured. We can apply the result in Eq. (3.41) by simply replacing $\mathcal{A}$ with $\dot{\mathcal{A}}$:

$$\langle \Delta \dot{\mathcal{A}}(t) \rangle = \int_0^t \chi_{\dot{A}B}(t-\tau)F(\tau)\,d\tau = \chi_{\dot{A}B}(t) * F(t). \tag{3.45}$$

As the trace operation is invariant under a cyclic permutation, it is possible to write $\chi_{\dot{A}B}(t)$ in two different but useful forms:

$$\chi_{\dot{A}B}(t) = \frac{1}{i\hbar} u(t) \langle [\mathcal{B}, \tilde{\dot{\mathcal{A}}}(t)] \rangle_{eq} = \frac{1}{i\hbar} u(t) \langle [\tilde{\dot{\mathcal{B}}}(-t), \dot{\mathcal{A}}] \rangle_{eq}. \tag{3.46}$$

Another useful form for $\chi_{\dot{A}\mathcal{B}}(t)$ is

$$\chi_{\dot{A}\mathcal{B}}(t) = \frac{i}{\hbar}u(t)\int_t^\infty \frac{d}{d\tau}\langle[\widetilde{\mathcal{B}}(-\tau),\dot{A}]\rangle_{eq}\,d\tau. \tag{3.47}$$

It requires that $\langle[\widetilde{\mathcal{B}}(-\tau),\dot{A}]\rangle_{eq}=0$ in the limit $\tau\to\infty$. This condition follows from the so-called *mixing property*, $\langle\widetilde{A}(\tau_1)\widetilde{\mathcal{B}}(\tau_2)\rangle_{eq}=\langle\widetilde{A}(\tau_1)\rangle_{eq}\langle\widetilde{\mathcal{B}}(\tau_2)\rangle_{eq}$ in the limit $|\tau_1-\tau_2|\to\infty$, which essentially says that there is no correlation between the two operators when the time difference is very large. Recall that the time derivative of $\mathcal{B}$ is obtained using $i\hbar\dot{\mathcal{B}}=[\mathcal{B},H_{eq}]$.

To prove the preceding result, we first note

$$\frac{d}{d\tau}\langle[\widetilde{\mathcal{B}}(-\tau),\dot{A}]\rangle_{eq}=-\langle[\dot{\widetilde{\mathcal{B}}}(-\tau),\dot{A}]\rangle_{eq}=-\langle[\dot{\mathcal{B}},\widetilde{A}(\tau)]\rangle_{eq}, \tag{3.48}$$

where we used the *Kubo identity*

$$\frac{d}{d\tau}\left[\exp(\tau H_{eq})B\exp(-\tau H_{eq})\right]=\exp(\tau H_{eq})[H_{eq},B]\exp(-\tau H_{eq}). \tag{3.49}$$

We then use Eq. (3.48) in Eq. (3.47) to show that

$$\begin{aligned}\chi_{\dot{A}\mathcal{B}}(t) &= \frac{i}{\hbar}u(t)\int_t^\infty \frac{d}{d\tau}\langle[\widetilde{\mathcal{B}}(-\tau),\dot{A}]\rangle_{eq}\,d\tau\\ &= \frac{1}{i\hbar}u(t)\int_t^\infty\langle[\dot{\mathcal{B}},\widetilde{A}(\tau)]\rangle_{eq}\,d\tau = \int_t^\infty\chi_{\dot{A}\dot{\mathcal{B}}}(\tau)\,d\tau.\end{aligned} \tag{3.50}$$

We can now calculate the Fourier transform of $\chi_{\dot{A}\mathcal{B}}$, which is known as the *generalized conductance*:

$$\begin{aligned}\sigma_{A\mathcal{B}}(\omega) &= \mathcal{F}\{\chi_{\dot{A}\mathcal{B}}\}(\omega) = \frac{1}{i\hbar}\int_{-\infty}^\infty e^{i\omega t}u(t)\int_t^\infty\langle[\dot{\mathcal{B}},\widetilde{A}(\tau)]\rangle_{eq}\,d\tau\,dt\\ &= \frac{1}{i\hbar}\int_0^\infty\langle[\dot{\mathcal{B}},\widetilde{A}(\tau)]\rangle_{eq}\int_0^\tau e^{i\omega t}dt\,d\tau = \int_0^\infty\frac{e^{i\omega\tau}-1}{i\omega}\chi_{\dot{A}\dot{\mathcal{B}}}(\tau)\,d\tau.\end{aligned} \tag{3.51}$$

As we shall see later, this expression is useful for applying the fluctuation-dissipation theorem to a physical system.

3.2.3 Generalized Susceptibility

In practice, the retarded linear response function can be constructed using the eigenstates of the Hamiltonian H_{eq} in Eq. (3.26). Let $|\Psi_n\rangle$ with $n=0,1,2,\dots$ form a complete set of the eigenstates of this Hamiltonian, with the corresponding eigenvalues E_n. Here, $|\Psi_0\rangle$ is the ground state with energy E_0, $|\Psi_1\rangle$ is the first excited state with energy E_1, and so on. The retarded linear response requires us to calculate an ensemble average over a chosen equilibrium distribution. It is common to choose the canonical or grand-canonical ensemble for this purpose. It is also possible to employ the microcanonical ensemble (see the derivation in Ref. [172]). However, as this choice does not involve any energy exchange, we can only establish a relation between the linear response function and the correlation function in terms of the total energy range of the ensemble, rather than its temperature. For this reason, we exclude the microcanonical ensemble in the following discussion.

Suppose the density operator ρ_{eq} represents one of the chosen ensembles and the corresponding partition function is given by $Z_{eq} = \text{Tr}\,(\rho_{eq})$. We know that the density operator is diagonal in the eigenstates of the equilibrium Hamiltonian, even though it would take the form of a symmetric density matrix ($\rho_{nm} = \rho_{mn}$) in other representations because of a *detailed balance principle* that is crucial for maintaining the equilibrium distribution within the ensemble [173]. Let $(\rho_{eq})_n$ be the nth diagonal element of ρ_{eq}. Adopting the notation, $\rho_{eq}\,|\Psi_n\rangle = [(\rho_{eq})_n/Z_{eq}]\,|\Psi_n\rangle$, for the action of the density operator on an eigenstate, we write the ensemble average for any operator O as

$$\langle O \rangle_{eq} = \frac{1}{Z_{eq}} \sum_n \langle \Psi_n | O | \Psi_n \rangle \, (\rho_{eq})_n. \tag{3.52}$$

We use this equation and the completeness relation $\sum_n |\Psi_n\rangle \langle \Psi_n| = 1$ in Eq. (3.33) to obtain

$$\mathcal{F}\{\chi_{AB}\}(\omega) = \frac{1}{i\hbar} \int_{-\infty}^{\infty} u(t)\, \langle [\mathcal{B}, \tilde{\mathcal{A}}(t)] \rangle_{eq}\, e^{-i\omega t} dt = \frac{1}{i\hbar} \sum_m \sum_n \frac{(\rho_{eq})_m}{Z_{eq}}$$

$$\int_{-\infty}^{\infty} u(t) e^{-i\omega t} \left[\langle \Psi_m | \mathcal{B} | \Psi_n \rangle \langle \Psi_n | \tilde{\mathcal{A}}(t) | \Psi_m \rangle - \langle \Psi_m | \tilde{\mathcal{A}}(t) | \Psi_n \rangle \langle \Psi_n | \mathcal{B} | \Psi_m \rangle \right] dt. \tag{3.53}$$

We can evaluate each term in the square bracket of the integral by invoking the definition $\tilde{\mathcal{A}}(t) = \exp(\frac{i}{\hbar} H_{eq} t)\, \mathcal{A} \exp(-\frac{i}{\hbar} H_{eq} t)$ in the interaction picture. The resulting values are

$$\langle \Psi_m | \mathcal{B} | \Psi_n \rangle \langle \Psi_n | \tilde{\mathcal{A}}(t) | \Psi_m \rangle = \langle \Psi_m | \mathcal{B} | \Psi_n \rangle \langle \Psi_n | \mathcal{A} | \Psi_m \rangle \exp\left[\frac{i}{\hbar}(E_n - E_m)t \right], \tag{3.54}$$

$$\langle \Psi_m | \tilde{\mathcal{A}}(t) | \Psi_n \rangle \langle \Psi_n | \mathcal{B} | \Psi_m \rangle = \langle \Psi_m | \mathcal{A} | \Psi_n \rangle \langle \Psi_n | \mathcal{B} | \Psi_m \rangle \exp\left[\frac{i}{\hbar}(E_n - E_m)t \right]. \tag{3.55}$$

To simplify further, we make use of the *Sokhotski–Plemelj formula*,

$$\lim_{\epsilon \to 0^+} \frac{1}{\omega \pm i\epsilon} = \text{P.V.} \frac{1}{\omega} \mp i\pi \delta(\omega). \tag{3.56}$$

Aside 3.3 shows the conditions under which this formula is valid and how it can be used in practice.

To ensure the convergence of Fourier integrals in Eq. (3.53), we introduce a small positive parameter ϵ through the factor $e^{-\epsilon t}$ and take the limit $\epsilon \to 0^+$ after performing the integration. The result equations are interpreted by invoking the Sokhotski–Plemelj formula. With this approach, the generalized susceptibility is found to be

$$\mathcal{F}\{\chi_{AB}\}(\omega) = \lim_{\epsilon \to 0^+} -\frac{i}{\hbar} \sum_m \sum_n \frac{(\rho_{eq})_m}{Z_{eq}} \frac{\langle \Psi_m | \mathcal{B} | \Psi_n \rangle \langle \Psi_n | \mathcal{A} | \Psi_m \rangle}{i\omega - i(E_n - E_m)/\hbar + \epsilon}$$

$$+ \lim_{\epsilon \to 0^+} \frac{i}{\hbar} \sum_m \sum_n \frac{(\rho_{eq})_m}{Z_{eq}} \frac{\langle \Psi_m | \mathcal{A} | \Psi_n \rangle \langle \Psi_n | \mathcal{B} | \Psi_m \rangle}{i\omega - i(E_m - E_n)/\hbar + \epsilon}. \tag{3.57}$$

This equation is known as the *Lehmann representation* (or spectral representation) of the generalized susceptibility. It is one of the most important results in linear response theory because it shows explicitly how a perturbation on a quantum device couples to its equilibrium energy spectrum before the perturbation is applied to the system.

Aside 3.3   Sokhotski–Plemelj Formula and Its Uses

To prove the Sokhotski–Plemelj formula, we choose a smooth function $f(\omega)$ that is non-singular in a neighborhood of $\omega = 0$ and consider the identity

$$\int_{-\infty}^{\infty} \frac{f(\omega)}{\omega + i\epsilon}\, d\omega = f(0) \int_{-\infty}^{\infty} \frac{1}{\omega + i\epsilon}\, d\omega + \int_{-\infty}^{\infty} \frac{f(\omega) - f(0)}{\omega + i\epsilon}\, d\omega.$$

Next we take the limit $\epsilon \to 0^+$. The first integral on the right side can be done using a contour in the upper-half complex plane that excludes the origin through a half-circle of radius ϵ centered at $\omega = 0$. Applying Cauchy's theorem, the result is

$$\lim_{\epsilon \to 0^+} \int_{-\infty}^{\infty} \frac{d\omega}{\omega + i\epsilon} = -\pi i.$$

The second integral can also be done using with the result

$$\lim_{\epsilon \to 0^+} \int_{-\infty}^{\infty} \frac{f(\omega) - f(0)}{\omega + i\epsilon}\, d\omega = \lim_{\epsilon \to 0^+} \int_{|\omega| > \epsilon} \frac{f(\omega)}{\omega + i\epsilon}\, d\omega = \text{P.V.} \int_{-\infty}^{+\infty} \frac{f(\omega)}{\omega}\, d\omega,$$

where we used $\lim_{\epsilon \to 0^+} \int_{|\omega| > \epsilon} \frac{f(0)}{\omega + i\epsilon}\, d\omega \to 0$ because it is an odd integral in that limit. The principle value (P.V.) notation indicates that the integral excludes the contribution at the singular point. Putting it together, we have proved the following result:

$$\int_{-\infty}^{\infty} \frac{f(\omega)}{\omega + i\epsilon}\, d\omega = \text{P.V.} \int_{-\infty}^{\infty} \frac{f(\omega)}{\omega}\, d\omega - i\pi f(0)$$

If we use the identity $f(0) = \int_{-\infty}^{\infty} f(\omega)\delta(\omega)\, d\omega$ and note that the preceding equations holds for any smooth function $f(\omega)$, we obtain the result in Eq. (3.56).

As an alternative proof, consider the logarithm function $\ln(z)$ of a complex number z. It is common to define its principal branch using the $\text{Ln}(z)$ notation as

$$\text{Ln}(z) = \ln(|z|) + i\,\text{Arg}(z), \qquad \text{Arg}(z) \in (-\pi, +\pi].$$

Using $z = \omega \pm i\epsilon$ where ω is real and taking the limit $\epsilon \to 0^+$, we obtain

$$\lim_{\epsilon \to 0^+} \text{Ln}(\omega \pm i\epsilon) = \ln(|\omega|) \pm i\pi u(\omega),$$

where $u(\omega)$ is the Heaviside step function. Differentiating this equation with respect to ω and taking the limit as $\epsilon \to 0^+$, we obtain the Sokhotski–Plemelj formula in Eq. (3.56) if we use the known relations:

$$\frac{d}{d\omega} u(\omega) = \delta(\omega), \qquad \frac{d}{d\omega} \ln(|\omega|) = \text{P.V.}\left(\frac{1}{\omega}\right).$$

Let us consider an application of the Sokhotski–Plemelj formula. When modeling quantum devices, one often has to evaluate a double integral of the form

$$I = \int_{-\infty}^{\infty} f(\omega)\, d\omega \int_{0}^{\infty} e^{i\omega t}\, dt.$$

This integral does not converge in the usual sense because the integrand $e^{i\omega t}$ does not vanish as $t \to \infty$. We can make the integrand vanish at infinity by multiplying it with $e^{-\epsilon t}$ and taking the limit $\epsilon \to 0^+$. With this modification, the integral becomes

$$I = \int_{-\infty}^{\infty} f(\omega)\, d\omega \int_0^{\infty} e^{-(\epsilon - i\omega)t} dt = \lim_{\epsilon \to 0^+} \int_{-\infty}^{\infty} \frac{f(\omega)}{\epsilon - i\omega}\, d\omega$$

$$= \pi f(0) + i\left(\text{P.V.} \int_{-\infty}^{\infty} \frac{f(\omega)}{\omega}\, d\omega \right),$$

where we used the Sokhotski–Plemelj formula in Eq. (3.56). This result makes sense only when $f(\omega)$ approaches zero for large ω to ensure the existence of the integral.

It is also useful to consider how the Fourier transform of the function $f(t) = 1/t$ behaves because it does not decay to zero fast enough for large values of t. We just write the results because a proper derivation requires the application of Cauchy's residue theorem with custom contours:

$$\text{P.V.} \int_{-\infty}^{\infty} \frac{\exp(i\omega t)}{t}\, dt == i\pi\, \text{sgn}(-\omega),$$

$$\lim_{\epsilon \to 0^+} \int_{-\infty}^{\infty} \frac{\exp(i\omega t)}{t \pm i\epsilon}\, dt = \pm 2\pi\, i u(\mp \omega).$$

These results will be useful later when we discuss the properties of certain quantum devices.

3.3 Fluctuation-Dissipation Theorem

The fluctuation-dissipation theorem provides a firm theoretical basis for the interaction of matter with its environment (surrounding fields) under the linear-response approximation. The underlying theory relates spontaneous fluctuations of microscopic variables to the kinetic coefficients that are responsible for energy dissipation.

3.3.1 Dynamic Correlation Function

We start by introducing two correlation functions, $C_{AB}(t)$ and $S_{AB}(t)$, defined as

$$C_{AB}(t) = \langle \widetilde{A}(t)B \rangle_{eq}, \qquad S_{AB}(t) = \frac{1}{2} \langle \widetilde{A}(t)B + B\widetilde{A}(t) \rangle_{eq}. \tag{3.58}$$

The function $C_{AB}(t)$ is called the *dynamic correlation function*, and it is a complex function even for Hermitian operators. The function $S_{AB}(t)$ is a real function and is known as the *symmetric correlation function*. By using the cyclic permutation property of the trace, we can show that $\langle \mathcal{B}\tilde{\mathcal{A}}(t)\rangle_{eq} = C_{\mathcal{B}\mathcal{A}}(-t)$ and write it in the form

$$S_{\mathcal{A}\mathcal{B}}(t) = \frac{1}{2}\left[C_{\mathcal{A}\mathcal{B}}(t) + C_{\mathcal{B}\mathcal{A}}(-t)\right]. \tag{3.59}$$

This function has real values if both $\mathcal{A}$ and $\mathcal{B}$ are Hermitian so that $C_{\mathcal{B}\mathcal{A}}(-t) = C^*_{\mathcal{A}\mathcal{B}}(t)$.

The linear response function $\mathcal{K}_{\mathcal{A}\mathcal{B}}(t)$ introduced in Eq. (3.43) is related to the dynamic correlation function as

$$\mathcal{K}_{\mathcal{A}\mathcal{B}}(t) = \frac{i}{\hbar}\left[C_{\mathcal{A}\mathcal{B}}(t) - C_{\mathcal{B}\mathcal{A}}(-t)\right]. \tag{3.60}$$

Taking the Fourier transform of this relation, we obtain

$$\mathcal{F}\{\mathcal{K}_{\mathcal{A}\mathcal{B}}\}(\omega) = \frac{i}{\hbar}\left[\mathcal{F}\{C_{\mathcal{A}\mathcal{B}}(t)\}(\omega) - \mathcal{F}\{C_{\mathcal{B}\mathcal{A}}(t)\}(-\omega)\right], \tag{3.61}$$

where we used the relation $\mathcal{F}\{C_{\mathcal{B}\mathcal{A}}(-t)\}(\omega) = \mathcal{F}\{C_{\mathcal{B}\mathcal{A}}(t)\}(-\omega)$. Thus, we only need to calculate the Fourier transform of the dynamic correlation function.

As the evaluation of this Fourier transform requires an ensemble average, we need to choose an equilibrium distribution. We choose the canonical ensemble as the relevant distribution. We may also choose the grand-canonical ensemble, for which the same results are obtained. We do not consider the microcanonical ensemble because it does not allow us to relate the temperature of the system to the derived quantities (see Ref. [172]). The use of the canonical equilibrium distribution provides us with $\rho_{eq} = \exp(-H_{eq}/k_BT)/Z_{eq}$. We use this form to obtain

$$C_{\mathcal{A}\mathcal{B}}(t) = \mathrm{Tr}\left[\rho_{eq}\tilde{\mathcal{A}}(t)\mathcal{B}\right] = \mathrm{Tr}\left[\rho_{eq}\exp(iH_{eq}t/\hbar)\mathcal{A}\exp(-iH_{eq}t/\hbar)\mathcal{B}\right]$$
$$= \mathrm{Tr}\left[\exp(iH_{eq}t'/\hbar)\mathcal{A}\exp(-iH_{eq}t'/\hbar)\rho_{eq}\mathcal{B}\right] = \mathrm{Tr}\left[\rho_{eq}\mathcal{B}\tilde{\mathcal{A}}(t')\right], \tag{3.62}$$

where we have introduced $t' = t + i\hbar/k_BT$ as a complex variable. Even though this concept has been used in many fields, including general relativity and quantum mechanics [174], no physical meaning can be assigned to t'.

We use the preceding expression to calculate the Fourier transform of $C_{\mathcal{A}\mathcal{B}}(t)$. The result can be simplified as follows by using invariance of the trace operation under cyclic permutations:

$$\mathcal{F}\{C_{\mathcal{A}\mathcal{B}}(t)\}(\omega) = \mathcal{F}\left\{\langle\mathcal{B}\tilde{\mathcal{A}}(t + i\hbar/k_BT)\rangle_{eq}\right\}(\omega) = \mathcal{F}\left\{\langle\tilde{\mathcal{B}}(-t - i\hbar/k_BT)\mathcal{A}\rangle_{eq}\right\}(\omega)$$

$$= \exp(-\hbar\omega/k_BT)\mathcal{F}\{C_{\mathcal{B}\mathcal{A}}(t)\}(-\omega). \tag{3.63}$$

The last result is known as the *detailed-balance condition* for the correlation function $C_{\mathcal{A}\mathcal{B}}(t)$. If we substitute this result in the symmetric correlation function $S_{\mathcal{A}\mathcal{B}}(t)$ defined in Eq. (3.58), we obtain

$$\mathcal{F}\{S_{\mathcal{A}\mathcal{B}}(t)\}(\omega) = \mathcal{F}\{S_{\mathcal{B}\mathcal{A}}(t)\}(-\omega) \rightarrow S_{\mathcal{A}\mathcal{B}}(t) = S_{\mathcal{B}\mathcal{A}}(-t). \tag{3.64}$$

The detailed-balanced condition is a generic property of all systems in thermodynamic equilibrium.

We use Eq. (3.63) to obtain the transform of the linear response function. Noting that $\mathcal{F}\{C_{BA}(-t)\}(\omega) = \mathcal{F}\{C_{BA}(t)\}(-\omega)$, we first obtain

$$\mathcal{F}\{S_{AB}(t)\}(\omega) = \frac{1}{2}\left[1 + \exp(\hbar\omega/k_B T)\right]\mathcal{F}\{C_{AB}(t)\}(\omega). \tag{3.65}$$

Then, using Eq. (3.61), we obtain the following two important relations:

$$\mathcal{F}\{\mathcal{K}_{AB}\}(\omega) = \frac{i}{\hbar}\left[1 - \exp(\hbar\omega/k_B T)\right]\mathcal{F}\{C_{AB}(t)\}(\omega), \tag{3.66}$$

$$\mathcal{F}\{\mathcal{K}_{AB}\}(\omega) = \frac{2i}{\hbar}\frac{\left[1 - \exp(\hbar\omega/k_B T)\right]}{\left[1 + \exp(\hbar\omega/k_B T)\right]}\mathcal{F}\{S_{AB}(t)\}(\omega). \tag{3.67}$$

We obtain the Fourier transform of $\mathcal{K}_{AB}(t)$ using the generalized susceptibility of the retarded linear response function [see Aside 3.2 and Eq. (3.44)]

$$\mathcal{F}\{\mathcal{K}_{AB}\}(\omega) = \int_{-\infty}^{\infty}\mathcal{K}_{AB}(t)e^{i\omega t}\,dt = \int_0^{\infty}\chi_{AB}(t)e^{i\omega t}\,dt - \int_{-\infty}^0\chi_{BA}(-t)e^{i\omega t}\,dt$$

$$= \mathcal{F}\{\chi_{AB}(t)\}(\omega) - \mathcal{F}\{\chi_{BA}(t)\}(-\omega). \tag{3.68}$$

This expression can be further simplified when both A and B are Hermitian operators. Given that the commutator of two Hermitian operators is anti-Hermitian, the expectation value of the commutator is a purely imaginary number, and $\chi_{AB}(t)$ is a real-valued function. Thus, $\mathcal{F}\{\chi_{AB}(t)\}(-\omega) = \mathcal{F}\{\chi_{AB}(t)\}(\omega))^*$. Using this relation in Eq. (3.68), we obtain

$$\mathcal{F}\{\mathcal{K}_{AB}\}(\omega) = 2i\Im\left[\mathcal{F}\{\chi_{AB}(t)\}(\omega)\right]. \tag{3.69}$$

Equating the right-hand sides of Eqs. (3.66) and Eq. (3.69) gives us the well-known *fluctuation-dissipation theorem* for both of the correlation functions [175]:

$$\mathcal{F}\{C_{AB}(t)\}(\omega) = \frac{2\hbar}{\left[1 - \exp(\hbar\omega/k_B T)\right]}\Im\left[\mathcal{F}\{\chi_{AB}(t)\}(\omega)\right], \tag{3.70}$$

$$\mathcal{F}\{S_{AB}(t)\}(\omega) = \frac{\hbar\left[1 + \exp(\hbar\omega/k_B T)\right]}{\left[1 - \exp(\hbar\omega/k_B T)\right]}\Im\left[\mathcal{F}\{\chi_{AB}(t)\}(\omega)\right]. \tag{3.71}$$

It is interesting to consider the classical limit in which $\hbar\omega \ll k_B T$. By approximating $\exp(\hbar\omega/k_B T)$ with $1 + (\hbar\omega/k_B T)$, the preceding two relations take the forms

$$i\omega\mathcal{F}\{C_{AB}(t)\}(\omega) = -k_B T \times 2i\Im\left[\mathcal{F}\{\chi_{AB}(t)\}(\omega)\right], \tag{3.72}$$

$$i\omega\mathcal{F}\{S_{AB}(t)\}(\omega) = -k_B T \times 2i\Im\left[\mathcal{F}\{\chi_{AB}(t)\}(\omega)\right]. \tag{3.73}$$

This result is expected because all operators commute with each other in the classical limit and lead to $C_{AB}(t) = S_{AB}(t)$. Using Eq. (3.69) and taking the inverse Fourier transform, we obtain the following time-domain result valid for $t \geq 0$:

$$\frac{\partial}{\partial t}C_{AB}(t) = -k_B T\mathcal{K}_{AB}(t). \tag{3.74}$$

This result represents the *classical* fluctuation-dissipation theorem.

The quantum fluctuation-dissipation theorem in Eq. (3.70) shows explicitly the close association between quantum fluctuations of a system, as described by the correlation functions, and the linear response of that system as governed by the generalized susceptibility $\mathcal{F}\{\chi_{AB}(t)\}(\omega)$. Aside 3.4 explains why these two quantities are identified as the "fluctuation" and "dissipation" terms by considering a simple system. We stress that, even though only the imaginary part of the generalized susceptibility appears explicitly in Eq. (3.70), the real and imaginary parts are related to each other through the Kramers–Kronig relations. The most general situation occurs when both $\mathcal{A}$ and $\mathcal{B}$ are not observable operators. The fluctuation-dissipation theorem in this case is obtained with the substitution

$$\Im\left[\mathcal{F}\{\chi_{AB}(t)\}(\omega)\right] = \frac{1}{2}\left[\mathcal{F}\{\chi_{AB}(t)\}(\omega) - \mathcal{F}\{\chi_{AB}(t)\}(-\omega)\right]. \qquad (3.75)$$

Aside 3.4 Understanding the Terminology behind the Fluctuation-Dissipation Theorem

To understand the terminology associated with the fluctuation-dissipation theorem, it is instructive to discuss the case of a periodic external perturbation using [see Eq. (3.26)]

$$H_{ext}(t) = -\mathcal{B}F(t) = -[\mathcal{B}^\dagger F_\omega \exp(-i\omega t - \epsilon t) + \mathcal{B}F_\omega^* \exp(i\omega t - \epsilon t)]$$

The infinitesimally small, positive parameter ϵ spreads the linear system's resonance over a finite range of frequencies, thus ensuring a periodic response with a finite amplitude. Its function is similar to that of a damping constant in a driven harmonic oscillator. Recalling that the Hamiltonian of a system represents its energy, the average energy at any instant is given by $\mathrm{Tr}\left(\rho(t)[H_{eq} + H_{ext}(t)]\right)$. The energy dissipation rate of the system can be calculated using the Hellmann–Feynman theorem (see Aside 3.5),

$$P_E(t) = \frac{d}{dt}\,\mathrm{Tr}\left[\rho(t)(H_{eq} + H_{ext})\right]$$

$$= \mathrm{Tr}\left[\frac{d\rho(t)}{dt}(H_{eq} + H_{ext})\right] + \mathrm{Tr}\left[\rho(t)\frac{d}{dt}H_{ext}(t)\right]. \qquad (3.76)$$

It follows from the Liouville equation that the first term vanishes. Using the preceding form of $H_{ext}(t)$, the second term can written as

$$\mathrm{Tr}\left[\rho(t)\frac{d}{dt}H_{ext}(t)\right] = (i\omega + \epsilon)\,\mathrm{Tr}[\rho(t)\mathcal{B}^\dagger]F_\omega \exp(-i\omega t - \epsilon t)$$

$$- (i\omega - \epsilon)\,\mathrm{Tr}[\rho(t)\mathcal{B}]F_\omega^* \exp(i\omega t - \epsilon t).$$

The average dissipated power $\overline{P_E(t)}$ is calculated by integrating Eq. (3.76) over one cycle of duration T in the limit $\epsilon \to 0$:

$$\overline{P_E(t)} = \lim_{\epsilon \to 0}\frac{1}{T}\int_0^T (i\omega + \epsilon)\,\langle\mathcal{B}^\dagger(t)\rangle_{eq}\, F_\omega \exp(-i\omega t - \epsilon t)\,dt$$

$$- \lim_{\epsilon \to 0}\frac{1}{T}\int_0^T (i\omega - \epsilon)\,\langle\mathcal{B}(t)\rangle_{eq}\, F_\omega^* \exp(i\omega t - \epsilon t)\,dt.$$

The linear response theory can be used to show that

$$\langle \mathcal{B}(t) \rangle_{eq} = \mathcal{F}\{\chi_{BB^\dagger}\}(\omega) F_\omega \exp(-i\omega t - \epsilon t),$$
$$\langle \mathcal{B}^\dagger(t) \rangle_{eq} = \mathcal{F}\{\chi_{B^\dagger B}\}(-\omega) F_\omega^* \exp(i\omega t - \epsilon t).$$

Using these results and noting that $\mathcal{F}\{\chi_{BB^\dagger}\}^*(\omega) = \mathcal{F}\{\chi_{B^\dagger B}\}(-\omega)$, we obtain

$$\overline{P_E(t)} = -2\omega \Im[\mathcal{F}\{\chi_{BB^\dagger}\}(\omega)]|F_\omega|^2. \tag{3.77}$$

This equation shows that, on average, the power dissipated by a system is equal to the power received by it.

The fluctuation-dissipation theorem is a powerful tool, both in classical and quantum mechanics, for predicting the behavior of systems that obey the detailed-balance condition. The theorem relies on the assumption that the response of a system in thermodynamic equilibrium to a small stimulus is the same as its response to a spontaneous fluctuation. A powerful consequence of this result is that it connects the relaxation of a linear system from a prepared nonequilibrium state to its quantum fluctuations occurring when the same system is in equilibrium.

Aside 3.5 Hellmann–Feynman Theorem

This theorem states that if $|\Psi\rangle$ is an energy eigenstate of a system with the eigenvalue E, then for any continuous parameter λ on which the Hamiltonian of the system depends, we have the relation

$$\frac{\partial E}{\partial \lambda}\langle \Psi | \Psi \rangle = \langle \Psi | \frac{\partial H}{\partial \lambda} | \Psi \rangle. \tag{3.78}$$

When the eigenstate is normalized ($\langle \Psi | \Psi \rangle = 1$), this theorem can be rephrased as the derivative of the energy with respect to a continuous parameter λ equals the expectation value of the derivative of the Hamiltonian with respect to that same parameter. Even though the theorem is attributed to Hellmann [176] and Feynman [177], it was proven independently by others including Güttinger [178] and Pauli [179].

The derivation starts from the eigenvalue equation $H|\Psi\rangle = E|\Psi\rangle$, which provides us with the identity $\langle \Psi | H | \Psi \rangle = E \langle \Psi | \Psi \rangle$. Differentiating this identity with respect to λ, we obtain

$$\frac{\partial E}{\partial \lambda}\langle \Psi | \Psi \rangle + E \langle \frac{\partial \Psi}{\partial \lambda} | \Psi \rangle + E \langle \Psi | \frac{\partial \Psi}{\partial \lambda} \rangle = \langle \frac{\partial \Psi}{\partial \lambda} | H | \Psi \rangle + \langle \Psi | \frac{\partial H}{\partial \lambda} | \Psi \rangle + \langle \Psi | H | \frac{\partial \Psi}{\partial \lambda} \rangle. \tag{3.79}$$

We can simplify this equation by noting that $\langle \Psi' | H | \Psi \rangle = E \langle \Psi' | \Psi \rangle$, where $\Psi' = \partial \Psi / \partial \lambda$. Also, the Hermitian nature of the Hamiltonian implies $\langle \Psi' | H | \Psi \rangle = \langle \Psi | H | \Psi' \rangle$. Using these two relations, we obtain the relation given in Eq. (3.78). It turns out that the theorem holds even for approximate energy eigenstates, making it quite useful for numerical analysis.

3.3.2 Johnson–Nyquist Noise

In 1928, Johnson discovered [180] and Nyquist explained [181] a source of noise in electrical conductors, now called the Johnson–Nyquist noise. It results from thermal motion of electrons inside an electrical conductor in equilibrium at some finite temperature and occurs regardless of any applied voltage. In 1951, Callen and Welton proved the Johnson–Nyquist result using a formulation based on the fluctuation-dissipation theorem [175]. Their analysis exploits the analogy of a physical system to an electrical circuit and is widely used in physics and engineering to describe the operation of lasers and transistors. It is also of utmost importance to characterize noise in quantum devices.

The fluctuation-dissipation theorem, when written in equivalent electrical terms, relates thermal fluctuations in the current or voltage response of a system to its linear response quantified by the admittance or impedance (through the Thevenin or Norton theorem for equivalent circuits). To understand this, consider a passive two-terminal electrical circuit containing linear components (resistors, capacitors, air-core inductors, etc.) and assume that the circuit is in thermal equilibrium with a reservoir at temperature T. The two terminals can be kept open, as shown in Figure 3.3. The dimensions of the circuit are assumed to be small compared to the wavelength (c/ω) so that the retardation effects can be neglected for quantities observed at its terminals. We connect an external battery to this circuit so that a current $I(t)$ flows through it. Using its Fourier transform $\mathcal{I}_E(\omega)$, the Ohm's law gives us the relation

$$\mathcal{I}_E(\omega) = \mathcal{E}_E(\omega)/Z_E(\omega), \tag{3.80}$$

where $Z_E(\omega)$ is the electrical impedance of the circuit.

To relate the current to our linear response theory, we need to map the operator $\mathcal{B}$ to $\mathcal{I}_E$. This mapping requires some thought about the function $F(t)$ that appears in the system Hamiltonian. Noting that average energy of the system at any instant is given by $\mathrm{Tr}\left(\rho(t)[H_{eq} - \mathcal{B}F(t)]\right)$, the energy-dissipation rate of the system is given by

$$
\begin{aligned}
P_E(t) &= \frac{d}{dt}\mathrm{Tr}\left[\rho(t)(H_{eq} - \mathcal{B}F(t))\right] \\
&= \mathrm{Tr}\left(\frac{d\rho(t)}{dt}[H_{eq} - \mathcal{B}F(t)]\right) - \frac{dF(t)}{dt}\mathrm{Tr}\left[\rho(t)\mathcal{B}\right].
\end{aligned}
\tag{3.81}
$$

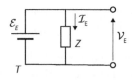

Fig. 3.3 A passive electrical circuit with impedance Z, in equilibrium with a thermal bath at temperature T. An electromotive force $\mathcal{E}_E$ is connected to the terminals and a current $\mathcal{I}_E$ flows through the element with a voltage drop $\mathcal{V}_E$ across it.

As before, the first term vanishes. Noting that $\text{Tr}\,[\rho(t)\mathcal{B}] = \langle\mathcal{B}(t)\rangle_{eq}$, we can identify $\mathcal{B}$ with $\mathcal{I}_E$ and get the relation

$$P_E = -\frac{dF(t)}{dt}\,\langle\mathcal{I}_E\rangle_{eq}. \tag{3.82}$$

As the rate of energy dissipation for a circuit is given by $\langle E_E(t)\mathcal{I}_E\rangle_{eq}$, we must relate the derivative dF/dt to the applied voltage $E_E(t)$ through the relation

$$\frac{dF(t)}{dt} = -E_E(t) \quad\rightarrow\quad i\omega\mathcal{F}\{F\}(\omega) = \mathcal{F}\{E_E\}(\omega), \tag{3.83}$$

where we used the Fourier transform to replace the time derivative with $-i\omega$.

We can now use the retarded linear transfer function, $\mathcal{F}\{\chi_{\mathcal{I}_E\mathcal{I}_E}\}(\omega)$, to obtain the relation

$$\mathcal{F}\{\chi_{\mathcal{I}_E\mathcal{I}_E}\}(\omega) = \frac{\mathcal{F}\{\langle\mathcal{I}_E\rangle_{eq}\}(\omega)}{\mathcal{F}\{F\}(\omega)} = \frac{i\omega}{Z(\omega)}. \tag{3.84}$$

The fluctuation-dissipation theorem requires the imaginary part of this quantity:

$$\Im\left[\mathcal{F}\{\chi_{\mathcal{I}_E\mathcal{I}_E}\}(\omega)\right] = \Im\left[\frac{i\omega}{Z(\omega)}\right] = \frac{\omega\Re\{Z(\omega)\}}{|Z(\omega)|^2}. \tag{3.85}$$

Using this result in Eq. (3.70), we obtain the relations

$$\mathcal{F}\{C_{\mathcal{I}_E\mathcal{I}_E}(t)\}(\omega) = -\frac{2\hbar\omega}{\left[1 - \exp(\hbar\omega/k_BT)\right]}\frac{\Re\{Z(\omega)\}}{|Z(\omega)|^2}, \tag{3.86}$$

$$\mathcal{F}\{S_{\mathcal{I}_E\mathcal{I}_E}(t)\}(\omega) = -\frac{\hbar\omega\left[1 + \exp(\hbar\omega/k_BT)\right]}{\left[1 - \exp(\hbar\omega/k_BT)\right]}\frac{\Re\{Z(\omega)\}}{|Z(\omega)|^2}. \tag{3.87}$$

In electrical literature, it is common to use the voltage V_E measured across the circuit shown in Figure 3.3. From the Ohm's law, we know that $\mathcal{F}\{V_E\}(\omega) = \mathcal{F}\{\mathcal{I}_E\}(\omega)Z(\omega)$. The corresponding voltage fluctuations are thus given by

$$\mathcal{F}\{C_{V_EV_E}(t)\}(\omega) = -\frac{2\hbar\omega\Re\{Z(\omega)\}}{\left[1 - \exp(\hbar\omega/k_BT)\right]}, \tag{3.88}$$

$$\mathcal{F}\{S_{V_EV_E}(t)\}(\omega) = -\hbar\omega\Re\{Z(\omega)\}\frac{\left[1 + \exp(\hbar\omega/k_BT)\right]}{\left[1 - \exp(\hbar\omega/k_BT)\right]}. \tag{3.89}$$

These results provide the spectral density of thermal noise at all frequencies. In the classical limit, we can approximate $\exp(\hbar\omega/k_BT)$ with $1+\hbar\omega/(k_BT)$. It is common to quote the noise spectral density using only positive frequencies, which requires multiplying the result with a factor of two. With these modifications, we obtain the well-known formula

$$\mathcal{F}\{C_{V_EV_E}\}(\omega) = 4K_BT\,\Re\{Z(\omega)\}. \tag{3.90}$$

Figure 3.4 shows the equivalent circuit representation of this result using a noise source with a bandwidth $\Delta f = \Delta\omega/(2\pi)$.

3.4 Dielectric Function

The dielectric function describes the reaction of a system to an external electromagnetic perturbation. As long as this reaction is linear, the response can be obtained by invoking the linear response theory of Section 3.3. As the linear response of a system is independent of the external perturbation, the dielectric function is a property of the system itself. In this section, we consider the electric susceptibility, which describes the polarization resulting from an incident electric field.

3.4.1 Calculation Using Linear Response Theory

The Gauss theorem provides a way to relate the charge density of a material to the scalar potential. As this potential and the electric field are unambiguously related to each other once the gauge is fixed, the dielectric function is most conveniently calculated by considering the charges and potentials involved. We employ the Coulomb gauge and account for the influence of the external field through the interaction Hamiltonian

$$H_{ext}(t) = \int \rho_e(\mathbf{r})\phi_{ext}(\mathbf{r}, t)\, d\mathbf{r}, \tag{3.91}$$

where the charge density $\rho_e(\mathbf{r})$ acts as the operator $\mathcal{B}$ in Eq. (3.26) and the scalar potential $\phi_{ext}(\mathbf{r}, t)$ plays the role of $F(t)$. In the quasi-static approximation, ϕ_{ext} is related to the external electric field through $E_{ext} = -\nabla\phi_{ext}$. The charge density ρ_e created by the perturbation generates the potential ϕ through the Poisson equation (see Section 2.3):

$$\nabla^2\phi(\mathbf{r}, t) = -\frac{1}{\epsilon_0}\rho_e(\mathbf{r}, t). \tag{3.92}$$

This equation can be solved using the Green's function to obtain

$$\phi(\mathbf{r}, t) = \frac{1}{4\pi\epsilon_0} \int \frac{\rho_e(\mathbf{r}', t)}{|\mathbf{r} - \mathbf{r}'|}\, d\mathbf{r}'. \tag{3.93}$$

We recall that $\langle\rho_e(\mathbf{r}, 0)\rangle_{eq} = 0$ as the medium is assumed to have no free charges before the perturbation is turned on. After the perturbation is applied, we can find $\langle\rho_e(\mathbf{r}), t\rangle_{eq}$ by

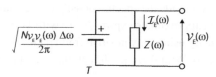

Fig. 3.4 Equivalent circuit for thermal noise generated by a passive electrical circuit that is in thermal equilibrium at temperature T. A noise source on the left generates the current $\mathcal{I}_E$ and the voltage drop $\mathcal{V}_E$ across an element with the impedance Z.

invoking the linear response theory of Section 3.3 (with ρ_e acting as both $\mathcal{A}$ and $\mathcal{B}$) and write the ensemble-averaged charge density as

$$\langle \rho_e(\mathbf{r},t) \rangle_{eq} = -\iint \chi_{\rho_e \rho_e}(\mathbf{r},t;\mathbf{r}_1,t_1) \phi_{ext}(\mathbf{r}_1,t_1) \, dt_1 \, d\mathbf{r}_1, \tag{3.94}$$

where $\chi_{\rho_e \rho_e}(\mathbf{r},t;\mathbf{r}_1,t_1)$ is the polarizability. As a charge density at one point creates a potential at another point, as indicated in Eq. (3.93), the total potential $\phi(\mathbf{r},t)$ can be written as the sum of the incident induced potentials:

$$\phi(\mathbf{r},t) = \phi_{ext}(\mathbf{r},t) - \iiint \frac{\chi_{\rho_e \rho_e}(\mathbf{r}_2,t;\mathbf{r}_1,t_1)}{4\pi \epsilon_0 |\mathbf{r} - \mathbf{r}_2|} \phi_{ext}(\mathbf{r}_1,t_1) \, dt_1 \, d\mathbf{r}_1 \, d\mathbf{r}_2. \tag{3.95}$$

We can rearrange this expression to show the input-output relationship explicitly as

$$\phi(\mathbf{r},t) = \iint \left[\delta(\mathbf{r} - \mathbf{r}_1)\delta(t - t_1) - \int \frac{\chi_{\rho_e \rho_e}(\mathbf{r}_2,t;\mathbf{r}_1,t_1)}{4\pi \epsilon_0 |\mathbf{r} - \mathbf{r}_2|} \, d\mathbf{r}_2 \right]$$
$$\times \, \phi_{ext}(\mathbf{r}_1,t_1) \, dt_1 \, d\mathbf{r}_1. \tag{3.96}$$

The *dielectric function* $\epsilon(\mathbf{r},t;\mathbf{r}_1,t_1)$ is defined to relate the field variables $\mathbf{E}$ and $\mathbf{D}$ as

$$\mathbf{E}(\mathbf{r},t) = \iint \epsilon^{-1}(\mathbf{r},t;\mathbf{r}_1,t_1)\mathbf{D}(\mathbf{r}_1,t_1) \, dt_1 \, d\mathbf{r}_1. \tag{3.97}$$

We can convert it to a relation between the two potentials by using

$$\mathbf{E}(\mathbf{r},t) = -\nabla\phi(\mathbf{r},t), \qquad \mathbf{D}(\mathbf{r},t) = -\nabla\phi_{ext}(\mathbf{r},t). \tag{3.98}$$

It is easy to show that the two potentials are also related as

$$\phi(\mathbf{r},t) = \iint \epsilon^{-1}(\mathbf{r},t;\mathbf{r}_1,t_1)\phi_{ext}(\mathbf{r}_1,t_1) \, dt_1 \, d\mathbf{r}_1. \tag{3.99}$$

A direct comparison of this equation with Eq. (3.96) provides us with the important result:

$$\epsilon^{-1}(\mathbf{r},t;\mathbf{r}_1,t_1) = \delta(\mathbf{r} - \mathbf{r}_1)(t - t_1) - \int \frac{\chi_{\rho_e \rho_e}(\mathbf{r}_2,t;\mathbf{r}_1,t_1)}{4\pi \epsilon_0 |\mathbf{r} - \mathbf{r}_2|} \, d\mathbf{r}_2. \tag{3.100}$$

If we assume that the medium has translational invariance in both space and time, its polarizability depends only on the differences $\mathbf{r}_2 - \mathbf{r}_1$ and $t - t_1$. The dielectric function for such a medium takes the form:

$$\epsilon^{-1}(\mathbf{r} - \mathbf{r}_1, t - t_1) = \delta(\mathbf{r} - \mathbf{r}_1)(t - t_1) - \int \frac{\chi_{\rho_e \rho_e}(\mathbf{r}_2 - \mathbf{r}_1, t - t_1)}{4\pi \epsilon_0 |\mathbf{r} - \mathbf{r}_2|} \, d\mathbf{r}_2, \tag{3.101}$$

As the last term is in the form of a convolution in both space and time, we can write ϵ^{-1} in the Fourier domain in the following compact form:

$$\mathcal{F}\{\epsilon^{-1}\}(\mathbf{k},\omega) = 1 - \frac{\mathcal{F}\{\chi_{\rho_e \rho_e}\}(\mathbf{k},\omega)}{\epsilon_0(\mathbf{k} \cdot \mathbf{k})}, \tag{3.102}$$

where the four-dimensional Fourier transform of a function $G(\mathbf{r},t)$ is defined as

$$\mathcal{F}\{G\}(\mathbf{k},\omega) = \iint G(\mathbf{r},t) \exp(i\mathbf{k} \cdot \mathbf{r} - i\omega t) \, d\mathbf{r} \, dt. \tag{3.103}$$

3.4.2 Relation to the Frequency-Dependent Conductivity

In practice, it is useful to write the dielectric function in terms of the material's conductivity that can be measured using a variety of techniques. Ohm's law gives us the relation

$$\mathcal{F}\{\mathbf{J}\}(\mathbf{k}, \omega) = \mathcal{F}\{\sigma\}(\mathbf{k}, \omega)\mathcal{F}\{\mathbf{E}\}(\mathbf{k}, \omega) = -i\mathbf{k}\mathcal{F}\{\sigma\}(\mathbf{k}, \omega)\mathcal{F}\{\phi\}(\mathbf{k}, \omega). \tag{3.104}$$

We can eliminate $\mathcal{F}\{\mathbf{J}\}(\mathbf{k}, \omega)$ using the continuity equation, $\partial\rho_e/\partial t + \nabla \cdot \mathbf{J} = 0$, which, in the Fourier domain, takes the form

$$\mathbf{k} \cdot \mathcal{F}\{\mathbf{J}\}(\mathbf{k}, \omega) = \omega\mathcal{F}\{\rho_e\}(\mathbf{k}, \omega) = -\omega\mathcal{F}\{\chi_{\rho_e\rho_e}\}(\mathbf{k}, \omega)\mathcal{F}\{\phi_{ext}\}(\mathbf{k}, \omega), \tag{3.105}$$

where we used Eq. (3.94) in the Fourier domain. Combining the two preceding equations, we obtain

$$\mathcal{F}\{\chi_{\rho_e\rho_e}\}(\mathbf{k}, \omega) = i\frac{\mathbf{k}^2}{\omega}\frac{\mathcal{F}\{\sigma\}(\mathbf{k}, \omega)}{1 + i\frac{\mathcal{F}\{\sigma\}(\mathbf{k},\omega)}{\epsilon_0\omega}}. \tag{3.106}$$

Using this form in Eq. (3.102), the dielectric function can be written in terms of the conductivity as

$$\mathcal{F}\{\epsilon\}(\mathbf{k}, \omega) = 1 + i\frac{\mathcal{F}\{\sigma\}(\mathbf{k}, \omega)}{\epsilon_0\omega}. \tag{3.107}$$

It is interesting to note that the frequency-dependent conductivity $\sigma(\mathbf{k}, \omega))$, which obeys the causality requirement, is proportional to $\mathcal{F}\{\epsilon^{-1}\} - 1$. This is the reason that the Kramers–Kronig relations must be applied to $\mathcal{F}\{\epsilon^{-1}\} - 1$, and not to the dielectric function itself.

The form of the dielectric function in Eq. (3.107) can be simplified further for materials whose linear response is spatially local (i.e., the response at any point depends only on the perturbation near that point; for details, see Refs. [182, 183]). In the Fourier domain, this amounts to setting $\mathbf{k} = 0$. It is common to drop the $\mathbf{k}$ dependence and write Eq. (3.107) as

$$\mathcal{F}\{\epsilon\}(\omega) = 1 + i\frac{\mathcal{F}\{\sigma\}(\omega)}{\epsilon_0\omega}. \tag{3.108}$$

This form is valid when the wavelength inside the material is significantly longer than other characteristic lengths such as the unit-cell size or the mean free path of electrons [184].

Equation (3.107) implicitly separates the charges into bound and free charges (via definitions of **E**, and **D**). At low frequencies, $\mathcal{F}\{\epsilon\}(\omega)$ is used for describing the response of bound charges to a driving field, leading to an electric polarization, while $\mathcal{F}\{\sigma\}(\omega)$ describes the contribution of free charges to the current flow. Even though such a clear separation is possible in certain spectral regions, the distinction between the bound and free charges is blurred at high frequencies, particularly at optical wavelengths. For example, in the case of highly doped semiconductors, the response of bound valence electrons can be lumped into a static dielectric constant ϵ_{sd}, while the response of conduction electrons is lumped into the conductivity $\mathcal{F}\{\sigma\}(\omega)$, leading to a dielectric function of the form [28]

$$\mathcal{F}\{\epsilon\}(\omega) = \epsilon_{sd} + i\frac{\mathcal{F}\{\sigma\}(\omega)}{\epsilon_0\omega}. \tag{3.109}$$

It is possible to rearrange the two terms without affecting the function value by replacing ϵ_{sd} with 1 and subtracting $i\epsilon_0\omega(\epsilon_{sd} - 1)$ from $\mathcal{F}\{\sigma\}(\omega)$, as suggested in Ref. [185].

We have chosen not to make such a change and write the dielectric function in its widely used form.

The preceding discussion shows that, in general, $\mathcal{F}\{\epsilon\}(\omega)$, and $\mathcal{F}\{\sigma\}(\omega)$ are complex functions that are often determined through experimental measurements. At optical frequencies, it is common to employ the concept of a *complex refractive index* using the definition,

$$n(\omega) = \sqrt{\mathcal{F}\{\epsilon\}(\omega)} = n_R(\omega) + in_I(\omega), \tag{3.110}$$

where the real part n_R is called the *refractive index* and the imaginary part n_I is related to the loss of electromagnetic power through absorption inside the medium.

3.4.3 Models Used for the Permittivity

Many models have been developed to find an analytic form $\mathcal{F}\{\epsilon\}(\omega)$ for different media. For example, a free-electron-gas model is often used for metals to calculate the response of electrons to a time-varying electric field. This model, known as the Drude model, provides the following expression [185]:

$$\mathcal{F}\{\epsilon_{\text{Drude}}\}(\omega) = \epsilon_0 - \frac{\epsilon_0\omega_p^2}{\omega^2 + i\gamma(\omega)\omega}, \tag{3.111}$$

where $\omega_p = \sqrt{q_e^2 N_0/(m_e\epsilon_0)}$ is the plasma frequency introduced in Section 1.3.1 (N_0 being the density of electrons) and γ is the collision rate of the free-electron gas [28]. This formula is valid if the smallest dimension of a metallic object is larger than the electron's mean free path l_e (defined in Section 1.2). If that is not the case, γ needs to be replaced by γ_n defined as [186]:

$$\gamma_n(\omega) = \gamma(\omega) + 2g_1(v_F/D) \tag{3.112}$$

where D is the enclosing diameter of the nanoscale object and

$$g_1 = \frac{E_F}{hbar\omega} \int_{1-\kappa}^{1} x^{3/2}(x+\kappa)^{1/2}\, dx. \tag{3.113}$$

The Fermi velocity v_F and the Fermi energy E_F have been defined in Section 1.2. The confinement of electrons to a nanoscale object provides the second contribution to γ_n.

To the lowest order, the linear response of an electron gas is governed by the plasma frequency that depends on the density N_0 of the electrons (see Section 1.3.1). Notice that this frequency does not depend on the size of the object that contains the electron gas [33]. At low frequencies such that $\omega \ll \gamma$, absorption is so large inside metals that any external electrical field decays from the metal surface as $e^{-z/\delta}$, where the skin depth δ is defined as [28]

$$\delta = \frac{c}{\omega_p}\sqrt{\frac{2\gamma}{\omega}}. \tag{3.114}$$

Owing to the frequency dependency of $\gamma(\omega)$, special care has to be taken to find the conductivity in the limit $\omega \to 0$ (see Ref. [187]).

The Drude model involves two time scales. The time scale for charge separation within the electron gas is governed by $1/\omega_p$; it restores the quasi-neutral behavior of the electron gas when subjected to disturbances. The second time scale is the relaxation time, defined as $\tau = 1/\gamma$ (see Ref. [188, 189, 190]). It describes the relaxation of excited electrons, and its magnitude depends on the frequency of the electromagnetic wave as $\gamma(\omega) = \gamma_0 + b\omega^2$, where γ_0 and b are two constants [191, 192]. The electron–phonon interaction is responsible for γ_0, while *Umklapp scattering* provides the constant b.

The treatment of a metal as a free-electron gas entirely ignores the fact that the motion of electrons is also affected by the metal's nuclei within the crystal lattice and other electrons bound to these nuclei. Their influence can be taken into account by introducing a background dielectric constant ϵ_b that is real, a reasonable approach in most cases of practical interest. This quantity can be calculated by considering the interband transitions from the fully occupied bands below the Fermi energy to the half-filled conduction band. The Drude's dielectric function given in Eq. (3.111) is then modified as

$$\mathcal{F}\{\epsilon\}(\omega) = \epsilon_b(\omega) - \frac{\epsilon_0 \omega_p^2}{\omega^2 + i\gamma(\omega)\omega}. \tag{3.115}$$

Even though the Drude model is a useful tool that can describe many significant features of a metal's permittivity, especially if we include the intraband contributions, it fails on several counts. First, because the model relies on the conduction-band electrons, any process that may excite electrons in the bands of lower energies (usually a d-band) can introduce significant deviations from its predictions. This is why the Drude model provides better predictions in the infrared region compared to the visible region. Second, the assumption that the conduction band is parabolic (through the effective-mass approximation) is a severe limitation for representing real metals. Third, the Drude model ignores the energy dependence of γ, even though significant energy dependence of this parameter is expected from a Boltzmann-type analysis. Fourth, a severe limitation results from the local-response assumption that completely ignores the wave-vector dependence of the permittivity. Such an assumption is valid only when the electric field varies so slowly that it is safe to discard spatial dispersion.

In reality, the permittivity has two length scales corresponding to two different phenomena. The first effect is the screening of the field at an interface of metals that can be described by classical electrostatic theory. The Drude model assumes a negligible screening length. However, at the atomic scale the screening process requires a certain distance to take place. The screening length is also known as the Thomas–Fermi length for metals [185]. The second length scale is associated with the nonlocal response of the metal. It corresponds to the distance traveled by an electron of velocity v_F over the duration of an optical cycle. When this length is much smaller than the wavelength of the external field, the optical properties are not affected much. This can occur when $v_F/\omega \ll \lambda$ and $\omega/\gamma \gg 1$. In contrast, non-local corrections are required when $v_F/\omega_p > \lambda/2\pi$. Also, when the electron velocity v_F equals the phase velocity ($v_p = \omega\lambda/2\pi$) of an electromagnetic field, the so-called Landau damping becomes important [193].

The Drude model is not suitable for modeling dielectrics or semiconductors. Reasonably accurate permittivity models that can describe dielectrics or semiconductors include the Sellmeier and Cauchy models for transparent materials with no conduction electrons [194], models for semiconductors based on the harmonic-oscillator approximation, the Lorentz model for dielectrics [195], and the Tauc–Lorentz model [196]. The Lorentz model assumes that every electron is bound to a positively charged atomic nucleus by a harmonic potential and that it exhibits damped harmonic oscillations under the influence of an external electric field. This model can be used to describe dielectrics that exhibit momentum-preserving intraband transitions. It is similar in form to the Drude model except for the addition of a resonance frequency ω_0 describing the oscillations of the bound electrons:

$$\mathcal{F}\{\epsilon_{\text{Lorentz}}\}(\omega) = \epsilon_b(\omega) + \frac{\epsilon_0 \omega_p^2}{\omega_0^2 - \omega^2 - i\omega\gamma}. \tag{3.116}$$

The Lorentz model is widely used for dielectric materials.

In recent years, new types of lasers, known as *spasers* because of the role played by surface plasmons, have been developed. They are based on 2D semimetals such as graphene and 2D semiconductors such as MoS_2 [197, 198, 199]. It is worth considering how the electrical behavior of these materials differs from the Drude or Lorentz-type models (see Refs. [200, 201]). Here we focus on the graphene, although a similar analysis can be adopted for other 2D materials. The main difference would be that some 2D materials have finite bandgaps (e.g., transition-metal dichalcogenides, hexagonal boron nitride, and silicene).

Experimental and theoretical results show that the permittivity of graphene is given by

$$\mathcal{F}\{\epsilon_{\text{graphene}}\}(\omega) = \epsilon_0 + \frac{i}{\omega d}\sigma_{\text{graphene}}(\omega), \tag{3.117}$$

where d is the thickness of graphene and $\sigma_{\text{graphene}}(\omega)$ is its conductivity. Early calculations of the graphene's conductivity were carried out using the Dirac Hamiltonian [202, 203, 204]. Most studies considered the effects of disorder in a phenomenological manner, but they were included self-consistently by Peres *et al.* [205]. Further improvements have incorporated electron–electron interactions [206, 207], a finite chemical potential, spatial dispersion [208, 209], a graphene bilayer [210, 211, 212, 213], the use of the Boltzmann distribution [214, 215], and the effects of temperature [216]. Based on these advances, the conductivity of graphene is found to be given by [216, 217, 218]

$$\mathcal{F}\{\sigma_{\text{graphene}}\}(\omega) = \frac{e^2}{8\hbar}\left[\tanh\left(\frac{\hbar\omega + 2E_F}{4k_BT}\right) + \tanh\left(\frac{\hbar\omega - 2E_F}{4k_BT}\right)\right]$$
$$- i\frac{e^2}{8\pi\hbar}\ln\left[\frac{(\hbar\omega + 2E_F)^2}{(\hbar\omega - 2E_F)^2 + (2k_BT)^2}\right] + i\frac{e^2}{\pi\hbar}\left(\frac{E_F}{\hbar\omega + i\hbar\gamma}\right), \tag{3.118}$$

where γ is the intraband scattering rate and E_F is the Fermi energy relative to the Dirac point obtained from $N_e = (E_F/\hbar v_F)^2/\pi$. The first two terms in Eq. (3.118) result from interband transitions, and the last term is due to intraband transitions. Owing to a strong dependence of the conductivity of graphene on the Fermi energy E_F, the permittivity of graphene can be easily controlled through electrostatic gating or chemical doping, both of which have been exploited in practice.

Consider the high-doping limit of the preceding equation. It is easy to see that graphene becomes lossy when $\hbar\omega > 2E_F$ (owing to the dominance of interband transitions), or when

$\hbar\omega < \hbar\gamma$ so that intraband free-carrier absorption dominates. If $E_F \gg k_BT$, Eq. (3.118) reduces to the standard Drude model because

$$\mathcal{F}\{\sigma_{\mathrm{graphene}}\}(\omega) = \frac{i\sigma_0}{\omega + i\gamma}, \quad \text{with} \quad \sigma_0 = \frac{e^2 E_F}{\pi \hbar^2}. \tag{3.119}$$

The parameter σ_0 is often used as a fitting parameter [219, 220]. For higher plasmon energies, we have to supplement this expression with an additional term that takes into account the interband contributions (for $E_F \gg k_BT$), resulting in

$$\mathcal{F}\{\sigma_{\mathrm{graphene}}\}(\omega) \approx \frac{i\sigma_0}{\omega + i\gamma} + \frac{e^2}{4\hbar}\left(u(\hbar\omega - 2E_F) - \frac{i}{\pi}\ln\left|\frac{\hbar\omega + 2E_F}{\hbar\omega - 2E_F}\right|\right), \tag{3.120}$$

where $u(x)$ is the Heaviside step function [169]. The logarithmic singularity in this expression at $\hbar\omega = E_F$ disappears at finite temperatures, as seen in Eq. (3.118).

3.5 Sum Rules

The *sum rules* of the retarded linear response function (see Section 3.3) are the relations in the form of an integral that any generalized susceptibility must satisfy. They stipulate that the Fourier transform $\mathcal{F}\{\chi_{AB}\}(\omega)$ of this function must be related to an equilibrium correlation function of certain derivatives of the operators $\mathcal{A}(t)$ and $\mathcal{B}(t)$ at the same moment. In practice, the sum rules are important to ensure the validity of any phenomenological model adopted for calculating the generalized susceptibility, especially at short wavelengths. When applied to the dielectric function of a medium (which is derived from a generalized susceptibility), the sum rules can be viewed as universal constraints on the frequency-domain results of the dielectric function of a medium. The existence of the integrals associated with the sum rules can be established using the superconvergence theorem given in Aside 3.6.

Our starting point is the fluctuation-dissipation theorem given in Eq. (3.69). Converting it to the time domain and using Eq. (3.43), this equation can be written as

$$\frac{i}{\pi}\int_{-\infty}^{\infty} \Im\left[\mathcal{F}\{\chi_{AB}\}(\omega)\right]\exp(-i\omega t)\,d\omega = \frac{i}{\hbar}\langle[\tilde{\mathcal{A}}(t), \mathcal{B}]\rangle_{eq}. \tag{3.121}$$

If we substitute $t = 0$ in this equation, we obtain the first sum rule

$$\frac{\hbar}{\pi}\int_{-\infty}^{\infty} \Im\left[\mathcal{F}\{\chi_{AB}\}(\omega)\right]\,d\omega = \langle[\mathcal{A}, \mathcal{B}]\rangle_{eq}, \tag{3.122}$$

where we used the relation $\tilde{\mathcal{A}}(t) = U^\dagger(t)\mathcal{A}U(t)$ with $U(t) = \exp(-\frac{i}{\hbar}H_{eq}t)$.

We can find other sum rules by taking the derivatives on both sides of Eq. (3.121) and using the relation

$$\frac{d^n}{dt^n}\tilde{\mathcal{A}}(t) = \left(\frac{i}{\hbar}\right)^n \mathcal{W}^n\tilde{\mathcal{A}}(t), \tag{3.123}$$

where the operator $\mathcal{W}$ is defined as $\mathcal{W}\widetilde{\mathcal{A}}(t) = [H_{eq}, \widetilde{\mathcal{A}}(t)]$. Taking the nth derivative, we obtain

$$\frac{\hbar}{\pi} \int_{-\infty}^{\infty} (i\omega)^n \Im\left[\mathcal{F}\{\chi_{\mathcal{AB}}\}(\omega)\right] \exp(-i\omega t)\, d\omega = \left(\frac{i}{\hbar}\right)^{n+1} \langle [\mathcal{W}^n \widetilde{\mathcal{A}}(t), \mathcal{B}]\rangle_{eq}. \qquad (3.124)$$

When this expression is evaluated at $t = 0$, we get multiple sum rules for different values of the integer n in the form

$$\frac{\hbar}{\pi} \int_{-\infty}^{\infty} (\hbar\omega)^n \Im\left[\mathcal{F}\{\chi_{\mathcal{AB}}\}(\omega)\right]\, d\omega = \langle [\mathcal{W}^n \mathcal{A}, \mathcal{B}]\rangle_{eq}. \qquad (3.125)$$

It is instructive to look at the simplest case for the generalized susceptibility associated with the operator $\mathcal{B}$ and given by $\mathcal{F}\{\chi_{\mathcal{BB}}\}(\omega)$. Using $\mathcal{A} = \mathcal{B}$ in the preceding equation, we obtain

$$\frac{\hbar}{\pi} \int_{-\infty}^{\infty} (\hbar\omega)^n \Im\left[\mathcal{F}\{\chi_{\mathcal{BB}}\}(\omega)\right]\, d\omega = \langle [\mathcal{W}^n \mathcal{B}, \mathcal{B}]\rangle_{eq}. \qquad (3.126)$$

Noting that $\Im\left[\mathcal{F}\{\chi_{\mathcal{AB}}\}(\omega)\right]$ is an odd function of frequency ω, the preceding integral vanishes for even values of n. Restricting to the odd powers by substituting $n = 2p + 1$, we get the relation

$$\frac{1}{\pi} \int_{-\infty}^{\infty} (\hbar\omega)^{2p+1} \Im\left[\mathcal{F}\{\chi_{\mathcal{BB}}\}(\omega)\right]\, d\omega = \langle [\mathcal{W}^{2p+1}\mathcal{B}, \mathcal{B}]\rangle_{eq}, \quad p = 0, 1, \ldots. \qquad (3.127)$$

The sum rule for $p = 0$ is known as the f-sum rule and is widely used for the complex refractive index of a medium. It is especially useful for checking the self-consistency of experimental or model-generated data. Also, as the dielectric function depends on atomic transitions, the sum rules can yield information about such transitions. It is interesting to note that the f-sum rule is analogous to the Thomas–Reiche–Kuhn sum rule used in quantum mechanics [221].

Aside 3.6 Superconvergence Theorem

The superconvergence theorem is central to the construction of sum rules. Here we just state this theorem without proof, as it is not hard to ascertain its validity by inspection. However, a rigorous proof requires an intricate analysis based on complex function theory. The theory is formulated for a function $f(x)$ that is continuously differentiable and asymptotically behaves as $f(x) = O[(x \ln x)^{-1}]$. This function is used to define a new function as $g(y) = \text{P.V.} \int_0^{\infty} [f(x)/(y^2 - x^2)])\, dx$. Then, the following asymptotic result holds for $y \gg x$ [170]:

$$g(y) = \frac{1}{y^2} \int_0^{\infty} f(x)\, dx + O(y^{-2}) \rightarrow \int_0^{\infty} f(x)\, dx = \lim_{y \to \infty} y^2 g(y).$$

We can find the f-sum rule for the complex refractive index $n(\omega)$ using the Kramers–Kronig relations given in Eq. (3.110):

$$
\begin{aligned}
n_R(\omega) - 1 &= \frac{2}{\pi} \, \text{P.V.} \int_0^\infty \frac{\omega' n_I(\omega')}{\omega' - \omega^2} \, d\omega' \\
n_I(\omega) &= -\frac{2\omega}{\pi} \, \text{P.V.} \int_0^\infty \frac{n_R(\omega') - 1}{\omega' - \omega^2} \, d\omega'.
\end{aligned}
\tag{3.128}
$$

The f-sum rule depends on the asymptotic behavior of $n(\omega)$ obtained from the Lorentz model in Eq. (3.116) using $\epsilon_b = 1$ and $n^2 = \epsilon/\epsilon_0$. In the limit $\omega \to \infty$, $n(\omega)$ behaves as

$$
n(\omega) = 1 - \frac{1}{2} \left(\frac{\omega_p}{\omega} \right)^2 + \dots,
\tag{3.129}
$$

where ω_p is the plasma frequency defined in Eq. (3.131). If this result is applied to the first Kramers–Kronig relation in Eq. (3.128), we get the *f-sum rule*:

$$
\int_0^\infty \omega n_I(\omega) \, d\omega = \frac{\pi}{4} \omega_p^2.
\tag{3.130}
$$

3.6 Surface Plasmons

Surface plasmons have attracted considerable attention in recent years in the context of spasers [184]. The properties of dielectric functions can be exploited to generate surface plasmons on the surface of a metal, both in passive and active structures. To describe their properties, we need to first understand how the electrons in a metal interact with an external electromagnetic field. A rigorous analysis normally impedes the insight gained from an approximate, back-of-the-envelope type analysis. It is worth carrying out such an analysis first to identify the relevant length, time, and energy scales.

A major simplification is realized by adopting the Born–Oppenheimer approximation [222] because it separates the vibrational dynamics of molecules from their electronic response. This approximation is based on the intuition that the motion of heavy nuclei can be separated from that of loosely bound electrons (mostly valence electrons) because electrons, being much lighter, can follow the motion of the nuclei almost instantaneously. As electrons move through a solid, they experience Coulomb forces from other electrons and atomic cores scattered throughout the solid. The net result is that the valence electrons experience a time-dependent potential, whose inclusion adds significant complexity and hinders our ability to build a simplified picture of the electron's response. The problem can be simplified by considering a single electron in an effective periodic but time-independent potential, produced by the stationary nuclei and other electrons in their equilibrium positions. It is common to neglect electron–electron interactions that cannot be represented as a local potential for the single electron under consideration (such as those arising from the exchange of two electrons). In this viewpoint, free electrons behave as a gas and move in a fixed distribution of positive charges, ensuring the electrical neutrality of the system. This model is known as the jellium model [33].

When subjected to an external electromagnetic field, the dynamics of the electron gas in the jellium model is characterized by collective oscillations of the electron gas. To the lowest order, the linear response of the electron gas is governed by the frequency of these oscillations. This frequency, known as the *plasma frequency*, has appeared in the Drude model discussed in Section 3.5.3. It depends on the density N_e of electrons as

$$\omega_p = \sqrt{q_e^2 N_e/(m_e \epsilon_0)}, \tag{3.131}$$

but it does not depend on the size of the object that contains the electron gas [33]. The plasma frequency marks the frequency beyond which a metal can no longer screen electric fields. The oscillations arise from the appearance of a restoring Coulomb force when electrons are displaced from their charge-neutral configuration (thus creating a net positive charge). Owing to their inertia, electrons do not replenish the positive region, but travel further away, thus creating an excess positive charge. This effect gives rise to coherent oscillations at the plasma frequency. The coherence of this collective motion is progressively destroyed by the Landau damping and by collisions of an electron with phonons or other electrons [34, 35, 36]. The Drude model discussed in Section 3.5.3 includes this damping through the parameter γ.

An intriguing feature, common to all metals that can be described by the Drude model, is the monotonous decrease of the real part of the permittivity. For example, $\Re[\epsilon(\omega)/\epsilon_0]$ for noble metals has a small positive value in the ultraviolet region, but it takes large negative values in the infrared region. A negative value of this quantity is essential to sustain the localized surface plasmons in nanostructures, and this is the main reason why metals are considered an essential ingredient for plasmonics. Indeed, many plasmonic applications require the range $-1 \geq \Re[\epsilon(\omega)/\epsilon_0] \geq -20$ in practice. To recognize the importance of the imaginary part of the permittivity, $\Im[\epsilon(\omega)/\epsilon_0]$, we need to consider the quality factor of localized plasmonic oscillations [223]. This quantity is responsible for losses inside the material. However, the real and imaginary parts are related through a quality factor Q that provides a figure of merit of the plasmons and is defined as

$$Q = \frac{\omega}{2[\Im(\epsilon(\omega))]^2} \frac{\mathrm{d}}{\mathrm{d}\omega} \Re(\epsilon(\omega)). \tag{3.132}$$

As discussed in Ref. [223], near the resonance frequency of a localized plasmon, Q depends solely on the complex dielectric function of the material and is independent of the geometry of the nanostructure. This result assumes the quasi-static limit in which dimensions of a nanoparticle are much smaller than the wavelength of the electromagnetic field. It also assumes that the dielectric part responsible for plasmonic resonance is lossless. The dielectric part, although essential for surface plasmons, plays a less significant role in determining the sharpness and quality of a plasmonic resonance. For typical metals, $\Im[\epsilon(\omega)/\epsilon_0]$ is in the range of 2–10. Surface plasmons become well pronounced only when $\Im[\epsilon(\omega)] \gg -\Re[\epsilon(\omega)]$, while losses remain relatively small [224].

Although our focus here is on the localized surface plasmons, it is instructive to also consider propagating surface plasmons, known as the surface-plasmon polaritons (SPPs) because of their coupling to an electromagnetic field. Their study provides useful length scales to estimate the behavior of localized surface plasmons. SPPs propagate along a metal–dielectric interface and suffer from losses in the directions both parallel and perpendicular to this interface. The conditions for the existence of SPPs can be written as [29]

$$\epsilon_d \epsilon_m < 0 \quad \text{and} \quad \epsilon_d + \epsilon_m < 0, \tag{3.133}$$

where ϵ_d and ϵ_m are the permittivities of the dielectric and the metal, respectively. The wavelength $\lambda_\parallel$ and the propagation distance $\delta_\parallel$ can be derived from the dispersion relation of SPPs:

$$k_\parallel = k_0 \sqrt{\frac{\epsilon_d \epsilon_m}{(\epsilon_d + \epsilon_m)}}. \tag{3.134}$$

where $k_0 = 2\pi/\lambda_0$ is the free-space wave vector. Taking the real and imaginary parts of $k_\parallel$, we obtain [225]

$$\lambda_\parallel = 2\pi/\Re(k_\parallel), \qquad \delta_\parallel = [2\Im(k_\parallel)]^{-1}. \tag{3.135}$$

The penetration depth of a SPP on the two sides of the metal–dielectric interface is calculated using $\delta_{d\perp} = [\Im(k_{d\perp})]^{-1}$ and $\delta_{m\perp} = [\Im(k_{m\perp})]^{-1}$. These two quantities are related to the permittivities of the metallic and dielectric materials as

$$k_{m\perp} = \sqrt{\epsilon_m/\epsilon_d}k_\parallel, \qquad k_{d\perp} = \sqrt{\epsilon_d/\epsilon_m}k_\parallel. \tag{3.136}$$

For a metallic nanoparticle with a size much smaller than the $\delta_{m\perp}$, the optical field penetrates the entire system and drives oscillations of the metal electrons.

The partition of energy in such a system into the electric and magnetic parts can be understood by the following intuitive argument [226]. If the particle is purely dielectric, the stored magnetic energy, $U_H \sim \frac{1}{2}H^2/c\sqrt{\epsilon_0}$, is much smaller than the electric energy, $U_E \sim \frac{1}{2}\epsilon E^2$, where E and H are the fields inside the particle. To sustain oscillations, each energy-storage mechanism must be able to fully store the energy in the other format. Owing to an energy-storage imbalance, a dielectric nanoparticle fails to function as a self-sustaining oscillator. However, if one introduces a free carrier (thus making the particle metallic), a part of the energy can also be stored as the kinetic energy of electrons given by $U_K \sim \frac{1}{2}\epsilon_0(\omega_p)(\omega)^2 E^2$. This opens the possibility of realizing energy balance at a frequency at which $U_H + U_K = U_E$ is satisfied. This frequency is precisely the plasma frequency of the nanoparticle. In this situation, the energy is mostly contained in the kinetic oscillations of electrons. It is the size of the nanoparticle and free-electron density that define the spatial scale of the localization of optical energy and govern it through the quality factor Q given in (3.132). As a result, optical fields can be confined to nanosize dimensions with their spatial distribution scaling with the system's size. This physical picture is at the heart of modern nanoplasmonics [184, 224].

The preceding analysis shows that the parameters that govern the formation of surface plasmons are fixed in the case of a metal–dielectric interface. The situation changes if we consider 2D materials such as graphene, for which a few key parameters that are responsible for sustaining plasmon oscillations can be tuned through chemical doping or by changing the electric and magnetic fields [227, 228]. Indeed, by varying such parameters, plasmons can be excited in 2D materials at frequencies ranging from microwave to the optical region, significantly surpassing the relatively low range of frequencies covered by the noble metals. Interestingly, it has also been observed that highly doped graphene exhibits lower losses and longer plasmon lifetimes compared with the noble metals [229, 230]. To understand how this is possible, we need to look at the features that make graphene a suitable candidate for sustaining surface plasmons.

Consider a graphene sheet (or another 2D material) surrounded by two dielectric media with permittivities ϵ_{1d} and ϵ_{2d}. We assume that the graphene sheet lies in the (x, y) plane, with the z-axis perpendicular to this plane. For both the TE and TM surface waves, we assume that the electric field has the form

$$\mathbf{E}_j^m(x, y, z, t) = \mathbf{E}_m \exp(ik_\parallel x - i\omega t - k_{j,\perp} z), \quad m = \text{TE or TM}, \tag{3.137}$$

where $j = 1, 2$ for the two dielectric materials surrounding the graphene sheet. The x and y axes are oriented such that the TE wave has its electric field oriented along the y direction. For TM waves, $\mathbf{E}^{TM}$ lies in the (x, z) plane and has nonzero components along both the x and z axes. The two surface waves propagate in the (x, y) plane such that $k_\parallel^2 = k_{j,\perp}^2 + k_0^2 \epsilon_{jd}$ for $j = 1, 2$. The general dispersion relations for the two surface waves in graphene are given by

$$\sigma_{\text{graphene}}(\omega) = \begin{cases} i\omega\epsilon_{1d}/k_{1,\perp} + i\omega\epsilon_{2d}/k_{2,\perp} & \text{for TM}, \\ (k_{1,\perp} + k_{2,\perp})/i\omega\mu_0 & \text{for TE}, \end{cases} \tag{3.138}$$

where $\sigma_{\text{graphene}}(\omega)$ is the conductivity of graphene at the frequency ω.

The preceding equation shows that the plasmonic features of graphene are mainly set by the imaginary part of its conductivity. It is possible to change the sign of the imaginary part by changing graphene's chemical potential [227]. When the imaginary part is positive, graphene shows metallic features and can support SPPs of the TM type under the right conditions [209, 220, 231, 232]. However, when the imaginary part is negative, graphene loses this ability but may be able to sustain a weak TE-type surface wave [220, 232]. It is instructive to look at the special case in which the same dielectric medium surrounds the graphene sheet on both sides (i.e., $\epsilon_{1d} = \epsilon_{2d} = \epsilon_d$). The dispersion relation in Eq. (3.138) in this case takes the form [220, 231, 233]

$$\sigma_{\text{graphene}}(\omega) = \begin{cases} 2i\omega\epsilon_d/k_\perp, & \text{for TM}, \\ 2k_\perp/i\omega\mu_0, & \text{for TE}, \end{cases} \tag{3.139}$$

where we used $k_\perp = k_{1,\perp} = k_{2,\perp}$. The propagation constant along the interface is given by $k_\parallel^2 = k_\perp^2 + \epsilon_d k_0^2$.

One can evaluate Eq. (3.139) approximately to gain some insight into the plasmonic behavior of graphene. Two figures of merit have been suggested to classify plasmonic materials [234]. The first one describes the confinement of the surface wave with respect to the vacuum wavelength and is given by $\text{FOM}_{\text{conf}} = (2\pi c/\omega)\Re(k_\perp)$. The second measure provides a normalized value for the SPP's propagation distance by introducing $\text{FOM}_{\text{prop}} = \Re(k_\parallel)/[2\pi\Im(k_\parallel)]$. We calculate these figures of merit for highly doped graphene using the conductivity given by the Drude model in Eq. (3.119). The dispersion relation in Eq. (3.139) then provides $k_\perp = 2\omega\epsilon_d(\omega + i\gamma)/\sigma_0$, resulting in

$$\text{FOM}_{\text{conf}} = \frac{4\pi c\epsilon_d}{\sigma_0}\omega, \qquad \text{FOM}_{\text{prop}} = sd. \tag{3.140}$$

These figures of merit tell us that the mode confinement for graphene is independent of loss, but depends inversely on the doping level through $\sigma_0 = e^2 E_F/\pi\hbar^2$ (see Table 1.2). In contrast, increased doping tends to boost the FOM_{prop}, but there is an upper bound on frequency, after which the propagation distance begins to decrease.

The preceding dielectric models can be refined using information gained through the ab initio methods that solve variants of the Schrödinger equation. A full quantum-mechanical treatment of the dielectric function would require the calculation of the material's wave function using multiple electrons and nuclei. However, considering the large mass of the nuclei relative to electrons, it is possible to use the Born–Oppenheimer approximation and decouple electron's motion from nuclei motion [222]. Even with this approximation, a full quantum-mechanical treatment remains difficult owing to the computational complexity, and further approximations are often made. In one approach, density functional theory is used with the local-density approximation for the exchange and correlation functions [235]. In another approach, a semianalytical dielectric function is constructed that depends on the jellium model for electrons that are confined by infinite potential barriers at the physical edges of the nanosize object [236].

4 Dissipation and Decoherence

I don't demand that a theory correspond to reality because I don't know what it is. Reality is not a quality you can test with litmus paper. All I am concerned with is that the theory should predict the results of measurements. Quantum theory does this very successfully. It predicts that the result of an observation is either that the cat is alive or that it is dead. It is like you can't be slightly pregnant: you either are or you aren't.

Stephen Hawking

4.1 Effect of Environment on a Quantum Device

All quantum devices operate at a finite temperature and are thus subject to thermodynamic laws. The second law of thermodynamics states that a state variable called entropy exists and its value changes with time because of external and internal perturbations such that [237]

$$dS = dS|_{\text{ext}} + dS|_{\text{int}}, \qquad (4.1)$$

where the internal change is always positive: $dS|_{\text{int}} \geq 0$. The external change in entropy depends on temperature as $dS|_{\text{ext}} = \delta Q/T$, where δQ is the amount of heat entering the closed system and T is the common temperature at the point where the heat transfer took place. This change vanishes, by definition, for a "thermodynamically closed" system for which $\delta Q = 0$. For a given physical process, entropy S of the whole system (including its environment) remains a constant if the process is reversible. An example of a reversible process is a *Qubit register*, which is a quantum circuit made of multiple reversible quantum gates (analogous to the registers used in electronic computers) (see Aside 2.10). Another example is the flow of electric current through a wire with zero resistance. In contrast, the combined entropy of a thermodynamically closed system (including its environment) can only increase when the underlying physical process is irreversible because $dS|_{\text{int}} > 0$. This is the reason why heat flows from the hot side to the cold side, never the reverse. Similarly, entropy increases if an electric current flows through a wire with finite resistance.

4.1.1 Entropy and Time Reversal

As entropy provides a way of measuring disorder in a system, the second law of thermodynamics can be restated as follows: The entropy of an isolated system will either increase (more disorder) or stay the same, but it can never decrease. Strikingly, this requirement imposes a direction on *time*, in contrast to other physical laws, which could possibly be *time reversed*. Most physical laws do have time-reversal symmetry built into them, that is, if one replaces t with $-t$ in their equations of motion, the transformed equations remain consistent with the physical law they represent. As a result, they can predict the state of a dynamical system in the past or in the future, just by knowing its present state, which provides the initial conditions for integrating the system's equations of motion. Clearly, the second law of thermodynamics breaks this feature. The British astronomer Arthur Eddington coined and popularized the term *time's arrow* to refer to the "one-way nature" of a thermodynamic system [238]. The "arrow of time" is one of the most salient features of our universe because it is intimately related to entropy, which demands this asymmetry. However, why this is so is one of the unsolved mysteries in modern science. The laws of physics have no control over the direction of time; they depend on it!

What we do know is that quantum mechanics provides accurate predictions at the microscopic level. However, governing physical laws at the microscopic level appear to be entirely symmetric in time, that is, if the direction of time were to reverse, the analysis and predictions based on quantum mechanics would remain true. This feature is obviously contradictory to our day-to-day perception because we perceive an obvious direction (or flow) of time.

Time reversal in the quantum regime requires careful implementation because it is not sufficient to just change the sign of the time variable. Wigner was the first person to consider how the Schrödinger equation behaves under time reversal [239]. He considered a Hamiltonian that was invariant under time reversal and found that the time-reversal symmetry was restored by complex conjugation of the wave function. The reason for this can be understood by using Wigner's antiunitary operator T defined as $\mathrm{T}\, c\, |q\rangle = c^*\, |q\rangle$ for any complex number c. Owing to this definition, when the operator T acts on two wave functions, say $|\phi\rangle$ and $|\psi\rangle$, it gives the result $\langle \mathrm{T}|\phi\rangle\,|\,\mathrm{T}|\psi\rangle\rangle = \langle \psi|\phi\rangle$. Recall that $\langle \psi|\phi\rangle$ is the conjugate of $\langle \phi|\psi\rangle$.

To see how the antiunitary operator leads to time symmetry, consider a state $|p\rangle$ of constant momentum and express it in the coordinate basis as $|p\rangle = \frac{1}{\sqrt{2\pi}} \int \exp(ipq) |q\rangle\, dq$. It follows that

$$\mathrm{T}\,|p\rangle = \frac{1}{\sqrt{2\pi}} \int \mathrm{T}\exp(ipq)\,|q\rangle\, dq = \frac{1}{\sqrt{2\pi}} \int \exp(-ipq)\,|q\rangle = |-p\rangle\,. \tag{4.2}$$

That is, the particle's momentum is flipped under the action of the T operator. In the Schrödinger picture, the Hamiltonian H remains invariant under time reversal because it is an even function of the momentum operator. However, we should consider how the state $\psi(t)$ is affected by the operator T. Using the time-evolution operator, $U(t) = \exp(-iHt/\hbar)$, it is easy to show that

$$\mathrm{T}\,\psi(t) = \mathrm{T}\,U(t)\psi(0) = U(-t)\psi^*(0). \tag{4.3}$$

Thus, the correct recipe for time reversal is to reverse the sign of t followed by complex conjugation of the wave function. For unhindered quantum flow, future and past are mere conventions resulting from the same governing equations of motion.

The next question is: what are the implications to the physical laws if time is not reversible? Intuitively, a process should be reversible if there is no loss of coherence or energy. Indeed, dissipation and decoherence are two irreversible processes common to any quantum system interacting with its environment. Here, dissipation refers to any process for which irreversible loss of energy happens. Decoherence is the phenomenon where a superposition of macroscopically distinct states loses its coherence because of phase fluctuations. Phase variations reduce the ability of different wave functions to interfere with each other because of the interaction of a quantum system with its surroundings. Even though these two processes are fundamentally different and occur on different time scales, the net effect of both on a quantum system is to produce an irreversible change in the measurable quantities of interest. In almost all cases, it is possible to separate a quantum system into a *relevant part* (representing a useful quantum device) and an *irrelevant part* (called the reservoir) that are coupled to each other. Often, the reservoir is a broadband system with a large number of degrees of freedom that is in thermal equilibrium with its environment. For this reason, the reservoir is also called a (heat) *bath*. In general, a quantum device has little influence on its reservoir because of the reservoir's many degrees of freedom, but the reservoir affects the device through dissipation (energy loss) and phase fluctuations. These fluctuations are included in the dynamics of the quantum device through an effective force whose characteristics depend on the reservoir and induce decoherence that increases the device's entropy through disorder.

4.1.2 Spontaneous and Stimulated Emissions

The simplest yet most instructive example of irreversible decay is the phenomenon of spontaneous emission from an excited atom or molecule. The irreversibility in this process can be traced back to a relatively short memory of the reservoir (the Markovian property) and the resulting exponential decay of the probability for the atom being in an excited state. This suggests that qualitatively different dynamics can be achieved if the conditions are changed. For example, it is possible for an atom coupled to a single mode of the electromagnetic field to undergo periodic exchange of energy between the atom and the field, resulting in the well-known Rabi oscillations. To understand the process of spontaneous emission, it is interesting to consider how the underlying theory evolved over time.

In the early days of quantum mechanics, scientists discovered that energy associated with an atom is quantized through discrete energy levels. Bohr proposed a model of atomic structure based on such discrete energy levels, connected through *emission* of photons of energy $\hbar\omega = E_m - E_n$ between two energy levels with $E_m > E_n$. However, this process is feasible only if there is a mechanism to excite a low-energy atom to a higher energy state. One mechanism is the *absorption* of photons that can raise an atom to a higher energy level. However, such an upward transition can occur only at discrete frequencies for which the energy of photon $\hbar\omega$ equals the energy difference $E_m - E_n$ of the two states involved.

As the absorption and emission processes seem to be connected with each other, Einstein made the first attempt in 1917 to find a relationship between them. Rather than considering a multilevel atom, he considered a simple system with only two energy levels: a ground level with the quantum state $|g\rangle$ and an excited level with the quantum state $|e\rangle$. The energy difference between these two states is $E_e - E_g = \hbar\omega_{eg}$. Einstein first considered only two processes: (1) absorption that can only occur in the presence of a radiation field and (2) spontaneous emission that can bring the atom to the ground state even without the presence of a field. If such a system is in thermal equilibrium at a fixed temperature T, it should satisfy the Boltzmann distribution, $\rho_e = \rho_g \exp(-\hbar\omega_{eg}/k_BT)$, where ρ_e and ρ_g are the population densities of the two states involved. Moreover, the radiation field in thermal equilibrium must also satisfy Planck's blackbody radiation formula (see Aside 3.1). Einstein realized that it was not possible to satisfy both the Boltzmann relation and Planck's blackbody radiation formula by limiting a two-level system to just absorption and spontaneous emission. His remedy was to postulate a third process known as *stimulated emission*, where a photon in the radiation field induces transition from the upper to lower energy level, followed up by the emission of a new photon. The emitted photon has exactly the same energy, momentum, and phase as the photon that induced the transition. Aside 4.1 provides details of Einstein's derivation based on the above argument.

Aside 4.1   Einstein's A and B Coefficients

Einstein's A and B coefficients represent the probability of absorption or emission of light from an atom [240]. The A coefficient is related to the rate of spontaneous emission and the B coefficient is related to the rate of absorption and stimulated emission. They are defined for a two-level atom with the lower energy level E_g in the quantum state $|g\rangle$ and the upper energy level E_e in the quantum state $|e\rangle$. The energy difference between them is written as $E_e - E_g = \hbar\omega_{eg} = h\nu_{eg}$. The system is in thermal equilibrium with the surrounding blackbody radiation. Transitions between the two energy states can occur in three different ways:

- spontaneous emission from $|e\rangle$ to $|g\rangle$ with probability A per unit time;
- absorption from $|g\rangle$ to $|e\rangle$ with probability $B_{ge}R_d$ per unit time;
- stimulated emission from $|e\rangle$ to $|g\rangle$ with probability $B_{eg}R_d$ per unit time;

where A, B_{eg}, and B_{ge} are constants and R_d is the spectral density of radiation at the frequency $\nu = \nu_{eg}$.

Given that the system is in thermal equilibrium, the occupation probabilities for the states $|e\rangle$ and $|g\rangle$ are $\exp(-E_e/k_BT)$ and $\exp(-E_g/k_BT)$, respectively. Also, the probability of upward transitions must exactly balance the probability of downward transitions in thermal equilibrium,

$$\left(A + B_{eg}R_d\right)\exp(-E_e/k_BT) = B_{ge}R_d\exp(-E_g/k_BT). \tag{4.4}$$

This equation can be rearranged to get

$$R_d = \frac{d\rho_\nu}{d\nu} = \frac{A}{B_{ge}\exp(h\nu_{eg}/k_BT) - B_{eg}}. \tag{4.5}$$

As this expression must coincide with the spectral density given in Eq. (3.25), we immediately obtain $B_{eg} = B_{ge} = B$, indicating that the absorption and stimulated emission processes are intimately connected. Further, the A and B coefficients are related as

$$\frac{A}{B} = \frac{8\pi h}{c^3} \nu^3. \tag{4.6}$$

This relation enables us to calculate Einstein's B coefficient once the spontaneous emission rate A is known. Although it is possible to use semiclassical arguments to estimate the value of A, its proper derivation requires quantum electrodynamics and is discussed later in this section.

Even though Einstein's treatment of the emission and absorption processes enabled him to establish relations between the A and B coefficients within a thermodynamic framework, the values of these coefficients cannot be calculated by resorting to semiclassical arguments. What one needs is an expression for spontaneous-emission rate A, because B can be obtained from it using Einstein's relation, given in Eq. (4.6). The hurdle for calculating A using a semiclassical approach stems from ignoring changes in the electromagnetic field induced by the emitted photon. Indeed, even after multiple attempts, no one has ever succeeded in deriving the rate of spontaneous emission solely using the classical description of an electromagnetic field [241]. Moreover, even the Schrödinger equation of quantum theory cannot be used to calculate the rate of spontaneous emission. To accomplish this task, one has to invoke the machinery of quantum electrodynamics.

Spontaneous emission, at the most basic level, is responsible for most of the light we see all around us. It is so omnipresent that there are many names associated with what is fundamentally the same thing. For example, if a material is excited by some means other than heating, spontaneous emission is called *luminescence*, which may also assume other names depending on the underlying process. Thus, *chemoluminescence* creates light through a chemical reaction. *Bioluminescence* is a type of chemoluminescence used by living creatures: Angler fish produce light lures to trap prey, and fireflies glow at night. However, if a material first absorbs light and then emits a part of it through spontaneous emission, the radiation is called *fluorescence*. If a material has a metastable state and continues to fluoresce long after the incident light causing excitation is turned off, we call it a *phosphorescent* material. Such materials are useful for making dials in clocks and watches. Even though we may not appreciate it, spontaneous emission is behind the multibillion-dollar industry of light-emitting diodes and lasers. Although lasers produce light mainly through stimulated emission, the trigger for stimulated emission comes from the photons emitted via spontaneous emission in these devices. Thus, not only does spontaneous emission set the scene for all fundamental radiative interactions, it also triggers these processes in many important quantum devices.

Ever since it was understood that absorption, spontaneous emission, and stimulated emission must coexist if matter and radiation were to achieve thermal equilibrium, it was realized that an atom in the excited state will inevitably radiate energy through spontaneous emission because it is the only process that does not requires any background radiation to be present. A consequence of this reasoning is that spontaneous emission must be an intrinsic property of matter [242]. This view, however, overlooks the fact that spontaneous emission is not a property of an isolated excited atom but a property of an atom coupled to its surroundings. The most distinctive feature of this emission, *irreversibility*, results from an infinite number of radiation modes (or states) available to the spontaneously emitted photon. If these states are modified, for example by placing the excited atom between mirrors of a cavity, spontaneous emission can be enhanced, or even inhibited. Purcell predicted in 1946 the ability to control the rate of spontaneous emission by changing the environment surrounding an excited system [243]. Since then, many theoretical and experimental studies have focused on various ways to influence the rate of spontaneous emission from an excited system [244, 245, 246].

Kleopnner extended Purcell's work in 1981, and this work [247] led to a new field called cavity quantum electrodynamics (or cavity QED). He showed that spontaneous emission is inhibited when the cavity's dimensions are small compared to the radiation wavelength and is enhanced if the cavity is resonant with the excited atomic system. In addition, the cavity changes the energy levels slightly in a manner analogous to the well-known Lamb shift. These properties opened up the possibility of tailoring spontaneous emission to suit applications. For example, in the case of a laser, it is desirable to limit spontaneous emission to modes that are lasing; in the case of solar cells, spontaneous emission is allowed in the detected modes [248]. An interesting example is a photonic crystal in which an excited quantum dot is placed inside a 3D periodic structure containing a photonic band gap that overlaps the electronic band edge of the quantum dot; spontaneous emission is found to be rigorously forbidden in such a system [248].

As we have mentioned, the ability to control spontaneous emission has applications for improving the efficiency of semiconductor devices such as solar cells, lasers, LEDs, spasers, and many other active quantum devices that rely on transitions between two excited states. Therefore, a fundamental understanding of the theory and techniques required to analyze such devices is critical. Dirac was the first person to derive an expression for Einstein's A coefficient for spontaneous emission based on quantum mechanics [249]. A more complete and insightful model was developed later in 1931 by Weisskopf and Wigner [250]. Their derivation, discussed next, has numerous applications beyond spontaneous emission because it is an example of a general class of problems that consider quantum-mechanical coupling of a small device to a large reservoir.

4.1.3 Weisskopf–Wigner Theory

A quantum system in an excited energy state cannot remain there indefinitely: it eventually makes a transition to a lower energy state by spontaneously emitting a photon. This transition is driven by the coupling of the excited atom to the reservoir surrounding it. Even when a vacuum acts as a reservoir for an excited atom, one must consider the atom's coupling to the radiation modes of the vacuum. Weisskopf and Wigner considered this special case to calculate the rate of spontaneous emission from an excited two-level atom (a small device) coupled to a continuum of electromagnetic modes of vacuum (a large reservoir).

Consider a quantum system with two energy states: the ground state $|g\rangle$ with energy E_g and an excited state $|e\rangle$ with energy E_e. The energy difference between these states can be written as $E_e - E_g = \hbar\omega_{eg}$. The vacuum surrounding this system acts as a reservoir.

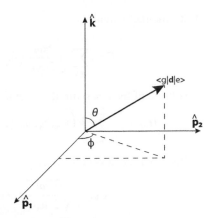

Fig. 4.1 Coordinate system used for representing a reservoir mode with the wave vector **k**. The associated plane wave is polarized in the transverse plane. The atomic dipole moment is oriented at an angle θ as shown.

Suppose the electromagnetic field in this reservoir is in its lowest energy state, denoted as $|\{0\}\rangle$. When this state is excited with one photon, the associated plane wave is polarized in a plane orthogonal to its direction of propagation governed by the vector **k**, whose magnitude is related to the frequency ω of the radiation as $k = \omega/c$. We denote unit vectors in this plane by $\hat{\mathbf{p}}_1$ or $\hat{\mathbf{p}}_2$, as shown in Figure 4.1.

We use the state vector $|\psi(0)\rangle$ to denote the initial state at $t = 0$ and define it to be a product state of the excited state $|e\rangle$ and the vacuum state with zero photons:

$$|\psi(0)\rangle = |e, \{0\}\rangle = |e\rangle \otimes |\{0\}\rangle. \tag{4.7}$$

When the atom makes a transition from this excited state to its ground state $|g\rangle$, a photon will be emitted in the direction **k** with polarization $\hat{\mathbf{p}}_s$ ($s = 1, 2$). The final state of the system can be written as $|g, 1_{\mathbf{k}s}\rangle = |g\rangle \otimes |1_{\mathbf{k}s}\rangle$, where we used the compact notation $|1_{\mathbf{k}s}\rangle$ to identify this single photon. The initial and final states form a complete basis for expanding the quantum state of the whole system at time t as

$$|\psi(t)\rangle = a(t)e^{-i\omega_{eg}t} |e, \{0\}\rangle + \sum_{\mathbf{k},s} b_{\mathbf{k}s}(t)e^{-i\omega_k t} |g, 1_{\mathbf{k}s}\rangle. \tag{4.8}$$

Note the sum over all possible directions and polarizations of the emitted photon.

We need to find how the coefficients $a(t)$ and $b_{\mathbf{k},s}(t)$ evolve with time, given that $a(0) = 1$ and $b_{\mathbf{k}s}(0) = 0$ initially. This can be done by using the Hamiltonian H of the whole system. We use an extension of the Jaynes–Cummings Hamiltonian (see Section 4.1.4) to describe the atom–field interaction in the form

$$H = \frac{1}{2}\hbar\omega_{eg}\sigma_3 + \sum_{\mathbf{k},s} \hbar\omega_k \hat{a}_{\mathbf{k}s}^{\dagger}\hat{a}_{\mathbf{k}s} + H_{int}, \tag{4.9}$$

where the first two terms represent the Hamiltonian of the atom and the field, respectively, and H_{int} describes the interaction between them. In the dipole approximation, $H_{int} = -\mathbf{d} \cdot \mathbf{E}$, where **d** is the dipole moment operator and **E** is the electric field. Using the quantized form of the electromagnetic field,

$$\mathbf{E} = \sum_{\mathbf{k},s} \sqrt{\frac{\hbar\omega_k}{2\epsilon_0 V}} \left(\hat{a}_{\mathbf{k}s}^\dagger + \hat{a}_{\mathbf{k}s}\right) \hat{\mathbf{p}}_s, \tag{4.10}$$

where V is the volume of the reservoir, the interaction Hamiltonian can be written as

$$H_{int} = -\sum_{\mathbf{k},s} \hbar \left(C_{\mathbf{k}s}\hat{a}_{\mathbf{k}s} |e\rangle \langle g| + C_{\mathbf{k}s}^*\hat{a}_{\mathbf{k}s}^\dagger |g\rangle \langle e|\right), \tag{4.11}$$

where the coupling coefficient is defined as

$$C_{\mathbf{k}s} = \sqrt{\frac{\omega_k}{2\hbar\epsilon_0 V}}(\langle e|\mathbf{d}|g\rangle \cdot \hat{\mathbf{p}}_s). \tag{4.12}$$

We can now use the Schrödinger equation, $i\hbar\frac{\partial}{\partial t} |\psi(t)\rangle = H |\psi(t)\rangle$, to obtain

$$i\hbar\left(\frac{da}{dt} - i\omega_{eg}a\right)e^{-i\omega_{eg}t} |e,\{0\}\rangle + i\hbar\sum_{\mathbf{k},s} \left(\frac{db_{\mathbf{k}s}}{dt} - i\omega_k b_{\mathbf{k}s}\right)e^{-i\omega_k t} |g, 1_{\mathbf{k}s}\rangle \tag{4.13}$$

$$= H\left[a(t)\exp(-i\omega_{eg}t) |e,\{0\}\rangle + \sum_{\mathbf{k},s} b_{\mathbf{k}s}(t)\exp(-i\omega_k t) |g, 1_{\mathbf{k}s}\rangle\right]. \tag{4.14}$$

We multiply the preceding equation first with $\langle e, \{0\}|$, then with $\langle g, 1_{\mathbf{k}s}|$, and use the orthonormal property of the quantum states. After some algebra, we obtain

$$\frac{da}{dt} = i\sum_{\mathbf{k},s} C_{\mathbf{k}s}\exp[-i(\omega_k - \omega_{eg})t]b_{\mathbf{k}s}(t) \tag{4.15}$$

$$\frac{db_{\mathbf{k}s}}{dt} = iC_{\mathbf{k}s}^*\exp[i(\omega_k - \omega_{eg})t]a(t). \tag{4.16}$$

To solve the preceding set of equations, we first formally integrate Eq. (4.16) to obtain

$$b_{\mathbf{k}s}(t) = iC_{\mathbf{k}s}^* \int_0^t \exp[i(\omega_k - \omega_{eg})t']a(t')\,dt', \tag{4.17}$$

and then substitute the result back into Eq. (4.15) to get

$$\frac{da(t)}{dt} = -\sum_{\mathbf{k},s} |C_{\mathbf{k}s}|^2 \int_0^t \exp[-i(\omega_k - \omega_{eg})(t - t')]a(t')\,dt'. \tag{4.18}$$

Equation (4.18) is an integral equation for $a(t)$ involving summation over all reservoir states. It can be solved only after making some reasonable approximations. Consider first the time integral:

$$I_a(t) = \int_0^t \exp[-i(\omega_k - \omega_{eg})(t - t')]a(t')\,dt'. \tag{4.19}$$

The exponential term in this equation oscillates rapidly unless $\omega_k \approx \omega_{eg}$ or t' is close to t. Physically, the excited-state amplitude $a(t)$ is expected to vary much slower than these rapid oscillations. As the largest contribution to the integral comes for t' close to t, we can replace $a(t')$ in the integrand by $a(t)$ and take it outside the integral. This is called the Weisskopf–Wigner approximation. It can also be recognized as the Markovian approximation because the dynamics of $a(t)$ does not depend on earlier times $t' < t$ (the system has no memory of the past). Using $\tau = t - t'$, the time integral becomes

$$I_a(t) \approx a(t) \int_0^t \exp[-i(\omega_k - \omega_{eg})\tau]\, d\tau. \qquad (4.20)$$

Since the largest contribution to this integral comes from the region near $\tau = 0$, we can extend the upper limit to ∞ without introducing much error. This integral then can be evaluated using the Sokhotski–Plemelj formula given in Aside 3.3. The result is

$$I_a(t) = a(t)\left[\pi\delta(\omega_k - \omega_{eg}) - i\text{P.V.}\left(\frac{1}{\omega_k - \omega_{eg}}\right)\right]. \qquad (4.21)$$

The sum on the right side of Eq. (4.18) can be done in the continuum limit by following the details in Aside 4.2.

Aside 4.2 The Continuum Limit

We calculate the sum $\sum_{\mathbf{k},s}|C_{\mathbf{k}s}|^2$ in the continuum limit (quantization volume $V \to \infty$) where the number of reservoir states becomes infinite. In this limit we replace the sum over k by an integral as

$$\sum_{\mathbf{k},s}|C_{\mathbf{k}s}|^2 \to \sum_{s=1}^{2}\int |C_{\mathbf{k}s}|^2\, D_{\text{os}}(k)d^3k, \qquad (4.22)$$

where $D_{\text{os}}(k) = V/(2\pi)^3$ is the density of states in the $\mathbf{k}$-space (see Aside 1.4 for details). The volume integration in the $\mathbf{k}$ space can be done using the spherical coordinates (k, θ, ϕ) to obtain (see Fig. 4.1)

$$\sum_{\mathbf{k},s}|C_{\mathbf{k}s}|^2 = \int \sum_{s=1}^{2}\frac{\omega_k}{2\hbar\epsilon_0 V}|\langle e|\mathbf{d}|g\rangle \cdot \hat{\mathbf{p}}_s|^2\, D_{\text{os}}(k)d^3k$$

$$= \int_0^{\infty}\frac{\omega_k k^2\, dk}{2(2\pi)^3\hbar\epsilon_0}\left[\int_0^{\pi}\int_0^{2\pi}\sum_{s=1}^{2}|\langle e|\mathbf{d}|g\rangle \cdot \hat{\mathbf{p}}_s|^2\sin\theta\, d\theta\, d\phi\right]. \qquad (4.23)$$

Consider the sum over s within the square brackets. Using

$$\sum_{s=1}^{2}|\langle e|\mathbf{d}|g\rangle \cdot \hat{\mathbf{p}}_s|^2 = [|\langle e|\mathbf{d}|g\rangle \cdot \hat{\mathbf{p}}_1|^2 + |\langle e|\mathbf{d}|g\rangle \cdot \hat{\mathbf{p}}_2|^2] \qquad (4.24)$$

and noting that $(\hat{\mathbf{p}}_1, \hat{\mathbf{p}}_2, \hat{\mathbf{k}})$ form a Cartesian coordinate system (see Fig. 4.1) we obtain

$$\sum_{s=1}^{2}|\langle e|\mathbf{d}|g\rangle \cdot \hat{\mathbf{p}}_s|^2 = |\langle e|\mathbf{d}|g\rangle|^2 - |\langle e|\mathbf{d}|g\rangle\cos\theta|^2 = |\langle e|\mathbf{d}|g\rangle|^2\sin^2\theta. \qquad (4.25)$$

Using this result and noting that

$$\int_0^{\pi}\int_0^{2\pi}\sin^3\theta\, d\theta\, d\phi = \frac{8\pi}{3}, \qquad (4.26)$$

the sum in Eq. (4.23) becomes

$$\sum_{\mathbf{k},s} |C_{\mathbf{k}s}|^2 = \frac{|\langle e|\mathbf{d}|g\rangle|^2}{6\pi^2\hbar\epsilon_0 c^3} \int_0^\infty \omega_k^3 \, d\omega_k, \tag{4.27}$$

where we have changed the integration variable from k to ω_k using $\omega_k = ck$.

We can now solve Eq. (4.18) approximately. Using the results in Eqs. (4.21) and (4.27), we obtain

$$\frac{da}{dt} = -\left(\frac{\gamma_a}{2} - i\Delta\omega_{LS}\right) a(t), \tag{4.28}$$

where the decay rate γ_a and the frequency shift $\Delta\omega_{LS}$ are defined as

$$\gamma_a = \frac{\omega_{eg}^3 |\langle e|\mathbf{d}|g\rangle|^2}{3\pi\epsilon_0\hbar c^3}, \tag{4.29}$$

$$\Delta\omega_{LS} = \frac{|\langle e|\mathbf{d}|g\rangle|^2}{6\pi^2\epsilon_0\hbar c^3} \text{P.V.} \int \frac{\omega_k^3 \, d\omega_k}{\omega_k - \omega_{eg}}. \tag{4.30}$$

Equation (4.28) can be easily integrated to obtain the final result

$$a(t) = a(0) \exp[-(\gamma_a/2 - i\Delta\omega_{LS})t)]. \tag{4.31}$$

It shows that the excited-state amplitude decays exponentially. Its phase also changes because of the frequency shift $\Delta\omega_{LS}$ known as the *Lamb shift*. The rate of spontaneous emission is related to the probability of the atom remaining in the excited state given by $|a(t)|^2 = \exp(-\gamma_a t)$. Indeed, γ_a is just Einstein's A coefficient.

As mentioned earlier, the Schrödinger equation cannot describe the process of spontaneous emission if we just focus on the two-level system and ignore its surroundings acting as a reservoir. The reason is that the Schrödinger equation describes processes that are *time reversible*. Spontaneous emission describes an *irreversible process* that transfers energy from the atom to its surroundings by emitting a photon into one of the reservoir states. The Weisskopf–Wigner theory introduces irreversibility into the Schrödinger equation by invoking two approximations: (1) the available reservoir modes cover a very broad spectrum and (2) the excited-state amplitude $a(t)$ depends only on t and not on earlier times (i.e., the system has no memory of the past).

The solution in Eq. (4.31) can be used to estimate the spectrum of spontaneously emitted emitted light. As the Lamb shift $\Delta\omega_{LS}$ is relatively small, we ignore it here. Using the result for $a(t)$ in Eq. (4.16), we obtain

$$\frac{db_{\mathbf{k}s}(t)}{dt} = iC_{\mathbf{k}s}^* \exp[i(\omega_k - \omega_{eg})t - \gamma_a t/2]a(0). \tag{4.32}$$

This equation can be easily integrated to obtain

$$b_{\mathbf{k}s}(t) = \frac{iC_{\mathbf{k}s}^* a(0)}{\gamma_a/2 - i(\omega_k - \omega_{eg})} \Big(\exp[i(\omega_k - \omega_{eg})t - \gamma_a t/2] - 1 \Big). \tag{4.33}$$

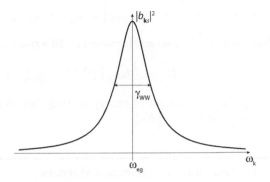

Fig. 4.2 Spontaneous-emission spectrum resulting from the Weisskopf–Wigner theory. It is centered at ω_{eg} and has a Lorentzian shape with the FWHM γ_a.

After a long time, the probability finding the system in the state $|g, 1_{\mathbf{k}s}\rangle$ is thus given by

$$\lim_{t \to \infty} |b_{\mathbf{k}s}(t)|^2 = \frac{|C_{\mathbf{k}s}|^2 |a(0)|^2}{\gamma_a^2/4 + (\omega_k - \omega_{eg})^2}. \tag{4.34}$$

This result shows that the spectrum of the emitted light has a Lorentzian shape and is centered at the frequency ω_{eg}. This spectrum is shown in Figure 4.2, and its full width at half-maximum (FWHM) is given by γ_a.

4.1.4 Jaynes–Cummings Hamiltonian

The Weisskopf–Wigner theory describes the evolution of a two-level system coupled to a large number of the radiation modes of a reservoir. The simplest example of this scenario is the Jaynes–Cummings model, where a two-level atom is coupled to a single electromagnetic mode of the reservoir. Here we discuss the Hamiltonian associated with this model.

As before, we consider an atom with two energy states: the ground state $|g\rangle$ with energy E_g and an excited state $|e\rangle$ with energy E_e. The transition frequency is related to the energy difference as $E_e - E_g = \hbar\omega_{eg}$. We assume that the frequency ω_k of the electromagnetic field is nearly resonant with ω_{eg}. The assumption $\omega_k \approx \omega_{eg}$ ensures that only one mode couples strongly to the two-level atom. The interaction of the atom with this mode is studied through a Hamiltonian containing three parts

$$H_{JC} = H_{eg} + H_f + H_{egf}, \tag{4.35}$$

where H_{eg} is the Hamiltonian of the two-level atom, H_f is the Hamiltonian of the electromagnetic field, and H_{egf} is the interaction Hamiltonian.

It is common to use the Pauli spin matrices to write the Hamiltonian of a two-level atom [251]. The raising and lowering operators, σ_+ and σ_-, introduced in Aside 4.3, can be used to obtain the following relations [252].

$$\sigma_+ |e\rangle = 0, \qquad \sigma_+ |g\rangle = |e\rangle, \qquad \sigma_- |g\rangle = 0, \qquad \sigma_- |e\rangle = |g\rangle, \tag{4.36}$$

$$[\sigma_3, \sigma_\pm] = 2\sigma_\pm, \qquad [\sigma_+, \sigma_-] = \sigma_3, \qquad \sigma_+^2 = \sigma_-^2 = 0, \qquad \sigma_+^\dagger = \sigma_-. \tag{4.37}$$

The Hamiltonian of a two-level system can be written in the form

$$H_{eg} = \frac{1}{2}\hbar\omega_{eg}\Big(|e\rangle\langle e| - |g\rangle\langle g|\Big) = \frac{1}{2}\hbar\omega_{eg}\sigma_3, \tag{4.38}$$

where we chose the origin of energy halfway between the two energy levels such that $E_e = \frac{1}{2}\hbar\omega_{eg}$ and $E_g = -\frac{1}{2}\hbar\omega_{eg}$.

Aside 4.3   Pauli Spin Operators and Matrices

A spin-$\frac{1}{2}$ system and a two-level atom have identical Hilbert spaces. The states $|g\rangle$ and $|e\rangle$ form an orthonormal basis such that $\langle g|g\rangle = \langle e|e\rangle = 1$ and $\langle g|e\rangle = \langle e|g\rangle = 0$. We define the Pauli spin operators in this basis as

$$\sigma_1 = |g\rangle\langle e| + |e\rangle\langle g|, \quad \sigma_2 = i(|g\rangle\langle e| - |e\rangle\langle g|), \quad \sigma_3 = |e\rangle\langle e| - |g\rangle\langle g| \tag{4.39}$$

$$\sigma_+ = |e\rangle\langle g| = (\sigma_1 + i\sigma_2)/2, \qquad \sigma_- = |g\rangle\langle e| = (\sigma_1 - i\sigma_2)/2. \tag{4.40}$$

With these definitions, it is easy to show that the Pauli spin operators $(\sigma_1, \sigma_2, \sigma_3)$ satisfy the following relations:

$$(\sigma_k)^2 = 1, \qquad \text{Tr}[\sigma_k] = 0, \qquad [\sigma_m, \sigma_n] = 2i\epsilon_{mnk}\sigma_k. \tag{4.41}$$

If we represent the states $|e\rangle$ and $|g\rangle$ in a matrix form as

$$|e\rangle = \begin{bmatrix} 1 \\ 0 \end{bmatrix}, \qquad |g\rangle = \begin{bmatrix} 0 \\ 1 \end{bmatrix}, \tag{4.42}$$

the Pauli spin operators have the following equivalent matrix representation:

$$\sigma_1 = \begin{bmatrix} 0 & 1 \\ 1 & 0 \end{bmatrix}, \ \sigma_2 = \begin{bmatrix} 0 & -i \\ i & 0 \end{bmatrix}, \ \sigma_3 = \begin{bmatrix} 1 & 0 \\ 0 & -1 \end{bmatrix}, \ \sigma_+ = \begin{bmatrix} 0 & 1 \\ 0 & 0 \end{bmatrix}, \ \sigma_- = \begin{bmatrix} 0 & 0 \\ 1 & 0 \end{bmatrix}. \tag{4.43}$$

The field Hamiltonian H_f requires the quantized form of the electric field,

$$\mathbf{E} = \sqrt{\frac{\hbar\omega_k}{2\epsilon_0 V}}\left(\hat{a}_{\mathbf{k}s}^\dagger + \hat{a}_{\mathbf{k}s}\right)\hat{\mathbf{p}}_s, \tag{4.44}$$

where $\hat{\mathbf{p}}_s$ is the polarization unit vector ($s = 1$ or 2) and $\hat{a}_{\mathbf{k}s}^\dagger$ and $\hat{a}_{\mathbf{k}s}$ are the creation and annihilation operators for the field at the frequency ω_k with $k = \omega_k/c$. The constant factor represents the average field per photon inside a cavity of volume V. In this representation, the Hamiltonian of the electromagnetic field is given by

$$H_f = \hbar\omega_k\hat{a}_{\mathbf{k}s}^\dagger\hat{a}_{\mathbf{k}s}, \tag{4.45}$$

where we have discarded the zero-point energy because it plays no role in the present situation.

The interaction energy in the dipole approximation is given as $V = -\mathbf{d} \cdot \mathbf{E}$, where $\mathbf{d}$ is the dipole-moment operator. In the current basis, we can write the interaction Hamiltonian as

$$H_{egf} = |e\rangle \langle e| V |g\rangle \langle g| + |g\rangle \langle g| V |e\rangle \langle e| . \tag{4.46}$$

Using $V = -\mathbf{d} \cdot \mathbf{E}$ and $\mathbf{E}$ from Eq. (4.44), it becomes [253]

$$H_{egf} = -\hbar \left(\hat{a}_{\mathbf{k}s}^\dagger + \hat{a}_{\mathbf{k}s} \right) \left[C_{\mathbf{k}s}\sigma_+ + C_{\mathbf{k}s}^*\sigma_- \right], \tag{4.47}$$

where we have defined $C_{\mathbf{k}s}$ as

$$C_{\mathbf{k}s} = \sqrt{\frac{\omega_k}{2\hbar\epsilon_0 V}} [\langle e|\mathbf{d}|g\rangle \cdot \hat{\mathbf{p}}_s]. \tag{4.48}$$

This quantity is often referred to as the *Rabi frequency* (especially when it is real). We keep the present notation to allow for its complex values.

We can further simplify the interaction Hamiltonian by invoking the rotating-wave approximation. In this approximation, the rapidly oscillating terms are neglected. More explicitly, terms oscillating at the low frequency $\omega_k - \omega_{eg}$ are kept, while terms oscillating at the high frequency $\omega_k + \omega_{eg}$ are neglected (as their contribution averages out to zero over a few cycles). To understand the origin of these frequencies, we need to consider the evolution of the operators appearing in Eq. (4.47) in the Heisenberg picture. It is easy to see from Eq. (4.45) that $\hat{a}_{\mathbf{k}s}$ evolves in time as $\exp(-i\omega_k t)$. Similarly, the atomic Hamiltonian H_{eg} can be used to show that the operators $\sigma_+ = |e\rangle \langle g|$ and $\sigma_- = |g\rangle \langle e|$ evolve in time as $\exp(i\omega_{eg}t)$ and $\exp(-i\omega_{eg}t)$, respectively. It follows that the combinations $\hat{a}_{\mathbf{k}s}\sigma_+$ and $\hat{a}_{\mathbf{k}s}^\dagger\sigma_-$ oscillate at the frequency $\omega_k - \omega_{eg}$. In contrast, the combinations $\hat{a}_{\mathbf{k}s}\sigma_-$ and $\hat{a}_{\mathbf{k}s}^\dagger\sigma_+$ oscillate at the frequency $\omega_k + \omega_{eg}$. In the rotating-wave approximation, we discard the later terms and write the Jaynes–Cummings Hamiltonian as

$$H_{JC} = \frac{1}{2}\hbar\omega_{eg}\sigma_3 + \hbar\omega_k\hat{a}_{\mathbf{k}s}^\dagger\hat{a}_{\mathbf{k}s} - \hbar \left(C_{\mathbf{k}s}\hat{a}_{\mathbf{k}s}\sigma_+ + C_{\mathbf{k}s}^*\hat{a}_{\mathbf{k}s}^\dagger\sigma_- \right). \tag{4.49}$$

Physically, $\hat{a}_{\mathbf{k}s}\sigma_+$ corresponds to the absorption of a photon that raises the atom to its excited state and $\hat{a}_{\mathbf{k}s}^\dagger\sigma_-$ corresponds to the emission of a photon that returns the atom to its ground state. These two terms conserve the total energy of the system. In contrast, the nonresonant terms that we dropped do not conserve energy. The accuracy of the rotating-wave approximation increases when the electromagnetic field is nearly resonant with the two-level atom and has a relatively small amplitude.

Several methods can be used to study the dynamics of the Jaynes–Cummings Hamiltonian [254, 255]. In a widely used method, we first find the stationary states of this Hamiltonian, known as the "dressed states," and then write the state of the system as a superposition of these dressed states [256]. In a variant of this method, called the Stenholm method [257], the problem is solved by noting that integer powers of the Hamiltonian can be calculated using analytical methods. Open quantum systems can be modeled by replacing the pure-state description of the system with a density-matrix description.

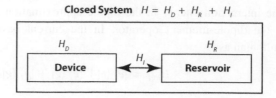

Fig. 4.3 The device plus reservoir ($D + R$) model commonly used for quantum devices. The combined Hamiltonian of the closed system includes a weak coupling term H_I. The reservoir has many degrees of freedom and is not affected by the device.

4.1.5 General Reservoir Model

The time evolution of any closed quantum system is governed by the Schrödinger equation. As discussed in Section 2.2, this equation shows that the quantum state $|\Psi(t)\rangle$ evolves from its initial state $|\Psi(t_0)\rangle$ as

$$|\Psi(t)\rangle = U(t, t_0) |\Psi(t_0)\rangle, \tag{4.50}$$

where $U(t, t_0) = \exp\left[-\frac{i}{\hbar}H(t - t_0)\right]$ is the unitary evolution operator. The Hamiltonian H of a closed system does not depend on time. Indeed, H cannot be a function of time, as this would imply an external mechanism driving the time dependence that does not exist for a closed system. As the Hamiltonian refers to total energy in the system, it follows that closed systems cannot dissipate energy as they evolve in time.

The key question that needs to be answered is: How can one incorporate dissipation and decoherence within the quantum-mechanics framework? There have been many attempts to incorporate dissipation into quantum mechanics. The most celebrated example is provided by lasers, which demand a satisfactory method to account for the lossy cavity that is required to provide optical feedback. Details about the methods used can be found in Refs. [258, 259, 260, 261, 262, 263].

It should be clear by now that we need to relax the requirement of the unitary evolution of a quantum system to incorporate irreversible changes. We have already seen a way of doing this in the Weisskopf–Wigner theory of spontaneous emission. In that case, the radiation modes of the vacuum surrounding a quantum system (a two-level atom) acted as a reservoir. We can extend this concept to all quantum devices through the scheme shown in Figure 4.3. In this scheme, the quantum device is coupled to a suitable reservoir, and it is this coupling that induces decoherence and dissipation. To make the problem tractable, we need to make three simplifying assumptions: (1) the quantum device couples to the reservoir weakly; (2) reservoir dynamics is not affected by this coupling; and (3) the reservoir is memoryless. In essence, interaction of the quantum device with the reservoir introduces randomness into the device dynamics. As a result, the device is better described using a density operator rather than its wave function.

The device plus reservoir model seen in Figure 4.3 solves the problem by using a Hamiltonian of the form $H = H_D + H_R + H_I$, where H_D is the device Hamiltonian, H_R is the reservoir Hamiltonian, and H_I describes the interaction between the two. However, it is important to separate the device dynamics from that of the reservoir to make the problem tractable. This can be done in two ways. One approach is based on the Schrödinger picture, and the dynamics is described through a reduced density matrix using a so-called master equation [264, 265, 266]. The second approach is based on the Heisenberg picture, where the problem is solved using the so-called Langevin equations (see Section 7.3) for a relevant set of noise operators of the reduced system [260, 267, 268].

Classical physics provides deterministic laws for describing observations and gives a sense of certainty even though many processes in nature are stochastic. Historically, the advent of quantum mechanics led to a prevailing view that nature is indeterministic. In 1926, Einstein made the following comment to Max Born on quantum mechanics: "The theory produces a good deal but hardly brings us closer to the secret of the Old One. I am at all events convinced that He does not play dice." The probabilistic nature of quantum mechanics is evident from the Heisenberg uncertainty principle (see Aside 2.11), which emphasizes that observable quantities do not have definite values in the quantum regime. Clearly, two basic concepts central to quantum phenomena, the quantum states and quantum observables, are intrinsically probabilistic. In contrast to this, classical physics considers probabilistic models only when one has an incomplete knowledge of the system.

4.2 Master-Equation Approach

The device plus reservoir model in Figure 4.3 requires the use of a density matrix. However, owing to the computational cost and the mere size of the full density matrix of a closed system, a numerical approach is often impractical. In this section we discuss two types of master equations that are used to reduce the complexity of the problem.

4.2.1 Master Equation for Occupation Probabilities

In both classical and quantum mechanics, probabilistic descriptions can be used to describe the past, present, and future states of a system. The dynamical equations describing the time evolution of these probabilities are called the *master equations*, which are first-order differential equations of the form

$$\frac{dP_m}{dt} = \sum_{n \neq m} (\Gamma_{m \leftarrow n} P_n - \Gamma_{n \leftarrow m} P_m), \qquad (4.51)$$

where $P_m(t)$ with $m = 1$ to M is the probability that the system is in the mth state and $\Gamma_{m \leftarrow n} \geq 0$ is the transition rate for moving the system from state n to state m ($\Gamma_{m \leftarrow m} = 0$). Since a master equation depends only on the transition rates, it can also be considered a *rate equation*. The first term containing $\Gamma_{m \leftarrow n}$ represents a gain of probability, while the second term with $\Gamma_{n \leftarrow m}$ corresponds to a loss of probability for the system to remain in the mth state. It is interesting to note that the underlying process can be time reversed if the transition matrix is symmetric.

If the solution of a set of master equations approaches a steady state after some time, the resulting probability $P_m(\infty)$ is said to satisfy the detailed-balance condition. In this situation, the relation $\Gamma_{n \leftarrow m} P_m(\infty) = \Gamma_{m \leftarrow n} P_n(\infty)$ holds for all states of the system. Even though all steady-state probabilities do not depend on time in the steady state, the transition rates Γ_{mn} need not be time independent, provided the detailed-balance condition holds for all times after the steady state has been reached. As we have seen in Chapter 3, this condition is sometimes used to study multiparticle quantum interactions in devices that are in thermal equilibrium with large reservoirs.

It is easy to show that the master equations conserve the total probability. Indeed if we sum the set in Eq. (4.51) over m, we find

$$\sum_m \frac{dP_m}{dt} = \sum_m \sum_{n \neq m} (\Gamma_{m \leftarrow n} P_n - \Gamma_{n \leftarrow m} P_m) = 0. \qquad (4.52)$$

This means that the sum $\sum_m P_m(t)$ remains constant over time. As $\sum_m P_m$ equals 1 at $t = 0$, the same value of this sum is maintained for all times. If we also assume that all transition rates must be positive in the master equation, the following result holds whenever the probability P_m becomes zero during its evolution:

$$\left.\frac{dP_m}{dt}\right|_{P_m=0} = \sum_n \Gamma_{m \leftarrow n} P_n > 0. \qquad (4.53)$$

As the derivative dP_m/dt is positive at the time P_m becomes zero, the momentary zero value of P_m quickly becomes positive again. This reasoning can be used to conclude that all probabilities remain positive provided they were initialized correctly at $t = 0$ [$0 \leq P_m(0) \leq 1$]. The requirement that all probabilities must remain below the limiting value of 1 is also satisfied because $\sum_m P_m(t) = 1$ must hold at all times, and thus nothing can have a value greater than 1 at any time. If that were the case, one or more probabilities must have negative values, which is not allowed.

The set of master equations is often put into a matrix form for computational purposes. As this set contains a first-order, linear differential equation, we can write it in the form

$$\frac{d}{dt} |P(t)\rangle = \Lambda(t) |P(t)\rangle, \qquad (4.54)$$

where $|P(t)\rangle$ is a column vector with elements $(P_1, P_2, \ldots, P_M)$ and the matrix $\Lambda(t)$ contains all transition rates. The matrix elements of $\Lambda(t)$ can be calculated by equating the terms in Eq. (4.54) with those in Eq. (4.51). It is easy to show that

$$\frac{dP_m}{dt} = \sum_n \Lambda_{mn} P_n = \sum_{n \neq m} \Lambda_{mn} P_n + \Lambda_{mm} P_m = \sum_{n \neq m} (\Lambda_{mn} P_n - \Lambda_{nm} P_m), \qquad (4.55)$$

where we used the relation $\Lambda_{mm} P_m = -\sum_{n \neq m} \Lambda_{nm} P_m$. This relation follows by noting that the preceding equation must hold in the special case $P_m = \delta_{nm}$, resulting in $\sum_n \Lambda_{mn} = 0$ for all n. By comparing this equation with Eq. (4.51), we obtain the relation

$$\Lambda_{mn} = \Gamma_{m \leftarrow n} - \delta_{mn} \sum_{m \neq n} \Gamma_{m \leftarrow n}. \qquad (4.56)$$

When evaluating this expression, it is important to remember that $\Gamma_{n \leftarrow n} = 0$ for all n. Noting that the master equation always has a steady-state solution, it follows that the matrix Λ must have a zero eigenvalue whose eigenvector represents the steady-state solution of the master equation.

It is common to resort to using Eq. (4.54) because one wants to focus on the quantum device without getting involved into the reservoir dynamics. The important question is how we can find the transition rates between different quantum states of the device. In practice, master equations are set up using phenomenological arguments and physically justified approximations. An insightful example of this approach is provided by the Pauli master equation. It is often used for modeling quantum devices because of its simplicity. It has also found applications outside quantum mechanics in areas such as biology and chemical kinetics.

The Pauli master equation considers all possible transitions between the energy states of a quantum device. As in Eq. (4.51), the occupation probability P_i of the device being in state $|i\rangle$ with energy E_i is written as

$$\frac{dP_i(t)}{dt} = \sum_{j \neq i} \left[w_{ij} P_j(t) - w_{ji} P_i(t) \right], \qquad (4.57)$$

where the transition rate w_{ij} for moving state j to i is calculated using the Fermi Golden Rule (see Section 2.2.6):

$$w_{ij} = \frac{2\pi}{\hbar} \eta^2 |\langle i|H_I|j\rangle|^2 \delta(E_i - E_j). \qquad (4.58)$$

It assumes that the device Hamiltonian is of the form $H = H_0 + \eta H_I$, where a small positive parameter η denotes the strength of the perturbation and both H_0 and H_I are time independent. The energies E_i and E_j are the eigenvalues of the unperturbed Hamiltonian H_0 (i.e., $H_0 |i\rangle = E_i |i\rangle$).

The Pauli master equation (4.57) is like a rate equation and tells us how the occupation probability of a specific state increases because of transitions that move the atom into that state (gain) or decreases because of transitions that move the atom out of that state (loss). It can be shown using the evolution operator, $U(t) = \exp\left[-\frac{i}{\hbar}(H_0 + \eta H_I)t\right]$, that the first-order perturbation term apples only for a relatively short duration [269]. It is necessary to include the second-order perturbation term ($\propto \eta^2$) for the device to establish equilibrium. The time needed to reach equilibrium is proportional to η^{-2}, if individual interactions are described within the Born approximation. This suggests that η must be small and t must be large such that the product $\eta^2 t$ remains finite. Therefore, we must keep the $\eta^2 t$ term in the perturbation expansion of the evolution operator but can neglect the terms containing $\eta^m t^n$ with $m \neq 2n$. As shown by van Hove [269], the resulting approximate expression for the evolution operator has a time dependence consistent with the Pauli master equation [270]. The main point to remember is that the Fermi golden rule is valid as long as η is small but t is large enough that $\eta^2 t$ is finite; this is sometimes referred to as the van Hove limit [269, 270].

4.2.2 Reduced Density Matrix

The usefulness of Eq. (4.57) is limited in practice because it only deals with the occupation probabilities that are related to the diagonal elements of a density matrix. The off-diagonal elements cannot be ignored if the reservoir-induced decoherence plays a significant role. In this section we focus on the technique used to obtain a reduced density matrix from the total density matrix of the combined system.

An important requirement for the master-equation approach is that the quantum device should couple to the reservoir weakly to ensure that the reservoir is not affected much by the device, and it is possible to employ the *Born approximation*. However, the device itself is profoundly affected by the reservoir and gets entangled with it. Owing to this entanglement, the device cannot be described using a "pure state" and requires a mixed-state description through a density operator (see Section 2.2.3). Also, owing to a relatively large size of the reservoir with a large number of degrees of freedom, the reservoir has closely spaced energy levels. In contrast, the device has a single or a few widely separated energy levels that interact with the reservoir. This allows one to make use of the *Markovian approximation*. This combination of the Born and Markovian approximations leads to a Markovian master equation that can often be solved analytically. Given the assumptions made in its derivation, such a master equation has the following properties: (1) it describes the dynamics of a quantum device on time scales larger than the correlation time of the reservoir; (2) its stationary solution corresponds to the state of thermal equilibrium with its reservoir; (3) it can be written in the Lindblad form (see Section 4.3).

To describe the weak interaction between the device and its reservoir, we invoke standard quantum mechanics and discuss how to account for dissipation and decoherence without violating its postulates covered in Section 2.2. The approach employs a mathematical technique to obtain a "reduced" density operator for the device alone starting from the density operator of the whole system (device plus reservoir). Even though the evolution of the total density matrix is always unitary, the reduced density matrix of the device can be nonunitary.

Let ρ_D, ρ_R, and ρ_{DR} denote the density operators of the device, the reservoir, and the composite system (device plus reservoir). As the composite system is a closed system, its density operator $\rho_{DR}(t)$ evolves as

$$\rho_{DR}(t) = U_{DR}(t,0)\rho_{DR}(0)U_{DR}^{\dagger}(t,0), \tag{4.59}$$

where $U_{DR}(t,0)$ is the evolution operator of the composite system (see Section 2.2.5 for details). We can recover ρ_D from ρ_{DR} by using the operation $\rho_D = \mathrm{Tr}_R[\rho_{DR}]$, where Tr_R denotes a partial trace over the reservoir. Sometimes this procedure is called *tracing out* over the reservoir; see Aside 4.4 for details. Further progress is made by assuming that there is no initial correlation between the device and the reservoir (i.e., $\rho_{DR}(0) = \rho_D(0) \otimes \rho_R(0)$; see Section 2.2.1 for details on the tensor product denoted by $\otimes$). This is a reasonable assumption when the device and reservoir are weakly coupled but may be inappropriate in some specific cases [271]. If we also assume that the device has negligible influence on the reservoir and use $\rho_R(t) \approx \rho_R(0)$, we arrive at the *Born approximation* and obtain

$$\rho_{DR}(t) \approx \rho_D(t) \otimes \rho_R(0). \tag{4.60}$$

These assumptions can be justified, noting that reservoir represents a very large environment relative to the device. Because of its many degrees of freedom and a relatively weak coupling to the device, a quantum device cannot influence the reservoir, even though it is affected considerably by the reservoir [272, 273, 274].

Aside 4.4   Partial Trace and Reduced Density Operators

Consider two systems, D (a quantum device) and R (a reservoir), with the Hilbert spaces H_D and H_R, respectively. Given any two states, $|d\rangle \in H_D$ and $|r\rangle \in H_R$, in these systems, their tensor product state $|d\rangle \otimes |r\rangle$ belongs to the composite system $H_D \otimes H_R$. We note that the composite state $|d\rangle \otimes |r\rangle$ is sometimes compactly written as $|d\rangle\,|r\rangle$, $|d, r\rangle$, or $|dr\rangle$. When one of these composite states acts on an operator of the composite state $\mathcal{O}_D \otimes \mathcal{O}_R$, where $\mathcal{O}_D \in H_D$ and $\mathcal{O}_R \in H_R$, we get

$$(\mathcal{O}_D \otimes \mathcal{O}_R)(|d\rangle \otimes |r\rangle) = (\mathcal{O}_D\,|d\rangle) \otimes (\mathcal{O}_R\,|r\rangle). \tag{4.61}$$

Refer to Sections 2.2.1 and 2.2.3 for details on the tensor product and the density operator, respectively.

Let ρ_D, ρ_R, and ρ_{DR} denote the density operators of the device, the reservoir, and the composite system. The concept of partial trace allows us to recover ρ_D from ρ_{DR} and we denote it as

$$\rho_D = \mathrm{Tr}_R[\rho_{DR}], \tag{4.62}$$

where $\mathrm{Tr}_R[\ ]$ is the partial trace taken in the state space of the reservoir. As a special case of Eq. (4.61), we obtain the relation

$$\mathrm{Tr}_R[|d_1\rangle\,\langle d_2| \otimes |r_1\rangle\,\langle r_2|] = |d_1\rangle\,\langle d_2|\,\mathrm{Tr}_R[|r_1\rangle\,\langle r_2|]. \tag{4.63}$$

Here $|d_1\rangle\,\langle d_2| \otimes |r_1\rangle\,\langle r_2|$ is an operator in the composite state space, but $|d_1\rangle\,\langle d_2|$ is an operator in the device space H_D alone. Also, $\mathrm{Tr}_R[|r_1\rangle\,\langle r_2|]$ is a complex number that can be evaluated by noting that trace is invariant under cyclic permutation, giving $\mathrm{Tr}_R[|r_1\rangle\,\langle r_2|] = \langle r_2|r_1\rangle$.

For easy reference, we list several key features of the partial trace. Like the standard density operator, the reduced density operator ρ_D satisfies the following five properties: (1) Hermitian nature; (2) trace invariance, (3) detailed balance, (4) translational invariance, and (5) positivity (only nonnegative eigenvalues, to comply with the probability interpretation). Note that for any operator, $\hat{O}$, the relation $\frac{d}{dt}\langle \hat{O} \rangle = \mathrm{Tr}\left[\hat{O} \frac{d}{dt} \rho_D(t) \right]$ does not depend on the coordinates.

Suppose the reservoir's density operator $\rho_R(0)$ corresponds to a mixture of states with the probability p_α for being in the α state such that $\sum_\alpha p_\alpha = 1$. We do not assume these states to be orthogonal, but each state is normalized such that $\langle \psi_{R\alpha}|\psi_{R\alpha}\rangle = 1$. The initial density operator of the reservoir can be written as a sum over all possible states:

$$\rho_R(0) = \sum_\alpha p_\alpha\,|\psi_{R\alpha}\rangle\,\langle \psi_{R\alpha}|. \tag{4.64}$$

We use this result to calculate the reduced density operator of the device as follows:

$$\rho_D(t) = \text{Tr}_R[\rho_{DR}(t)] = \text{Tr}_R[U_{DR}(t,0)\rho_D(0) \otimes \rho_R(0)U_{DR}^\dagger(t,0)]$$

$$= \text{Tr}_R\left[U_{DR}(t,0)\rho_D(0) \otimes \left(\sum_\alpha p_\alpha |\psi_{R\alpha}\rangle \langle\psi_{R\alpha}|\right) U_{DR}^\dagger(t,0)\right]. \quad (4.65)$$

To make further progress, we make use of an orthonormal basis for the reservoir. Let the set $\{|\phi_{R\beta}\rangle\}$ represent this basis and let I_D and I_R be the identity operators in the Hilbert spaces of the device and the reservoir, respectively. Noting that trace is cycle-invariant under any orthonormal basis, we write $\rho_D(t)$ using the preceding relation as

$$\rho_D(t) = \sum_\beta (I_D \otimes \langle\phi_{R\beta}|)U_{DR}(t,0)\rho_D(0)$$

$$\otimes \left(\sum_\alpha p_\alpha |\psi_{R\alpha}\rangle \langle\psi_{R\alpha}|\right)U_{DR}^\dagger(t,0)(I_D \otimes |\phi_{R\beta}\rangle)$$

$$= \sum_\beta \sum_\alpha W_{\alpha\beta}(t)[\rho_D(0) \otimes 1]W_{\alpha\beta}^\dagger(t), \quad (4.66)$$

where we collected various terms to introduce the operator $W_{\alpha\beta}(t)$ as [272, 273, 274]

$$W_{\alpha\beta}(t) = \left(I_D \otimes \langle\phi_{R\beta}|\right) U_{DR}(t,0) \left(I_D \otimes \sqrt{p_\alpha}\,|\psi_{R\alpha}\rangle\right). \quad (4.67)$$

It is important to check whether $\text{Tr}_D[\rho_D(t)] = 1$, as expected. It is easy to see that this requirement will be satisfied if the double sum $S = \sum_\alpha \sum_\beta W_{\alpha\beta}^\dagger(t)W_{\alpha\beta}(t) = I_D$:

$$\text{Tr}_D\left[\rho_D(t)\right] = \text{Tr}_D\left[\sum_\alpha \sum_\beta W_{\alpha\beta}(t)(\rho_D(0) \otimes 1)W_{\alpha\beta}^\dagger(t)\right]$$

$$= \text{Tr}_D\left[(\rho_D(0) \otimes 1)\sum_{\alpha,\beta} W_{\alpha\beta}^\dagger(t)W_{\alpha\beta}(t)\right]$$

$$= \text{Tr}_D\left[(\rho_D(0) \otimes 1)(I_D \otimes 1)\right] = 1. \quad (4.68)$$

We verify the double sum as follows:

$$S = \sum_\alpha p_\alpha (I_D \otimes \langle\psi_{R\alpha}|) U_{DR}^\dagger(t,0)\left(I_D \otimes \sum_\beta |\phi_{R\beta}\rangle \langle\phi_{R\beta}|\right)U_{DR}(t,0)(I_D \otimes |\psi_{R\alpha}\rangle)$$

$$= \sum_\alpha p_\alpha (I_D \otimes \langle\psi_{R\alpha}|) U_{DR}^\dagger(t,0)(I_D \otimes I_R) U_{DR}(t,0)(I_D \otimes |\psi_{R\alpha}\rangle)$$

$$= \sum_\alpha p_\alpha (I_D \otimes \langle\psi_{R\alpha}|)(I_D \otimes |\psi_{R\alpha}\rangle) = \sum_\alpha p_\alpha (I_D \otimes \langle\psi_{R\alpha}|\psi_{R\alpha}\rangle) = I_D \otimes 1. \quad (4.69)$$

4.2.3 Markovian Master Equation

The preceding section has shown that it is possible to define an operator mapping $V(t)$ such that

$$\rho_D(t) = V(t)\rho_D(0) = \sum_{\alpha,\beta} W_{\alpha\beta}(t)[\rho_D(0) \otimes 1]W_{\alpha\beta}^\dagger(t), \quad (4.70)$$

Even though we treated time t as a fixed parameter up to now, if we allow it to vary keeping $t \geq 0$, we obtain a one-parameter family of dynamical maps that are completely positive and trace preserving. We now show that this family belongs to a semigroup, which will allow us to use known results in this area. The *semigroup condition* requires that $V(t_1)V(t_2) = V(t_1 + t_2)$ holds for $V(t)$; this is also known as the *Markovian property*. To prove it, we use Eq. (4.67) and the result $U_{DR}(t_1 + t_2, 0) = U_{DR}(t_1, 0)U_{DR}(t_2, 0)$:

$$V(t_2)\left[V(t_1)\rho_D(0)\right] = \sum_{\alpha,\beta} W_{\alpha\beta}(t_2)\left[\left(\sum_{\alpha',\beta'} W_{\alpha\beta}(t_1)[\rho_D(0) \otimes 1]W_{\alpha\beta}^{\dagger}(t_1)\right) \otimes 1\right]W_{\alpha\beta}^{\dagger}(t_2)$$

$$= \sum_{\alpha,\beta} W_{\alpha\beta}(t_1 + t_2)[\rho_D(0) \otimes 1]W_{\alpha\beta}^{\dagger}(t_1 + t_2) = V(t_1 + t_2)\rho_D(0). \qquad (4.71)$$

It is important to note that the constraint of time being positive $(t_1, t_2 \geq 0)$ implies that we can only propagate the system forward in time (i.e., this dynamical map does not have an inverse). This behavior is quite different from the coherent evolution of closed systems, where an inverse operation with a negative time argument always exists. Mathematically, even though the dynamical map of a closed quantum system forms a group, the corresponding map for an open quantum system can only form a semigroup. The generator of the semigroup is called the Liouvillian $\mathcal{L}$ defined through $V(t) = \exp(\mathcal{L}t)$. The Liouvillian operator is a generalization of the concept of a superoperator. One important consequence of this generalization is that the von Neumann entropy [see Eq. (2.50)] defined for our open system is not a conserved quantity. As the von Neumann entropy is essentially a quantum analog of the well-known Shannon entropy used in communication theory, this result suggests that the system loses coherence (or information) as it evolves. We can find the Liouville equation using a small time step (Δt) such that

$$\rho_D(t + \Delta t) = V(\Delta t)\rho_D(t) = \exp(\mathcal{L}\Delta t)\rho_D(t) = (1 + \mathcal{L}\Delta t)\rho_D(t) + O(\Delta t^2). \qquad (4.72)$$

If we take the limit $\Delta t \to 0$, we arrive at the Liouville equation

$$\frac{d}{dt}\rho_D(t) = \mathcal{L}\rho_D(t). \qquad (4.73)$$

The final task is to find an explicit form of the Liouvillian $\mathcal{L}$ in terms of the device and reservoir parameters. For this purpose, we need to find a basis to represent $\mathcal{L}$ in the Liouville space, which has a dimension of N^2 in contrast with the dimension N of the device's Hilbert space. This is done through a set of N^2 operators denoted by $\{F_i\}$ with i varying from 1 to N^2 such that $F_1 = I_D$ [272]. The inner product for this basis is defined as

$$F_i \cdot F_j = \frac{1}{N}\text{Tr}_D[F_i^{\dagger}F_j], \qquad (4.74)$$

so that the orthonormality condition, $F_i \cdot F_j = \delta_{ij}$, holds for any pair of basis elements. Using $F_1 = I_D$, we can show that all other basis elements are traceless:

$$F_1 \cdot F_j = \frac{1}{N}\text{Tr}_D[I_D^{\dagger}F_j] = \text{Tr}_D[F_j] = 0 \quad (j \neq 1). \qquad (4.75)$$

We use the orthonormal basis $\{F_i\}$ to expand the operator $W_{\alpha\beta}(t)$ as

$$W_{\alpha\beta}(t) = \sum_{i=1}^{N^2} [W_{\alpha\beta}(t) \cdot F_i] F_i. \tag{4.76}$$

Substitution of this result in Eq. (4.70) gives us

$$
\begin{aligned}
V(t)\rho_D(0) &= \sum_{\alpha} \sum_{\beta} \left[\sum_{i=1}^{N^2} [W_{\alpha\beta}(t) \cdot F_i] F_i \right] [\rho_D(0) \otimes 1] \left[\sum_{j=1}^{N^2} [W_{\alpha\beta}(t) \cdot F_j]^* F_j^\dagger \right] \\
&= \sum_{i=1}^{N^2} \sum_{j=1}^{N^2} \sum_{\alpha} \sum_{\beta} [W_{\alpha\beta}(t) \cdot F_i][W_{\alpha\beta}(t) \cdot F_j]^* F_i [\rho_D(0) \otimes 1] F_j^\dagger \\
&= \sum_{i=1}^{N^2} \sum_{j=1}^{N^2} c_{ij}(t) F_i [\rho_D(0) \otimes 1] F_j^\dagger,
\end{aligned} \tag{4.77}
$$

where we defined the time-dependent coefficients $c_{ij}(t)$ as

$$c_{ij}(t) = \sum_{\alpha} \sum_{\beta} [W_{\alpha\beta}(t) \cdot F_i][W_{\alpha\beta}(t) \cdot F_j]^*. \tag{4.78}$$

It follows that the matrix c formed using these coefficients is Hermitian ($c_{ij} = c_{ji}^\dagger$). The matrix is also positive because, for any complex vector v, we have the relation [272]

$$\sum_{i=1}^{N^2} \sum_{j=1}^{N^2} c_{ij} v_i^* v_j = \sum_{\alpha} \sum_{\beta} \left| \left(\sum_i v_i F_i \right) \cdot W_{\alpha\beta}(t) \right|^2 \geq 0. \tag{4.79}$$

Thus, all eigenvalues of the matrix c are real and positive. We use this property later.

As $V(t)\rho_D(0) = \exp(\mathcal{L}t)\rho_D(0)$, we can find $\mathcal{L}\rho_D(0)$ by evaluating $\frac{dV}{dt}\rho_D(0)$ at $t = 0$. Differentiating Eq. (4.77) with respect to t and setting $t = 0$, we obtain

$$\mathcal{L}\rho_D(0) = \frac{d}{dt} V(t)\rho_D(0) \bigg|_{t=0} = \sum_{i=1}^{N^2} \sum_{j=1}^{N^2} a_{ij} F_i [\rho_D(0) \otimes 1] F_j^\dagger, \tag{4.80}$$

where we introduced the time-independent coefficients a_{ij} as

$$a_{ij} = \frac{dc_{ij}(t)}{dt} \bigg|_{t=0}. \tag{4.81}$$

We can simplify further by using $F_1 = I_D$. Separating the $i = j = 1$ term from the double sum in the preceding expression, we obtain

$$\mathcal{L}\rho_D(0) = a_{11}I_D[\rho_D(0)\otimes 1]I_D^\dagger + \sum_{i=2}^{N^2} a_{i1}F_i[\rho_D(0)\otimes 1]I_D^\dagger$$

$$+ \sum_{j=2}^{N^2} a_{1j}I_D[\rho_D(0)\otimes 1]F_j^\dagger + \sum_{i=2}^{N^2}\sum_{j=2}^{N^2} a_{ij}F_i[\rho_D(0)\otimes 1]F_j^\dagger. \tag{4.82}$$

We simplify this expression by introducing a new operator F as

$$F = \sum_{i=2}^{N^2} a_{i1}F_i, \qquad F^\dagger = \sum_{i=2}^{N^2} a_{1i}F_i^\dagger, \tag{4.83}$$

where we used the fact that the matrix $a = \{a_{ij}\}$ is Hermitian because $c = \{c_{ij}\}$ is Hermitian. In terms of F, we can write $\mathcal{L}\rho_D(0)$ as

$$\mathcal{L}\rho_D(0) = a_{11}[\rho_D(0)\otimes 1] + F[\rho_D(0)\otimes 1] + [\rho_D(0)\otimes 1]F^\dagger +$$

$$\sum_{i=2}^{N^2}\sum_{j=2}^{N^2} a_{ij}F_i[\rho_D(0)\otimes 1]F_j^\dagger. \tag{4.84}$$

Noting that the trace of the reduced density operator does not change as it evolves in time, the relation $\mathrm{Tr}_D[\mathcal{L}\rho_D(0)] = 0$ is satisfied for any $\rho_D(0)$. Using it, we obtain

$$\mathrm{Tr}_D\left[\left(a_{11}I_D + F + F^\dagger + \sum_{i=2}^{N^2}\sum_{j=2}^{N^2} a_{i,j}F_j^\dagger F_i\right)[\rho_D(0)\otimes 1]\right] = 0. \tag{4.85}$$

It follows from this equation that

$$a_{11}I_D + F + F^\dagger + \sum_{i=2}^{N^2}\sum_{j=2}^{N^2} a_{ij}F_j^\dagger F_i = 0. \tag{4.86}$$

Using this result, we finally obtain the quantum master equation in a form referred to as the "first standard form" [272]:

$$\frac{d\rho_D(t)}{dt} = -\frac{i}{\hbar}[H_F, \rho_D(t)] + \mathcal{D}(\rho_D(t)), \tag{4.87}$$

where $H_F = \frac{i}{2}(F - F^\dagger)$ and the functional $\mathcal{D}(X)$ is defined as

$$\mathcal{D}(X) = \sum_{i=2}^{N^2}\sum_{j=2}^{N^2} a_{ij}\left(F_iXF_j^\dagger - F_j^\dagger F_iX\right) = \frac{1}{2}\sum_{i=2}^{N^2}\sum_{j=2}^{N^2} a_{ij}\left([F_iX, F_j^\dagger] + [F_i, XF_j^\dagger]\right). \tag{4.88}$$

When $\mathcal{D}(\rho_D(t)) = 0$, Eq. (4.87) reduces to the standard Liouville equation for a closed system. Thus, we can immediately identify H_F as the Hamiltonian that enables unitary evolution of a quantum device when dissipation and decoherence are absent; H_F is referred to as the *effective Hamiltonian of the device*. One effect of the reservoir is to change the original device Hamiltonian H_D to H_F, which results in shifting of energy levels of the device (the Lamb shift). The second effect of the reservoir comes through the operator $\mathcal{D}(\rho_D(t))$, which introduces decoherence and dissipation because of the device's interaction with the reservoir. This operator is sometimes referred to as the *dissipator* in literature. We stress that Eq. (4.87) is local in time because it depends on ρ_D at time t, but not on ρ_D values at earlier times (the Markovian property). Even though the derivation of this equation was presented systematically in 1976 in Refs. [275, 276], similar equations appeared earlier in the context of spin relaxation dynamics and laser theory [277, 278, 279].

4.3 Lindblad Equation

The Lindblad equation [276] is a variant of the Markovian quantum master equation derived in Section 4.2. Even though it bears Lindblad's name, equivalent results were obtained by Gorini et al. [275, 280]. The Lindblad equation makes use of the concept of *positive mapping* that was first introduced by Stinespring [281] and later studied by Kraus [282].

4.3.1 Derivation of the Lindblad Equation

The use of the orthonormal basis $\{F_i\}$ in Section 4.2.3 led to the coefficients c_{ij} in Eq. (4.78) that were used to define a_{ij} in Eq. (4.81). However, even though the $N^2 \times N^2$ matrix of a_{ij} coefficients is Hermitian, it is not necessarily diagonal in the $\{F_i\}$ basis. Clearly, the dissipator will have a less complicated structure in a basis in which the coefficients a_{ij} form a diagonal matrix. To realize this simplification, we diagonalize the a matrix using a unitary matrix U such that $a = UDU^\dagger$, where D is a diagonal matrix with the elements $(\gamma_{a1}, \gamma_{a2}, \ldots, \gamma_{aN^2})$. These eigenvalues are real and positive because the matrix a is Hermitian and positive definite. We use the matrix U to introduce a new set of operators A_k as

$$F_i = \sum_{k=2}^{N^2} U_{ki} A_k, \tag{4.89}$$

where $k = 2, 3, \ldots, N^2$; F_1 is not expanded because $F_1 = I_D$. Writing the matrix relation $a = UDU^\dagger$ in its explicit form, we have

$$a_{ij} = \sum_{p=1}^{N^2} \sum_{q=1}^{N^2} U_{ip} D_{pq} U_{jq}^* = \sum_p \gamma_{ap} U_{ip} U_{jp}^*. \tag{4.90}$$

We substitute Fi and a_{ij} in Eq. (4.87) to simplify the dissipator $\mathcal{D}(\rho_D(t))$:

$$\mathcal{D}(\rho_D(t)) = \frac{1}{2} \sum_{i=2}^{N^2} \sum_{p=2}^{N^2} \sum_{j=2}^{N^2} \gamma_{ap} U_{ip} U_{jp}^* \left[\sum_{k=2}^{N^2} U_{ki} A_k \rho_D(t), \sum_{k=2}^{N^2} A_k^\dagger U_{kj}^\dagger \right]$$

$$+ \frac{1}{2} \sum_{i=2}^{N^2} \sum_{p=2}^{N^2} \sum_{j=2}^{N^2} \gamma_{ap} U_{ip} U_{jp}^* \left[\sum_{k=2}^{N^2} U_{ki} A_k, \rho_D(t) \sum_{k=2}^{N^2} A_k^\dagger U_{kj}^\dagger \right]. \tag{4.91}$$

This equation can be simplified considerably because U is a unitary matrix. The final result is given by

$$\mathcal{D}(\rho_D(t)) = \frac{1}{2} \sum_{k=2}^{N^2} \gamma_{ak} \left([A_k \rho_D(t), A_k^\dagger] + [A_k, \rho_D(t) A_k^\dagger] \right). \tag{4.92}$$

When we substitute this result in Eq. (4.87), we obtain the Lindblad equation [276]

$$\frac{d\rho_D(t)}{dt} = -\frac{i}{\hbar} [H_F, \rho_D(t)] + \frac{1}{2} \sum_{k=2}^{N^2} \gamma_{ak} \left([A_k \rho_D(t), A_k^\dagger] + [A_k, \rho_D(t) A_k^\dagger] \right). \tag{4.93}$$

Even though more general evolution equations are conceivable using the Kraus map [283], the Lindblad equation is the most widely used quantum master equation. It has the following properties: (1) It is local in time (Markovian property), (2) has constant coefficients, and (3) preserves the Hermitian and positivity properties of the density matrix. The A_k operators are referred to as the Lindblad operators, or as jump operators of the system. It is important to note that all coefficients γ_{ak} represent rates (unit s^{-1}) because A_k operators are dimensionless. In practice, these rates correspond to the relaxation rates of the device.

Apart from quantum optics and quantum computing, the Lindblad equation has found applications in other branches of physics and chemistry. For example, it has been used to model the effect of environment on a ultrafast predissociation process [284]. It has also been used to model the dynamics of laser-induced nonthermal desorption of neutral molecules from metal surfaces [285]. In nuclear physics, a semigroup formalism similar to that used for the Lindblad equation is applied to model giant resonances in the nuclear spectra above the neutron-emission threshold [286].

It is also useful to consider how the Lindblad formalism relates to the Pauli master equation in Eq. (4.57). For this purpose, we first note that the diagonal elements of the density matrix correspond to populations of the associated energy levels. Second, we choose a basis $\{|m\rangle\}$ of the energy eigenstates by diagonalizing the Hamiltonian of the device. The Lindblad equation for the diagonal elements in this basis takes the form

$$\frac{d}{dt}(\rho_D)_{mm}(t) = \sum_{j \neq m} \left[(A_j)_{mn_j} (\rho_D)_{n_j n_j} (A_j^\dagger)_{n_j m} - |(A_j)_{mn_j}|^2 (\rho_D)_{mm} \right]$$

$$= \sum_{j \neq m} |(A_j)_{mn_j}|^2 \left[(\rho_D)_{n_j n_j} - (\rho_D)_{mm} \right], \tag{4.94}$$

where we assumed that the Lindblad operator A_j couples the state $|m\rangle$ only to the state $|n_j\rangle$. We now make the assignment $(\rho_D)_{mm}(t) \rightarrow P_m(t)$. The resulting equation can be written in the form of the Pauli master equation as

$$\frac{d}{dt}P_m(t) = \sum_{j \neq m} \left[w_{mj}P_j(t) - w_{jm}P_m(t) \right], \tag{4.95}$$

where $w_{mj} = \sum_{j \neq m} |(A_j)_{mn_j}|^2 \delta_{m,n_j}$. The important point to note is that this equation is based solely on the diagonal elements of the density matrix. It may be valid in some limiting cases, but off-diagonal matrix elements of the density matrix also need to be considered in the quantum-coherence regime. Even though one may circumvent these elements by adopting a formulation of coherent dynamics based on the diagonal elements, such an approach is nonintuitive and computationally hard because one needs to deal with the nonlocal time derivatives [273].

Because the Lindblad equation is a linear equation, it can be written in the form $\frac{d\rho_D(t)}{dt} = \mathcal{L}\rho_D(t)$. However, this format does not allow one to make use of readily available computational techniques because the density operator $\rho_D(t)$ is in the form of a matrix. Most computational schemes are designed for equations of the form $dV/dt = MV$, where V is a vector and M is a matrix. In practice, it is required to map the density operator $\rho_d(t)$ to a column vector using a strategy known as vec-ing [287, 288]. In some cases, the Lindblad equation can be solved analytically. The simplest example of such a dissipative quantum system is a damped harmonic oscillator. We discuss it next because this example provides insight that is lost in describing complex systems.

4.3.2 A Damped Harmonic Oscillator

Consider a harmonic oscillator oscillating at frequency ω_0. The Hamiltonian of an undamped harmonic oscillator has the form

$$H_D = \hbar\omega_0(\hat{a}^\dagger\hat{a} + 1/2), \tag{4.96}$$

where $\hat{a}$ and $\hat{a}^\dagger$ are the annihilation and creation operators that satisfy the commutation relation $[\hat{a}, \hat{a}^\dagger] = 1$. To account for dissipation induced by the oscillator's coupling to a reservoir, we invoke the Lindblad equation given in Eq. (4.93). Given that the Hilbert space is two-dimensional ($N = 2$), the Liouville space has dimensions $N^2 = 4$. Here we consider only three Lindblad operators and identify them as $\{A_1 = I_D, A_2 = \hat{a}, A_3 = \hat{a}^\dagger\}$, where I_D is the identity element. Introducing two positive eigenvalues γ_{a2} and γ_{a3}, we can write the Lindblad equation in the form

$$\frac{d\rho_D}{dt} = -i\omega_0[\hat{a}^\dagger\hat{a}, \rho_D(t)] + \frac{1}{2}\gamma_{a2}\left([\hat{a}\rho_D(t), \hat{a}^\dagger] + [\hat{a}, \rho_D(t)\hat{a}^\dagger]\right)$$
$$+ \frac{1}{2}\gamma_{a3}\left([\hat{a}^\dagger\rho_D(t), \hat{a}] + [\hat{a}^\dagger, \rho_D(t)\hat{a}]\right). \tag{4.97}$$

It follows from the cyclic-invariance property of the trace operator that $\text{Tr}[\rho_D(t)]$ maintains its initial value, as it should, because the trace of each commutator on the right side of the preceding equation is zero.

To find the equation of motion for averaged quantities, we use the identity $\text{Tr}[A[B,C]] = \text{Tr}[[A,B]C]$ that can be proven using the cyclic-invariance property of trace. Using $\langle \hat{a}(t)\rangle = \text{Tr}[\rho(t)\hat{a}]$, we obtain

$$\frac{d}{dt}\langle \hat{a}\rangle = \text{Tr}\left[\frac{d\rho}{dt}\hat{a}\right] = -i\omega\langle \hat{a}\rangle - \frac{1}{2}(\gamma_{a2} - \gamma_{a3})\langle \hat{a}\rangle . \tag{4.98}$$

We can integrate this equation to find the following analytical solution:

$$\langle \hat{a}(t)\rangle = \exp\left[-i\omega t - \frac{1}{2}(\gamma_{a2} - \gamma_{a3})t\right]\langle \hat{a}(0)\rangle . \tag{4.99}$$

This result shows that, as long as $\gamma_{a2} > \gamma_{a3}$, the amplitude of a quantum harmonic oscillator is damped in the same way as expected for a classical oscillator. It is important to emphasize that this decay occurs for the expectation value of the $\hat{a}$ operator, which itself cannot decay because that would violate the commutation $[\hat{a},\hat{a}^\dagger] = 1$. If $\gamma_{a3} > \gamma_{a2}$, the preceding equation describes a quantum amplifier. However, the saturation effects must be included because an exponential growth cannot continue forever.

Another quantity of interest is $\langle N_p\rangle = \text{Tr}[\hat{a}^\dagger\hat{a}\rho_D(t)]$, representing average population of the harmonic oscillator. Using the same procedure, we obtain the following rate equation:

$$\frac{d}{dt}\langle N_p\rangle = -\gamma_{a2}\langle N_p\rangle + \gamma_{a3}(\langle N_p\rangle + 1). \tag{4.100}$$

It follows from this equation that, when $\gamma_{a2} > \gamma_{a3}$ and the harmonic oscillator is damped, the population reaches a steady state value of

$$\langle N_p\rangle_{ss} = \frac{\gamma_{a3}}{\gamma_{a2} - \gamma_{a3}}. \tag{4.101}$$

If the harmonic oscillator is in thermal equilibrium, its average energy, $E_{av} = \langle N_p\rangle_{ss}\hbar\omega$, can be calculated from the Boltzmann distribution at a specific temperature T. Using

$$E_{av} = \langle N_p\rangle_{ss}\hbar\omega = \frac{\hbar\omega}{\exp\left(\frac{\hbar\omega}{k_BT}\right) - 1}, \tag{4.102}$$

we obtain the relation

$$\frac{\gamma_{a2}}{\gamma_{a3}} = \exp\left(\frac{\hbar\omega}{k_BT}\right) = \frac{\langle N_p\rangle_{ss} + 1}{\langle N_p\rangle_{ss}}. \tag{4.103}$$

Here $\langle N_p\rangle_{ss}$ is the number of thermal photons at the frequency ω, calculated using the Planck distribution. As both γ_{a2} and γ_{a3} are temperature dependent, we can introduce a positive real number $\gamma_0(T)$. In terms of this quantity, $\gamma_{a2} = \gamma_0(T)[\langle N_p\rangle_{ss} + 1]$ and $\gamma_{a3} = \gamma_0(T)\langle N_p\rangle_{ss}$.

4.3.3 A Damped Two-Level Atom

In this section we apply the Lindblad equation to a two-level atom to see how its coupling to a reservoir leads to its damping. We have considered this problem from a different angle when we used the Weisskopf–Wigner theory to derive the rate of spontaneous emission for an excited atom. We adopt the notation used in Section 4.1.4 and make use of Pauli spin operators introduced in Aside 4.3.

As before, the two-level atom has a ground state $|g\rangle$ with energy E_g and an excited state $|e\rangle$ with energy E_e. The energy difference between these states is written as $E_e - E_g = \hbar\omega_{eg}$. The device Hamiltonian takes the form $H_D = \frac{1}{2}\hbar\omega_{eg}\sigma_3$. The Pauli operators acting on this Hamiltonian either reduce or increase energy by $\hbar\omega_{eg}$ because $[H_D, \sigma_\pm] = \pm\hbar\omega_0\sigma_\pm$. The appropriate basis in the Liouville space for this system is the set $\{A_1 = I_D, A_2 = \sigma_3, A_3 = \sigma_-, A_4 = \sigma_+\}$, where I_D is the identity element. Therefore, we introduce three positive eigenvalues $\{\gamma_{a2}, \gamma_{a3}, \gamma_{a4}\}$ and obtain the Lindblad equation (4.93) in the following form:

$$\frac{d\rho_D}{dt} = -i\omega_{eg}[\sigma_3, \rho_D(t)] + \frac{1}{2}\gamma_{a2}\left([\sigma_3\rho_D(t), \sigma_3] + [\sigma_3, \rho_D(t)\sigma_3]\right)$$

$$+ \frac{1}{2}\gamma_{a3}\left([\sigma_-\rho_D(t), \sigma_+] + [\sigma_-, \rho_D(t)\sigma_+]\right) + \frac{1}{2}\gamma_{a4}\left([\sigma_+\rho_D(t), \sigma_-] + [\sigma_+, \rho_D(t)\sigma_-]\right).$$

(4.104)

This equation can be simplified using the commutation relations given in Aside 4.3 to obtain

$$\frac{d\rho_D}{dt} = -i\omega_{eg}[\sigma_3, \rho_D(t)] + \gamma_{a2}\left[\sigma_3\rho_D(t)\sigma_3 - \rho_D(t)\right]$$

$$+ \frac{1}{2}\gamma_{a3}\left[2\sigma_-\rho_D(t)\sigma_+ - \sigma_+\sigma_-\rho_D(t) - \rho_D(t)\sigma_+\sigma_-\right]$$

$$+ \frac{1}{2}\gamma_{a4}\left[2\sigma_+\rho_D(t)\sigma_- - \sigma_-\sigma_+\rho_D(t) - \rho_D(t)\sigma_-\sigma_+\right].$$ (4.105)

It is possible to make connections with the analysis for a damped harmonic oscillator (in Section 4.3.2) by identifying $\hat{a} \to \sigma_-$ and $\hat{a}^\dagger \to \sigma_+$. If we discard the γ_{a2} term leading to the Lamb shift in Eq. (4.105), we can use Eq. (4.102) and write the remaining decay rates as $\gamma_{a3} = \gamma_e[\langle N_p\rangle_{ss} + 1]$, and $\gamma_{a4} = \gamma_e\langle N_p\rangle_{ss}$, where γ_e is a constant that may depend on temperature and $\langle N_p\rangle_{ss}$ is the number of thermal photons at the frequency ω_{eg} (calculated using the Planck distribution). If we now compare these coefficients with the Weisskopf–Wigner theory of spontaneous emission in Section 4.1.3, we find that γ_e is just the spontaneous-emission rate given by

$$\gamma_e = \frac{\omega_{eg}^3 |\langle e|\mathbf{d}|g\rangle|^2}{3\pi\epsilon_0\hbar c^3}.$$ (4.106)

We stress that $\gamma_{a3} = \gamma_e[\langle N_p\rangle_{ss} + 1]$ contains two emission terms: $\gamma_e\langle N_p\rangle_{ss}$ representing the rate of stimulated emission and γ_e representing the rate of spontaneous emission. In contrast, γ_{a4} is responsible for absorption.

4.4 Redfield Equation

As we have seen in the preceding section, the Lindblad equation provides a way to track evolution of the density matrix of a device coupled to a reservoir. However, the map describing this evolution is not always unitary. Indeed, it is this nonunitary feature that introduces dissipation and decoherence in the standard formulation of quantum mechanics. The *Krauss operator-sum representation* provides such a map, although the resulting description is not unique [283, 282]. We can view the *Redfield equation* [289] as a generalization of the Lindblad equation because the latter can be derived from the former. However, this process cannot be reversed because not all resonant interactions with the environment are retained in the so-called secular approximation that converts the Redfield equation into the Lindblad equation. What is very important to note is that both the Redfield and Lindblad equations are Markovian master equations that are valid only for quantum devices that are weakly coupled to a large reservoir. Even though the Redfield equation is trace-preserving (just like the Lindblad equation), it does not guarantee a positive time-evolution of the density matrix, which opens the possibility for having negative populations (clearly an unphysical situation). In spite of this, the Redfield equation converges asymptotically to the thermal equilibrium distribution set by the reservoir.

4.4.1 Derivation of the Redfield Equation

We again consider a quantum device, coupled weakly to a large reservoir with the total Hamiltonian $H_{DR} = H_D + H_R + H_I$, where H_I is the interaction part. As the combination $D + R$ is a closed system, its density matrix $\rho_{DR}(t)$ obeys the standard Liouville equation

$$\frac{d}{dt}\rho_{DR}(t) = -\frac{i}{\hbar}[H_{DR}, \rho_{DR}(t)]. \tag{4.107}$$

It is useful to transform this equation to the interaction picture. For this transformation, we adopt the notation of Section 2.2.5 and represent the operators in the interaction picture with a tilde on top. In terms of the evolution operator $U_{DR}(t) = \exp[-\frac{i}{\hbar}(H_D + H_R)t]$, we have

$$\rho_{DR}(t) = U_{DR}(t)\widetilde{\rho}_{DR}(t)U_{DR}^{\dagger}(t), \qquad H_I(t) = U_{DR}(t)\widetilde{H}_I(t)U_{DR}^{\dagger}(t). \tag{4.108}$$

The resulting Liouville equation in the interaction picture has the form

$$\frac{d}{dt}\widetilde{\rho}_{DR}(t) = -\frac{i}{\hbar}[\widetilde{H}_I(t), \widetilde{\rho}_{DR}(t)]. \tag{4.109}$$

This equation can be solved implicitly to obtain

$$\widetilde{\rho}_{DR}(t) = \widetilde{\rho}_{DR}(0) - \frac{i}{\hbar}\int_0^t [\widetilde{H}_I(t'), \widetilde{\rho}_{DR}(t')]\, dt', \tag{4.110}$$

where $\widetilde{\rho}_{DR}(0)$ is the initial value of the density operator. It is not possible to find an exact solution of this integral equation. However, it can be solved approximately as a series expansion by replacing $\widetilde{\rho}_{DR}(t)$ on the right side with its value at $t = 0$, and repeating this procedure multiple times.

As our focus is on the quantum device, in Eq. (4.109) we trace over the reservoir to obtain the reduced density operator of the device. Using a series expansion and retaining terms up to second order, the final result is given by

$$\frac{d}{dt}\widetilde{\rho}_D(t) = -\frac{i}{\hbar}\operatorname{Tr}_R[\widetilde{H}_I(t), \widetilde{\rho}_{DR}(0)] - \frac{1}{\hbar^2}\int_0^t \operatorname{Tr}_R[\widetilde{H}_I(t), [\widetilde{H}_I(t'), \widetilde{\rho}_{DR}(t')]]\, dt'. \qquad (4.111)$$

As before, it is reasonable to assume that no initial correlation exists between the device and the reservoir (i.e., $\widetilde{\rho}_{DR}(0) = \rho_D(0) \otimes \rho_R(0)$). Although his assumption may be unreasonable in some specific situations [271], we assume that it holds for our quantum device. If we further assume that the evolution of the device has negligible influence on the reservoir ($\widetilde{\rho}_R(t) \approx \rho_R(0)$), we arrive at the Born approximation: $\widetilde{\rho}_{DR}(t) \approx \widetilde{\rho}_D(t) \otimes \rho_R(0)$. As a result, the composite system remains in an approximate product state at all times, and temporal changes in the density matrix of the environment can be neglected. This is justified by noting that the reservoir represents a large environment with many degrees of freedom, and it remains in thermal equilibrium because it interacts with the device weakly [272, 273, 274]. Under these conditions,

$$\operatorname{Tr}_R[\widetilde{H}_I(t), \widetilde{\rho}_{DR}(0)] = \operatorname{Tr}_R[\widetilde{H}_I(t), \rho_D(0) \otimes \rho_R(0)] \to 0, \qquad (4.112)$$

because when taking the trace over the reservoir using the eigenstates of H_R, the commutator bracket vanishes owing to the diagonal representation of $\rho_R(0)$. As trace is invariant, the above result holds for any other basis for the reservoir as well.

With these simplifications, Eq. (4.111) reduces to

$$\frac{d}{dt}\widetilde{\rho}_D(t) = -\frac{1}{\hbar^2}\int_0^t \operatorname{Tr}_R[\widetilde{H}_I(t), [\widetilde{H}_I(t'), \widetilde{\rho}_D(t') \otimes \rho_R(0)]]\, dt'. \qquad (4.113)$$

This is still an integro-differential equation and it cannot be solved easily. To simplify it, we make use of the Markovian approximation, which amounts to assuming that the quantum device has no memory of its past. We thus replace $\widetilde{\rho}_D(t')$ with $\widetilde{\rho}_D(t)$ in Eq. (4.113) and obtain the Redfield equation in the form [289, 272]

$$\frac{d}{dt}\widetilde{\rho}_D(t) = -\frac{1}{\hbar^2}\int_0^t \operatorname{Tr}_R[\widetilde{H}_I(t), [\widetilde{H}_I(t'), \widetilde{\rho}_D(t) \otimes \rho_R(0)]]\, dt'. \qquad (4.114)$$

The Redfield equation is useful in quantum optics but has found applications in many other fields including magnetic resonance [290, 291] and optical spectroscopy [292].

Even though the Redfield equation is local in time for the reduced density matrix, it is still not fully Markovian as it contains an implicit dependence on the initial value at $t = 0$ of the reduced density operator. By making the substitution $t' \to t - t_1$ and extending the upper integration limit to infinity, we obtain a fully Markovian master equation in the form

$$\frac{d}{dt}\widetilde{\rho}_D(t) = -\frac{1}{\hbar^2}\int_0^\infty \operatorname{Tr}_R[\widetilde{H}_I(t), [\widetilde{H}_I(t - t_1), \widetilde{\rho}_D(t) \otimes \rho_R(0)]]\, dt_1. \qquad (4.115)$$

This equation is valid for reservoirs with a "short memory" and is justified when the time scale over which the quantum device evolves can be considered large compared to the time scale over which correlations in the reservoir decay. As a result, the Markovian version of the Redfield equation cannot resolve device dynamics over time scales comparable to or shorter than the correlation time of the reservoir. Thus, even though the Markovian Redfield equation is a differential equation, it can only describe device evolution over a coarse-grained time scale [272]. Also, there is no guarantee that Eq. (4.115) describes a generator of a dynamical semigroup [293, 294, 272]. Another issue is that Eq. (4.115) contains rapidly oscillating terms that are problematic for its numerical implementation. A way to remove these rapidly oscillating terms is provided by the rotating-wave approximation that we consider next.

4.4.2 Rotating-Wave Approximation

To implement the rotating-wave approximation, we write the interaction Hamiltonian H_I in the form [272]:

$$H_I = \sum_\alpha \Upsilon_\alpha \otimes \Lambda_\alpha, \tag{4.116}$$

where the Hermitian operators Υ_α and Λ_α act respectively on the quantum device and the reservoir. We denote eigenvalues of the Hamiltonian H_D of the quantum device by ϵ, and the corresponding eigenstates by $|\epsilon\rangle$. For any two eigenstates such that $\epsilon' - \epsilon = \hbar\omega$, we define a new operator at the frequency ω as

$$\Upsilon_\alpha(\omega) = \sum_{\epsilon,\epsilon'} \delta_{\epsilon'-\epsilon,\hbar\omega} |\epsilon\rangle \langle\epsilon|\Upsilon_\alpha|\epsilon'\rangle \langle\epsilon'|. \tag{4.117}$$

We calculate the commutator $[H_D, \Upsilon_\alpha(\omega)]$ as

$$[H_D, \Upsilon_\alpha(\omega)] = \sum_{\epsilon,\epsilon'} \delta_{\epsilon'-\epsilon,\hbar\omega}\epsilon |\epsilon\rangle \langle\epsilon|\Upsilon_\alpha|\epsilon'\rangle \langle\epsilon'| - \sum_{\epsilon,\epsilon'} \delta_{\epsilon'-\epsilon,\hbar\omega} |\epsilon\rangle \langle\epsilon|\Upsilon_\alpha|\epsilon'\rangle \epsilon' \langle\epsilon'|$$

$$= (\epsilon - \epsilon') \sum_{\epsilon,\epsilon'} \delta_{\epsilon'-\epsilon,\hbar\omega} |\epsilon\rangle \langle\epsilon|\Upsilon_\alpha|\epsilon'\rangle \langle\epsilon'| = -\hbar\omega \Upsilon_\alpha(\omega). \tag{4.118}$$

Similarly, we can show that $[H_D, \Upsilon_\alpha^\dagger(\omega)] = \hbar\omega \Upsilon_\alpha^\dagger(\omega)$. It follows that the relation $\Upsilon_\alpha^\dagger(\omega) = \Upsilon_\alpha(-\omega)$ holds. We can also show that $[H_D, \Upsilon_\alpha^\dagger(\omega)\Upsilon_{\alpha'}(\omega)] = 0$. As the energy eigenstates form a complete basis in the associated Hilbert space, the sum over all energy levels such that $\epsilon' - \epsilon = \hbar\omega$ amounts to summing over all frequencies. Thus, the original operator can be written as

$$\Upsilon_\alpha = \sum_\omega \Upsilon_\alpha(\omega). \tag{4.119}$$

As the Redfield equation (4.115) is written in the interaction picture, we need to write the operator Υ_α in the interaction picture using

$$\widetilde{\Upsilon}_\alpha = \exp\left(\frac{i}{\hbar}H_D t\right) \left(\sum_\omega \Upsilon_\alpha(\omega)\right) \exp\left(-\frac{i}{\hbar}H_D t\right), \tag{4.120}$$

We make use of the Baker–Hausdorff formula for any two operators A and B:

$$\exp(iBt)A\exp(-iBt) = A + it[B, A] + \left(\frac{i^2 t^2}{2!}\right)[B, [B, A]] + \dots. \tag{4.121}$$

Using it, we find the relations

$$\widetilde{\Upsilon}_\alpha = \sum_\omega \exp(-i\omega t)\Upsilon_\alpha(\omega), \qquad \widetilde{\Upsilon}_\alpha^\dagger = \sum_\omega \exp(i\omega t)\Upsilon_\alpha^\dagger(\omega). \tag{4.122}$$

Combining these results, the interaction Hamiltonian can be written as

$$\widetilde{H}_I(t) = \sum_\alpha \sum_\omega \exp(-i\omega t)\Upsilon_\alpha(\omega) \otimes \widetilde{\Lambda}_\alpha(t), \tag{4.123}$$

where $\widetilde{\Lambda}_\alpha(t)$ is given by

$$\widetilde{\Lambda}_\alpha(t) = \exp\left(\frac{i}{\hbar}H_R t\right)\Lambda_\alpha \exp\left(-\frac{i}{\hbar}H_R t\right). \tag{4.124}$$

It follows from Eq. (4.112) that the average value of $\widetilde{\Lambda}_\alpha(t)$ is zero.

Before using these results in Eq. (4.115), we rewrite them in the following equivalent form:

$$\frac{d}{dt}\widetilde{\rho}_D(t) = \frac{1}{\hbar^2}\int_0^\infty \mathrm{Tr}_R\Big[\widetilde{H}_I(t - t')\widetilde{\rho}_D(t) \otimes \rho_R(0)\widetilde{H}_I(t)$$
$$- \widetilde{H}_I(t)\widetilde{H}_I(t - t')\widetilde{\rho}_D(t) \otimes \rho_R(0)\Big]\,dt' + \mathrm{H.C}, \tag{4.125}$$

where H.C denotes the Hermitian conjugate of the previous expression. Substituting the interaction Hamiltonian from Eq. (4.123) and simplifying by rearranging terms, we obtain

$$\frac{d}{dt}\widetilde{\rho}_D(t) = \sum_{\alpha,\alpha'}\sum_{\omega,\omega'}\Gamma_{\alpha\alpha'}(\omega)\exp[i(\omega' - \omega)t]$$
$$\times \left(\Upsilon_{\alpha'}(\omega)\widetilde{\rho}_D(t)\Upsilon_\alpha^\dagger(\omega') - \Upsilon_\alpha^\dagger(\omega')\Upsilon_{\alpha'}(\omega)\widetilde{\rho}_D(t)\right) + \mathrm{H.C}, \tag{4.126}$$

where we have introduced

$$\Gamma_{\alpha\alpha'}(\omega) = \frac{1}{\hbar^2}\int_0^\infty \mathrm{Tr}_R\left[\widetilde{\Lambda}_\alpha^\dagger(t)\widetilde{\Lambda}_{\alpha'}(t - t')\rho_R(0)\right]\exp(i\omega t')\,dt'. \tag{4.127}$$

The trace over reservoir represents an averaging procedure. Thus, we can write $\Gamma_{\alpha\alpha'}(\omega)$ in terms of the reservoir's correlation function as

$$\Gamma_{\alpha\alpha'}(\omega) = \frac{1}{\hbar^2}\int_0^\infty \langle\widetilde{\Lambda}_\alpha^\dagger(t)\widetilde{\Lambda}_{\alpha'}(t - t')\rangle\exp(i\omega t')\,dt'. \tag{4.128}$$

Because the reservoir is large and remains in thermal equilibrium ($[H_R, \rho_R(0)] = 0$), the correlation function depends only on the time difference [272]:

$$\langle\widetilde{\Lambda}_\alpha^\dagger(t)\widetilde{\Lambda}_{\alpha'}(t - t')\rangle = \langle\widetilde{\Lambda}_\alpha^\dagger(t')\widetilde{\Lambda}_{\alpha'}(0)\rangle. \tag{4.129}$$

Equation (4.126) is still too complicated. We can simplify it by noting that the term $\exp[i(\omega' - \omega)t]$ oscillates rapidly when $\omega \neq \omega'$ and $t \gg (\omega' - \omega)^{-1}$. We can eliminate such terms by making the *rotating-wave approximation* and setting $\omega' = \omega$, resulting in

$$\frac{d}{dt}\widetilde{\rho}_D(t) = \sum_{\alpha,\alpha'} \sum_{\omega} \Gamma_{\alpha\alpha'}(\omega) \left[\Upsilon_{\alpha'}(\omega)\widetilde{\rho}_D(t)\Upsilon_{\alpha}^{\dagger}(\omega) - \Upsilon_{\alpha}^{\dagger}(\omega)\Upsilon_{\alpha'}(\omega)\widetilde{\rho}_D(t) \right] + \text{H.C.} \quad (4.130)$$

To simplify it further, we introduce the Fourier transform of the reservoir's correlation function as

$$\gamma_{\alpha\alpha'}(\omega) = \frac{1}{\hbar^2} \int_{-\infty}^{\infty} \langle \widetilde{\Lambda}_{\alpha}^{\dagger}(t)\widetilde{\Lambda}_{\alpha'}(t - t') \rangle \exp(i\omega t')\, dt', \quad (4.131)$$

and note that $\Gamma_{\alpha\alpha'}(\omega)$ is related to this quantity as [272]

$$\Gamma_{\alpha\alpha'}(\omega) = \frac{1}{2}\gamma_{\alpha\alpha'}(\omega) + iS_{\alpha\alpha'}(\omega). \quad (4.132)$$

Using these results in Eq. (4.130), we obtain the Redfield equation in the rotating-wave approximation:

$$\frac{d}{dt}\widetilde{\rho}_D(t) = -\frac{i}{\hbar}[H_{LS}, \widetilde{\rho}_D(t)] + \mathcal{D}\big(\widetilde{\rho}_D(t)\big), \quad (4.133)$$

where the dissipator is defined as

$$\mathcal{D}\big(\widetilde{\rho}_D(t)\big) = \sum_{\alpha,\alpha'} \sum_{\omega} \frac{1}{2}\gamma_{\alpha\alpha'}(\omega) \left([\Upsilon_{\alpha'}(\omega)\widetilde{\rho}_D(t), \Upsilon_{\alpha}^{\dagger}(\omega)] + [\Upsilon_{\alpha'}(\omega), \widetilde{\rho}_D(t)\Upsilon_{\alpha}^{\dagger}(\omega)] \right). \quad (4.134)$$

The Hamiltonian H_{LS} in Eq. (4.133) is called the Lamb-shift Hamiltonian because it causes a Lamb-type shift of the unperturbed energy levels. It is defined as

$$H_{LS} = \hbar \sum_{\alpha,\alpha'} \sum_{\omega} S_{\alpha\alpha'} \Upsilon_{\alpha}^{\dagger}(\omega)\Upsilon_{\alpha'}(\omega). \quad (4.135)$$

It is easy to show that the Lamb-shift Hamiltonian commutes with the device Hamiltonian (i.e., $[H_{LS}, H_D] = 0$).

4.5 Quantum-Optics Master Equation

A typical situation found in quantum optics is that many electromagnetic modes of the reservoir interact weakly with a quantum device. These interactions enable the flow of energy from the device to these modes. However, owing to the weak nature of the interaction, it takes much longer for energy to flow back to the device once it has left. The reason for this behavior is related to different Rabi frequencies associated with different radiation modes. In essence, because of a continuum of radiation modes, energy transfer to the reservoir is an irreversible process that destroys the coherence of the atomic state. It is important to stress that this energy transfer satisfies well both the Markovian and Born approximations [295, 296, 297]. Thus, we can apply the Markovian version of the Redfield equation (4.133) to describe this situation and understand how a quantum device interacts with the radiation modes. The resulting equation is the quantum-optics master equation.

We use the dipole approximation to write the interaction Hamiltonian as

$$H_I = -\mathbf{d} \cdot \mathbf{E} = -\sum_{\alpha=1}^{3} d_\alpha E_\alpha, \tag{4.136}$$

where $\alpha = 1, 2, 3$ represent three Cartesian coordinates, $\mathbf{d}$ is the dipole-moment operator, and $\mathbf{E}$ is the interacting electric field, assumed to be a plane wave polarized along the direction $\hat{\mathbf{p}}_s$. This unit vector is always orthogonal to the propagation vector $\mathbf{k}$ (i.e., $\mathbf{k} \cdot \hat{\mathbf{p}}_s = 0$). The frequency of the plane wave is given by $\omega_k = ck$. The quantized form of the electric field is given by

$$\mathbf{E} = \sum_{ks} \sqrt{\frac{\hbar \omega_k}{2\epsilon_0 V}} \left(\hat{a}_{\mathbf{k}s}^\dagger + \hat{a}_{\mathbf{k}s} \right) \hat{\mathbf{p}}_s, \tag{4.137}$$

where V is the reservoir volume. The total Hamiltonian of the device plus reservoir is thus given by

$$H = H_D + \sum_{\mathbf{k},s} \hbar \omega_k \hat{a}_{\mathbf{k}s}^\dagger \hat{a}_{\mathbf{k}s} - \sum_\alpha \mathbf{d}_\alpha \otimes \mathbf{E}_\alpha, \tag{4.138}$$

where H_D is device Hamiltonian and $[\hat{a}_{\mathbf{k}s}, \hat{a}_{\mathbf{k}'s'}^\dagger] = \delta_{\mathbf{k}\mathbf{k}'}\delta_{ss'}$.

We assume that the electromagnetic radiation interacting with the quantum device is in thermal equilibrium. In this situation, the density operator of a mode with frequency ω_{kj} has the form

$$\rho_R(\omega_{kj}) = \frac{1}{1 - \exp(-\hbar \omega_{kj}/k_B T)} \sum_{n_{kj}=0}^{\infty} \exp\left[-n_{kj}(\hbar \omega_{kj}/k_B T)\right] |n_{kj}\rangle \langle n_{kj}|, \tag{4.139}$$

where n_{kj} represents the number of photons in the state $|n_{kj}\rangle$. Using this expression, we can write the equilibrium density operator $\rho_R(0)$ in Eq. (4.115) as

$$\rho_R(0) = \rho_R(\omega_{k1}) \otimes \rho_R(\omega_{k2}) \otimes \ldots \otimes \rho_R(\omega_{km}) \otimes \ldots . \tag{4.140}$$

The utility of this expression is that we can we can calculate averages using the relation $\langle O \rangle = \text{Tr}[O\rho_R(0)]$, where O is any operator. It is easy to show that

$$\langle \hat{a}_{\mathbf{k}s} \hat{a}_{\mathbf{k}'s'} \rangle = 0, \qquad \langle \hat{a}_{\mathbf{k}s}^\dagger \hat{a}_{\mathbf{k}'s'}^\dagger \rangle = 0, \tag{4.141}$$

$$\langle \hat{a}_{\mathbf{k}s} \hat{a}_{\mathbf{k}'s'}^\dagger \rangle = \delta_{\mathbf{k}s,\mathbf{k}'s'}[1 + N_{BI}(\omega_k)], \tag{4.142}$$

$$\langle \hat{a}_{\mathbf{k}s}^\dagger \hat{a}_{\mathbf{k}'s'} \rangle = \delta_{\mathbf{k}s,\mathbf{k}'s'} N_{BI}(\omega_k), \tag{4.143}$$

where the *Bose–Einstein occupation number* is given by

$$N_{BI}(\omega_k) = \frac{1}{1 - \exp\left(-\hbar \omega_k/k_B T\right)}. \tag{4.144}$$

If the equilibrium condition of the reservoir is different (a squeezed state rather than thermal equilibrium), the correlation averages may have different values, but the following analysis remains valid.

We can use the preceding results to calculate the most important quantity in the Redfield equation (4.133), namely the reservoir correlation function $\Gamma_{\alpha\alpha'}(\omega)$. Using Eq. (4.127), it can be written as

$$\Gamma_{\alpha\alpha'}(\omega) = \frac{1}{\hbar^2} \int_0^\infty \langle \mathbf{E}_\alpha^\dagger(t)\mathbf{E}_{\alpha'}(t-t')\rangle \exp(i\omega t')\, dt'. \tag{4.145}$$

Substituting $\mathbf{E}$ from Eq. (4.137), we obtain

$$\Gamma_{\alpha\alpha'}(\omega) = \frac{1}{\hbar^2} \int_0^\infty \sum_{ks}\sum_{k's'} \sqrt{\frac{\hbar\omega_k}{2\epsilon_0 V}}\sqrt{\frac{\hbar\omega_{k'}}{2\epsilon_0 V}}\, \hat{\mathbf{p}}_{s\alpha}\hat{\mathbf{p}}_{s'\alpha'} \exp(i\omega t')\times$$

$$\Big(\langle \hat{a}_{ks}^\dagger \hat{a}_{k's'}^\dagger\rangle \exp[i\omega_k t + i\omega_{k'}(t-t')] + \langle \hat{a}_{ks}\hat{a}_{k's'}\rangle \exp[-i\omega_k t - i\omega_{k'}(t-t')]+$$

$$\langle \hat{a}_{ks}\hat{a}_{k's'}^\dagger\rangle \exp[-i\omega_k t + i\omega_{k'}(t-t')] + \langle \hat{a}_{ks}^\dagger \hat{a}_{k's'}\rangle \exp[i\omega_k t - i\omega_{k'}(t-t')] \Big) dt'. \tag{4.146}$$

If we now apply the averages given in Eq. (4.141) and note that $\hat{\mathbf{p}}_{s\alpha}\hat{\mathbf{p}}_{s'\alpha'} = (\hat{\mathbf{p}}_{s\alpha})^2\delta_{\alpha\alpha'}\delta_{ss'}$, we obtain

$$\Gamma_{\alpha\alpha'}(\omega) = \frac{1}{\hbar^2} \sum_{k,s} \frac{\hbar\omega_k}{2\epsilon_0 V}\, \langle\hat{\mathbf{p}}_{s\alpha}\rangle^2 \delta_{\alpha\alpha'}\delta_{ss'}\Big([1+N_{BI}(\omega_k)]$$

$$\times \int_0^\infty \exp[-i(\omega_k-\omega)t']\, dt' + N_{BI}(\omega_k)\exp[i(\omega_k+\omega)t']\, dt'\Big). \tag{4.147}$$

Finally, using the Sokhotski–Plemelj formula given in Aside 3.3, $\Gamma_{\alpha\alpha'}(\omega)$ can be written in the form

$$\Gamma_{\alpha\alpha'}(\omega) = \delta_{\alpha,\alpha'}\left(\frac{1}{2}\gamma(\omega) + iS(\omega)\right), \tag{4.148}$$

where $\gamma(\omega) = (\omega^3 \langle p_\alpha\rangle^2 /3\pi\epsilon_0\hbar c^3)b(\omega)$ and $b(\omega)$ is defined as

$$b(\omega) = \begin{cases} 1+N_{BI}(\omega), & \omega \geq 0 \\ -N_{BI}(-\omega), & \omega < 0. \end{cases} \tag{4.149}$$

The corresponding Lamb shift of energy levels is given by

$$S(\omega) = \frac{\langle p_\alpha\rangle^2}{6\pi^2\hbar\epsilon_0 c^3}\text{P.V.}\int_0^\infty \left(\frac{1+N_{BI}(\omega')}{\omega-\omega'} + \frac{N_{BI}(\omega')}{\omega+\omega'}\right)\omega'^3\, d\omega'. \tag{4.150}$$

When these quantities are used in Eq. (4.133), the resulting equation is called the quantum-optics master equation.

5 Quantum Current Flow

What we observe is not nature itself, but nature exposed to our method of questioning.

Werner Heisenberg

5.1 Quantum Transport

Ohm's law is an empirical law that is known to hold across a wide range of length scales. It states that the local current density of a material is proportional to the electric field strength at that point ($J = \sigma E$). The proportionality constant σ is known as the conductivity of the material, and it can even be defined using the same exact ratio when this law breaks down. Indeed, Ohm's law is known to break down when the electric field is very strong or very weak. As a result, it is natural to assume that it may not hold for nanoscale conductors where the discrete nature of electric charges cannot be ignored, but experiments have not met this expectation. It was observed in 2012 that Ohm's law holds for silicon wires as small as four atoms wide and one atom high [298]. What this observation tells us is that we need to revise our understanding of conductivity, especially at the nanoscale, and develop a more accurate model of charge transport in nanoscale conductors. In this chapter, we focus on different ways of characterizing charge transfer through nanostructures, taking into account the quantum nature of charge carriers and that of the surrounding medium.

Historically, charge transport in conductors was studied by Drude and Lorentz. The Drude theory predicts that the conductivity of a metal is given by $\sigma = N_e q_e^2 \tau_e / m_e$, where N_e is the density of electrons and τ_e is the mean collision time of electrons related to their mean free path. In this model, conductivity of a metal is calculated by extending the local conductivity associated with an infinitesimal length of the conductor to a finite size using the well-established machinery of calculus. The model also assumes that charge transport through a conductor is essentially diffusive in nature with isotropic relaxation.

It was realized in the 1950s that the Drude theory has a limited validity for quantum devices with nanoscale conductive channels. Rather than integrating local conductivity over the size of a conductor, another approach emerged where the conductance of the whole channel was found directly. This approach is known as the *scattering method*. The advantage of this method is that both the coherent and incoherent processes occurring within a quantum device can be accounted for when charge carriers are driven far from thermal equilibrium. By coherent processes we mean processes such as tunneling and ballistic transport; incoherent processes include scattering via phonons and charge transport via hopping mechanism (in inorganic semiconductors). The scattering method also abandons the notion that charge carriers behave like classical particles inside a conductor and enable one to invoke the full machinery of quantum mechanics.

In electromagnetic theory, the "current" (measured in amperes) is defined as the amount of charge (in Coulombs) transferred through a given cross section of a conductor per unit time (measured in seconds). This definition was also officially adopted by NIST (National Institute of Standards and Technology, USA) in May 20, 2019, along with three other SI base units: the kilogram (mass), kelvin (temperature), and mole (amount of substance). One promising way to quantify current with high precision makes use of a nanoscale technique called single-electron transport (SET) pumping [299]. This technique is used at NIST for measuring currents. It involves applying a gate voltage to a transistor-like device, which ejects one electron through a high-resistance tunneling junction into a quantum island made using a microscopic quantum dot. As discussed in Aside 5.1, this electron is removed from the island via another tunneling junction.

Aside 5.1   Single-Electron Transport (SET) Pumping

A highly precise current source can be built using a SET pump. As shown in Figure 5.1, it is constructed by connecting two tunneling junctions and a gate electrode to a quantum island capable of holding single electrons. The structure resembles a metal-oxide semiconductor field-effect transistor (MOSFET) used in conventional electronic circuits. The gate electrode controls flow of electrons from the source to the drain. However, these electrons need to cross two tunneling junctions that isolate the source and the gate from the quantum

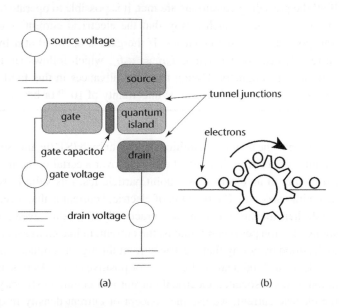

(a)   (b)

Fig. 5.1   (a) Schematic diagram of a single-electron transistor (SET). Current flows from the source to the drain, as in a conventional MOSFET. (b) Schematic showing how the SET pumps individual electrons.

island in the middle. The quantum island plays a key role in passing single electrons from the source to the drain. The principle exploited is called the Coulomb blockade effect, a well-understood phenomenon discussed in Section 1.5. This effect can be understood by noting that, whenever an electron is tunneled to the quantum island, the electrostatic energy of the system increases by an amount E_C given by

$$E_C = \frac{q_e^2}{2C_{SET}}, \tag{5.1}$$

where $C_{SET} = C_S + C_D + C_G$ is the total capacitance of the device, C_S, C_D, and C_G being capacitances between the quantum island and the source, the drain, or the gate, respectively. Often, E_C is referred to as the charging energy because it is precisely the energy gained with the injection of a single electron.

The tunneling of an electron can only occur if an energy greater than E_C is externally supplied to the device. However, if thermal energy $K_B T$ is larger than E_C, the process is hampered by the operating temperature of the device. The charging energy can be increased beyond thermal energy by reducing the total capacitance C_{SET}, which amounts to reducing the physical dimensions of tunneling junctions to near 100 nm. The characteristic time to tunnel an electron through such a tunnel junction is given by $t_T = C_T R_T$, where C_T can be C_S, C_D, or C_G. However, the uncertainty principle induces an energy uncertainty of $\frac{h}{2t_T}$ to the required charging energy. To guarantee tunneling events, we need to ensure $E_C \gg \frac{h}{2t_T}$. This condition requires that the resistance of the tunnel junction be large enough to satisfy $R_T \gg h/q_e^2$. The ratio h/q_e^2 is known as the *von Klitzing constant*; it has a numerical value of 25.9 kΩ.

If all the preceding conditions are met, it is possible to operate the SET pump by sweeping the gate voltage in such a way that the electrical current is generated through clocked transport of individual electrons. If the gate is operated at a frequency f_G, the generated current I_{SD} can be written as $I_{SD} = q_e f_G$, which follows straight from the definition of current in a conductor. Owing to recent advances in this field, SET pumps can generate currents ~ 100 pA with a relative uncertainty of 10^{-6} or better.

Owing to the adopted definition of current by NIST, the measurement of current becomes a matter of counting individual electrons over a certain time interval. However, the classical picture of an electron as a point particle loses its validity when its quantum nature is taken into account. The paradox of electrical current is that, even though charges are quantized, their transfer rate can have practically any value, even a fraction of the charge of a single electron per second; that is, the current, unlike charges, is not quantized. This can be understood by noting that electric current through a conductor is merely a displacement of the electron cloud against the lattice of positive cores. As this displacement can be by any amount, the associated electrical current is a continuously varying quantity. To calculate the electric current, we use the concept of current density in quantum mechanics. If the motion of a quantized charge carrier of mass m_e and charge q_e is governed by the wave function Ψ_e, the resulting current density is given by

$$\mathbf{J} = -i\frac{q_e\hbar}{2m_e}\left(\Psi_e^*\nabla\Psi_e - \Psi_e\nabla\Psi_e^*\right) = \frac{q_e\hbar}{m_e}\Im\left[\Psi_e^*\nabla\Psi_e\right], \qquad (5.2)$$

where $\Im[\ldots]$ stands for the imaginary part. The conductance of a quantum channel can be used to characterize how this current interacts with other parts of the quantum system.

According to the semiclassical theory, there are no fundamental restrictions on the conductivity of a material, and it can have all continuous values in a range that depends on the intrinsic properties of a specific material. However, following the discovery of the quantum Hall effect, it became clear that conductance is not a continuous variable but changes in steps of a basic quantum unit, the so-called *conductance quantum* given by

$$G_Q = q_e^2/h. \qquad (5.3)$$

The existence of the conductance quantum was first observed in a 2D electron gas formed between GaAs and AlGaAs semiconducting layers [59, 60]. This discovery opened up a new area of research on quantum localization. A very useful relationship, due to Landauer, states that conductance of any material can be calculated by multiplying the conductance quantum G_Q with the quantum-mechanical transmission coefficient of that material. The important question that seems like a paradox is how dissipation occurs as a result of current flowing through a quantum conductor. Whenever a current I flows through a material with conductance G, it also dissipates energy at the rate I^2/G. As elastic scattering is the only process responsible for the appearance of the conductance, there is no dynamical mechanism that can cause energy dissipation. The answer comes from the appearance of a contact resistance when a quantum conductor is connected to a reservoir. The reasoning is based on the fluctuation-dissipation theorem discussed in Section 3.3 that predicts the behavior of systems obeying detailed balance; electrical resistances in quantum systems are also covered by this theorem.

A typical quantum transport system is depicted in Figure 5.2. It consists of a nanoscale quantum device that can either hold or transport charge carriers to the electrodes (which could be more than two in multilead scenarios). If an electrode supplies charge carriers, it is called a "source." If an electrode removes charge carriers, it is termed a "drain." Depending on details of the sources, drains, and the quantum device, such a system can be analyzed using several approaches differing in details of how the dynamics of charge carriers through the quantum device is treated. The first approach we consider is known as the Landauer–Büttiker method, and it provides a simple and intuitive description for the majority of quantum systems found in practice. In this approach, charge dynamics is considered as ballistic transport (pure elastic scattering) near thermal equilibrium. This

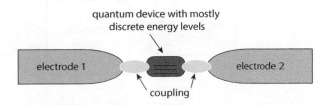

Fig. 5.2   A typical quantum transport setup where a quantum device is connected to two electrodes.

method completely obscures the quantum features of the device and treats it like a potential barrier, which is represented by a scattering matrix. Such a passive approach may not be suitable in cases where the quantum device (1) is nonlinear and its response covers a broad frequency range, (2) functions far from equilibrium with a high source-drain voltage, or (3) contains charge carriers that interact with each other as they move inside the device. To handle such scenarios, it is common to invoke transport theory, especially tailored for interacting particles under nonequilibrium conditions. The underlying method is known as the nonequilibrium Green's function method.

5.2 Landauer–Büttiker Method

The Landauer–Büttiker method maps quantum transport to an equivalent scattering problem and establishes a relation between the scattering amplitude of a charge carrier in a quantum device and its conducting properties (see Fig. 5.3). The beginnings of this method can be traced back to the pioneering work in 1957 by Rolf Landauer [63, 64], who heuristically derived an expression for the electric current using an approach based on scattering theory. This work was refined in 1986 by Büttiker, who laid the foundation of the formalism presented here [300]. This method is general to the extent that it can be applied to any quantum channel containing noninteracting charge carriers. The absence of inelastic scattering can be relaxed, provided the mean-field description remains valid. In this method, interactions among charge carriers are accounted for through changes in the charge distribution that modify the scattering potential. This type of charge transport is called coherent because quantum coherence properties are preserved across the scattering region.

5.2.1 Scattering Matrix Representation

The scattering matrix S provides all the information needed to find the outputs at the ports of a quantum device, given any inputs at these ports (see Fig. 5.3). This matrix takes into account only linear operations taking place within the quantum device. Such an approach is also called the *S-parameter* method. The *X-parameter* method is a generalization of the S-parameter method and is used for characterizing quantum devices subject to large input

Fig. 5.3   Model of quantum transport based on scattering theory. Coupling of the quantum device to the electrodes is represented by a scattering matrix.

power levels that drive them into the nonlinear regime [301]. Here we focus on the S-parameter method and consider the simplest case of a quantum device with only two ports. Generalization to three or more ports is not difficult but is not considered in this book. The utility of the scattering matrix is that it fully characterizes linear operations of a quantum device without requiring a detailed knowledge of all components and processes inside the quantum device. This feature makes it useful in practice, especially for devices that have complex internal setup.

Figure 5.4 shows a two-port quantum device, each port supporting a single channel (or mode). The dynamics of the device is governed by a potential barrier situated at its center. One can view the ports as transmission lines bringing the wave function to this potential barrier for scattering. We employ a local coordinate system, with z_η and $\mathbf{r}_{\perp\eta}$ denoting the longitudinal and transverse coordinates at the port η with $\eta = 1, 2$. The origin of these coordinates lies at the device's center. Assuming that the current flows along the longitudinal direction, we ignore the transverse coordinates in the following discussion.

All charge particles (electrons) move freely on both sides of the potential barrier and are affected by this barrier only at $z_\eta = 0$. Thus, the wave function of each particle is in the form of a plane wave. Consider one such particle moving in the forward direction. Its associated plane wave arrives at the potential barrier located at $z_\eta = 0$ from the left. As a result of scattering, this plane wave is split into reflected and transmitted waves. However, as the scattering process is assumed to be elastic, the number of particles as well as their total energy and momentum must be conserved before and after the scattering event. Assuming that the ports of the device are symmetric, the velocity of the particle at each port is given by $v_\eta(E) = \hbar k_\eta(E)/m$, where m is the particle's mass, $\hbar k$ is its momentum, and $E = \hbar^2 k_\eta^2/2m$ is its energy. Since velocities are equal for a given energy, k_η is the same at both ports. In general, both E and v_η can vary over a wide range.

As a result of scattering, there will be reflected and transmitted waves at both ports of the device. At any given point along either of the ports, the total wave function will be a superposition of these two waves. As the scattering matrix is defined using the amplitudes of the forward and backward propagating plane waves (denoted by a_η and b_η), the flux density corresponds to $|a_\eta|^2$ and $|b_\eta|^2$. We normalize these amplitudes such that the particle number is preserved. In mathematical terms, this normalization amounts to writing the incident plane wave in the form $\psi(z_\eta, k_\eta) = (2\pi\hbar v_\eta)^{-1/2}\exp(ik_\eta z_\eta)$. This form ensures that the following relation denoting the conservation of particles holds:

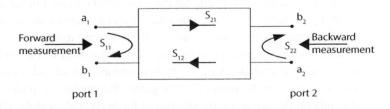

A two-port quantum device and scattering parameters associated with it. The same dS parameters are also used for commercial vector-network analyzers.

$$\int_E \int_{z_\eta} \psi(z_\eta, k_\eta)\psi^*(z_\eta, k'_\eta)\, dz_\eta\, dE = \frac{1}{2\pi} \int_{k_\eta} \int_{z_\eta} e^{i(k_\eta - k'_\eta)z_\eta}\, dz_\eta\, dk_\eta$$

$$= \int_{k_\eta} \delta(k_\eta - k'_\eta)\, dk_\eta = 1, \qquad (5.4)$$

where we used the relation $dE = \hbar v_\eta\, dk_\eta$.

With the preceding normalization, we can write the wave functions on two sides of the potential barrier in the form

$$\Psi(z_\eta) = \begin{cases} (2\pi\hbar v_1)^{-1/2}[a_1 \exp(ik_1 z_\eta) + b_1 \exp(-ik_1 z_\eta)] & \text{if } z_\eta < 0, \\ (2\pi\hbar v_2)^{-1/2}[b_2 \exp(ik_2 z_\eta) + a_2 \exp(-ik_2 z_\eta)] & \text{if } z_\eta > 0, \end{cases} \qquad (5.5)$$

where a_1 and a_2 are the incident wave amplitudes and b_1 and b_2 are the scattered wave amplitudes at ports 1 and 2, respectively. Also, we can set $k_1 = k_2$ and $v_1 = v_2$ owing to the symmetry assumption. The scattering matrix is defined through the relation

$$\begin{bmatrix} b_1 \\ b_2 \end{bmatrix} = \begin{bmatrix} S_{11} & S_{12} \\ S_{21} & S_{22} \end{bmatrix} \begin{bmatrix} a_1 \\ a_2 \end{bmatrix}. \qquad (5.6)$$

Transmission and reflection coefficients of the forward wave at the potential barrier are found by setting $a_2 = 0$ and are given by

$$t_{21} = \left.\frac{b_2}{a_1}\right|_{a_2=0}, \qquad r_{11} = \left.\frac{b_1}{a_1}\right|_{a_2=0}. \qquad (5.7)$$

If the input comes from the other port, the transmission and reflection coefficients, t_{12} and r_{22}, are obtained by setting $a_1 = 0$. Using them, it possible to write the scattering matrix in the form

$$\mathbf{S} = \begin{bmatrix} r_{11} & t_{12} \\ t_{21} & r_{22} \end{bmatrix}. \qquad (5.8)$$

As any wave function satisfies the Schrödinger equation, these four coefficients can be found by solving this equation using the form of the wave function given in Eq. (5.5). The resulting scattering matrix $\mathbf{S}$ is unitary owing to the flux conservation (no particles are lost during the scattering process),

$$\mathbf{S}^\dagger \mathbf{S} = \mathbf{S}\mathbf{S}^\dagger = 1. \qquad (5.9)$$

The situation changes if particles of the same energy have different velocities at the two ports. In this case, we can still use the wave function given in Eq. (5.5) with $k_1 \neq k_2$ and $v1 \neq v_2$. Following the same procedure, we can calculate the matrix elements as before, while maintaining the unitary nature of the scattering matrix.

It is possible to generalize the preceding formalism to the multichannel case, where the particles at each port can belong to different channels (also called transverse modes). When a port is in the form of a waveguide, the number of modes supported by that port depends on the geometric dimensions of the waveguide and the effective mass of charge carriers. In this case, channels are analogous to the electromagnetic modes of a microwave or optical waveguide found by solving Maxwell's equations. The only difference is that the eigenfunctions of the Schrödinger equation provide the modes for the channels.

If N_1 and N_2 denote the numbers of channels at the two ports, the incoming and outgoing amplitudes can be written in a vector form using the following two column vectors:

$$
\mathbf{a} = \begin{bmatrix} a_{11} \\ \vdots \\ a_{1N_1} \\ a_{21} \\ \vdots \\ a_{2N_2} \end{bmatrix} \quad ; \quad \mathbf{b} = \begin{bmatrix} b_{11} \\ \vdots \\ b_{1N_1} \\ b_{21} \\ \vdots \\ b_{2N_2} \end{bmatrix},
\tag{5.10}
$$

where the first subscript denotes the port and the second subscript denotes the channel of that port. For the two-port case, the resulting scattering matrix is a square matrix with $N_1 + N_2$ rows and columns. It is sometimes possible to partition this scattering matrix into submatrices corresponding to reflected and transmitted wave functions, analogous to the 2×2 scattering matrix in the single-channel case:

$$
\mathbf{S} = \begin{bmatrix} \mathbf{s}_{11} & \mathbf{s}_{12} \\ \mathbf{s}_{21} & \mathbf{s}_{22} \end{bmatrix} \rightarrow \mathbf{S} = \begin{bmatrix} \mathbf{r}_{11} & \mathbf{t}_{12} \\ \mathbf{t}_{21} & \mathbf{r}_{22} \end{bmatrix}.
\tag{5.11}
$$

Here, the transmission matrix $\mathbf{t}_{21}$ has dimensions of $N_1 \times N_2$, and the reflection matrix $\mathbf{r}_{11}$ has dimensions of $N_1 \times N_1$.

As the complexity of the system grows, it is useful to specify the notation clearly to identify individual components in the scattering matrix. We employ the standard notation used by equipment manufacturers such as Keysight™. For a two-port passive linear device, let the subscripts p and q denote the ports, and m and n denote the channels at these ports. The elements of the scattering matrix are then written as

$$
\mathbf{S} = \left[S_{(pm)(qn)} \right].
\tag{5.12}
$$

However, when each port has a single channel, we can ignore the trivial mode index 1 and use the compact notation $\mathbf{S} = \left[S_{(p1)(q1)} \right] = \left[S_{pq} \right]$. We always ensure that a scattering matrix is unitary. If a nonunitary scattering matrix is encountered, we can make it unitary with the following scaling:

$$
S_{(pm)(qn)} \rightarrow \sqrt{\frac{v_{qn}}{v_{pm}}} S_{(pm)(qn)},
\tag{5.13}
$$

where v_{pm} and v_{qn} are the velocities of a charge carrier in the designated channels. As this rescaling can always be performed, we only consider unitary scattering matrices in this book.

5.2.2 Charge Transport in Two-Port Devices

Even though the Landauer–Büttiker method can be used for multiport devices, we consider the simplest case seen in Figure 5.3, where two electrodes (acting as reservoirs in thermal equilibrium) are connected to a quantum device (mesoscopic scatterer). Owing to the equilibrium state of the two electrodes, electrons inside them behave incoherently. However, if an electron leaves an electrode and enters the quantum device, its transport through the device preserves its coherence by undergoing only elastic scattering events. This type of transport is referred to as *phase-coherent transport*. As particle flux is conserved in this kind of transport, the scattering matrix provides a complete description of the passage through a quantum device. The Landauer–Büttiker method determines the current flowing through this device using its scattering matrix, which depends on the device's geometry. The reservoirs (contacts) come into this description through their equilibrium distributions, which depend on the chemical potential and temperature.

To calculate the scattering matrix, we begin with the Hamiltonian of an electron with the effective mass m_e written as

$$H_\eta = \frac{p_{z_\eta}^2}{2m_e} + \frac{p_{\perp\eta}^2}{2m_e} + V(\mathbf{r}_{\perp\eta}), \tag{5.14}$$

where the kinetic energy of the particle is separated into its longitudinal and transverse parts, using z_η and $\mathbf{r}_{\perp\eta}$ as the local coordinates in the two directions. The motion of the particle along the z_η direction is not constrained, but it is quantized in the transverse plane owing to the confinement potential $V(\mathbf{r}_{\perp\eta})$. One can find the "transverse" channels by solving the eigenvalue equation

$$\left[\frac{p_{\perp\eta}^2}{2m_e} + V(\mathbf{r}_{\perp\eta})\right] \psi_{\eta n}(\mathbf{r}_{\perp\eta}) = \epsilon_{\eta n}\psi_{\eta n}(\mathbf{r}_{\perp\eta}). \tag{5.15}$$

As usual, the transverse profiles of different modes are orthogonal and are normalized such that

$$\int \psi_{\eta n}^*(\mathbf{r}_{\perp\eta})\psi_{\zeta m}(\mathbf{r}_{\perp\zeta})\,d\mathbf{r}_{\perp\eta} = \delta_{\eta\zeta}\delta_{mn}. \tag{5.16}$$

Since the particle moves freely in the z_η direction, its energy (dispersion relation) can be written as

$$E_{\eta n}(k_{\eta n}) = \frac{\hbar^2 k_{\eta n}^2}{2m_e} + \epsilon_{\eta n}. \tag{5.17}$$

The incoming and outgoing wave functions for the nth channel can now be expressed as

$$a_{\eta n}(z_\eta, \mathbf{r}_{\perp\eta}) = \psi_{\eta n}(\mathbf{r}_{\perp\eta})\exp(+ik_{\eta n}z_\eta), \tag{5.18}$$

$$b_{\eta n}(z_\eta, \mathbf{r}_{\perp\eta}) = \psi_{\eta n}(\mathbf{r}_{\perp\eta})\exp(-ik_{\eta n}z_\eta). \tag{5.19}$$

With these incoming and outgoing wave functions, we can use the scattering-matrix description to obtain the outgoing wave function for the nth channel at the port ζ in the form

$$b_{\zeta n} = \sum_\eta \sum_{m\in\{\eta\}} S_{(\zeta n)(\eta m)} a_{\eta m}. \tag{5.20}$$

We quantize the incoming and outgoing amplitudes using the second-quantization formulation and denote the corresponding annihilation operators by $\hat{a}_{\eta m}$ and $\hat{b}_{\eta m}$ and the creation operators by $\hat{a}_{\eta m}^{\dagger}$ and $\hat{b}_{\eta m}^{\dagger}$, respectively. With these definitions, the operator form of Eq. (5.20) can be written as

$$\hat{b}_{\zeta n}(E) = \sum_{\eta} \sum_{m \in \{\eta\}} S_{(\zeta n)(\eta m)}(E)\hat{a}_{\eta m}(E), \tag{5.21}$$

$$\hat{b}_{\zeta n}^{\dagger}(E) = \sum_{\eta} \sum_{m \in \{\eta\}} (S_{(\zeta n)(\eta m)})^{*}(E)\hat{a}_{\eta m}^{\dagger}(E). \tag{5.22}$$

Armed with these expressions, we can construct the wave function in the k-space associated with the particles traveling through each port:

$$\Psi_{\eta} = \sum_{n \in \{\eta\}} \Psi_{\eta}(\mathbf{r}_{\eta}, k_{\eta n}), \tag{5.23}$$

where the wave function for the nth channel is given by

$$\Psi_{\eta}(\mathbf{r}_{\eta}, k_{\eta n}) = \psi_{\eta n}(\mathbf{r}_{\perp \eta}) \left[\hat{a}_{\eta n}(k_{\eta n})e^{ik_{\eta n}z_{\eta}} + \hat{b}_{\eta n}(k_{\eta n})e^{-ik_{\eta n}z_{\eta}} \right]. \tag{5.24}$$

The definition of current density given in Eq. (5.2) demands a wave function $\Psi_{\eta}(\mathbf{r}_{\eta}, t)$ in physical space and time. We can obtain it by taking the inverse Fourier transform of the k-space wave function as

$$\Psi_{\eta}(\mathbf{r}_{\eta}, t) = \sum_{n \in \{\eta\}} \int_{-\infty}^{\infty} \Psi_{\eta}(\mathbf{r}_{\eta}, k_{\eta n}) \exp\left[-\frac{i}{\hbar}E_{\eta n}(k_{\eta n})t\right] dk_{\eta n}. \tag{5.25}$$

We substitute this result in Eq. (5.2) to obtain the current density and integrate across the transverse plane of the quantum device to calculate the current. The result is given by

$$\hat{I}_{\eta}(z_{\eta}, t) = \frac{\hbar q_e}{2im_e}\left[\int \Psi_{\eta}^{\dagger}(\mathbf{r}_{\eta}, t)\frac{\partial}{\partial z_{\eta}}\Psi_{\eta}(\mathbf{r}_{\eta}, t)\, d\mathbf{r}_{\perp \eta} - \int \left(\frac{\partial}{\partial z_{\eta}}\Psi_{\eta}^{\dagger}(\mathbf{r}_{\eta}, t)\right)\Psi_{\eta}(\mathbf{r}_{\eta}, t)\, d\mathbf{r}_{\perp \eta} \right]. \tag{5.26}$$

In practice, the complexity of the problem can be significantly reduced by changing the integration from the k space to the energy domain while taking the inverse Fourier transform in Eq. (5.25). This can be done by using the dispersion relation in Eq. (5.17) to get $dE_{\eta n} = (\hbar^2 k_{\eta n}/m_e)dk_{\eta n} = \hbar v_{\eta n}dk_{\eta n}$. However, we also need to map the annihilation and creation operators to their energy-space representation. We recall from Chapter 2 that we need to ensure the invariance of the commutator or anticommutator relations, regardless of the representation used. We are dealing with electrons here, which are fermions. Thus, we need to ensure that the anticommutator relation is invariant in both representations. In the k space, this relation has the form

$$\hat{a}_{\eta m}(k_{\eta m})\hat{a}_{\zeta n}^{\dagger}(k_{\zeta n}) + \hat{a}_{\zeta n}^{\dagger}(k_{\zeta n})\hat{a}_{\eta m}(k_{\eta m}) = \delta_{\eta \zeta}\delta_{mn}\delta(k_{\eta m} - k_{\zeta n}). \tag{5.27}$$

In the energy domain, this relation is preserved if we use

$$\hat{a}_{\eta m}(E_{\eta m})\hat{a}_{\zeta n}^{\dagger}(E_{\zeta n}) + \hat{a}_{\zeta n}^{\dagger}(E_{\zeta n})\hat{a}_{\eta m}(E_{\eta m}) = \frac{\delta_{\eta \zeta}\delta_{mn}}{\hbar v_{\eta m}(E_{\eta m})}\delta(E_{\eta m} - E_{\zeta n}). \tag{5.28}$$

The preceding equation shows that we can transform all creation and annihilation operators from the k space to the energy domain with a simple mapping of the form

$$\hat{\Upsilon}_{\eta n}(k_{\eta n}) \to \frac{1}{\sqrt{\hbar v_{\eta n}(E_{\eta n})}} \hat{\Upsilon}_{\eta n}(E_{\eta n}), \quad \text{where} \quad \Upsilon \in \{\hat{a}, \hat{a}^\dagger, \hat{b}, \hat{b}^\dagger\}. \tag{5.29}$$

Even though this transformation looks complex because the prefactor depends on the energy, it turns out that its use cancels out certain terms in the Fourier integral, simplifying it considerably. Using the preceding relations, we can write the wave function in Eq. (5.24) in the energy basis as

$$\Psi_\eta(\mathbf{r}_\eta, E_{\eta n}) = \frac{\psi_{\eta n}(\mathbf{r}_{\perp \eta})}{\sqrt{\hbar v_{\eta n}}} \Big[\hat{a}_{\eta n}(E_{\eta n}) \exp(ik_{\eta n} z_\eta) + \hat{b}_{\eta n}(E_{\eta n}) \exp(-ik_{\eta n} z_\eta) \Big]. \tag{5.30}$$

We can now calculate the current given in Eq. (5.26). In view of the orthogonal nature of different transverse modes, when we multiply and integrate across the transverse area of the quantum device, only terms with the matching port and channel indices survive in the final expression. Consider the first integral in Eq. (5.26) with the wave function given in Eq. (5.25). Taking the z derivative and using the mode-orthogonality relation, we obtain

$$\int \Psi_\eta^\dagger(\mathbf{r}_\eta, t) \frac{\partial}{\partial z_\eta} \Psi_\eta(\mathbf{r}_\eta, t)\, d\mathbf{r}_{\perp \eta} = \sum_{n \in \{\eta\}} \iint ik_{\eta n}(E'_{\eta n}) \frac{\kappa_{\eta n 1}(E_{\eta n})}{\sqrt{\hbar v_{\eta n}(E_{\eta n})}} \frac{\kappa_{\eta n 2}(E'_{\eta n})}{\sqrt{\hbar v_{\eta n}(E'_{\eta n})}}$$

$$\times \exp\left[-\frac{i}{\hbar}(E'_{\eta n} - E_{\eta n})t \right] dE_{\eta n}\, dE'_{\eta n}, \tag{5.31}$$

where we have defined the following two relations:

$$\kappa_{\eta n 1}(E_{\eta n}) = \Big[\exp[-ik_{\eta n}(E_{\eta n}) z_\eta] \hat{a}_{\eta n}^\dagger(E_{\eta n}) + \exp[ik_{\eta n}(E_{\eta n}) z_\eta] \hat{b}_{\eta n}^\dagger(E_{\eta n}) \Big] \tag{5.32}$$

$$\kappa_{\eta n 2}(E'_{\eta n}) = \Big[\exp[ik_{\eta n}(E'_{\eta n}) z_\eta] \hat{a}_{\eta n}(E'_{\eta n}) - \exp[-ik_{\eta n}(E'_{\eta n}) z_\eta] \hat{b}_{\eta n}(E'_{\eta n}) \Big]. \tag{5.33}$$

We can simplify further by noting that speeds of electrons of different energies are not that different because speed is a slowly varying function of energy. If we assume that they are all the same in the energy range of interest, we can use $v_{\eta n}(E_{\eta n}) \approx v_{\eta n}(E'_{\eta n}) = \frac{\hbar k_{\eta n}}{m_e}$. Using this approximation in Eq. (5.31), we obtain

$$\int \Psi_\eta^\dagger(\mathbf{r}_\eta, t) \frac{\partial}{\partial z_\eta} \Psi_\eta(\mathbf{r}_\eta, t)\, d\mathbf{r}_{\perp \eta} = \frac{im_e}{h^2} \sum_{n \in \{\eta\}} \iint \kappa_{\eta n 1}(E_{\eta n}) \kappa_{\eta n 2}(E'_{\eta n})$$

$$\times \exp\left[-\frac{i}{\hbar}(E'_{\eta n} - E_{\eta n})t \right] dE_{\eta n}\, dE'_{\eta n}. \tag{5.34}$$

Using the same procedure, the second term of Eq. (5.26) is found to be

$$\int \left(\frac{\partial}{\partial z_\eta} \Psi_\eta^\dagger(\mathbf{r}_\eta, t) \right) \Psi_\eta(\mathbf{r}_\eta, t)\, d\mathbf{r}_{\perp \eta} = \frac{im_e}{h^2} \sum_{n \in \{\eta\}} \iint \kappa_{\eta n 2}(E_{\eta n}) \kappa_{\eta n 1}(E'_{\eta n})$$

$$\times \exp\left[-\frac{i}{\hbar}(E'_{\eta n} - E_{\eta n})t \right] dE_{\eta n}\, dE'_{\eta n}. \tag{5.35}$$

Using these results in Eq. (5.26), we finally obtain the current in the form

$$\hat{I}_\eta(t) = \frac{q_e}{2h} \iint \sum_{n\in\{\eta\}} \left[\kappa_{\eta n1}(E_{\eta n})\kappa_{\eta n2}(E'_{\eta n}) - \kappa_{\eta n2}(E_{\eta n})\kappa_{\eta n1}(E'_{\eta n}) \right]$$
$$\times \exp\left[-\frac{i}{\hbar}(E'_{\eta n} - E_{\eta n})t \right] dE_{\eta n}\, dE'_{\eta n}. \qquad (5.36)$$

This result can be simplified by invoking the anticommutator relations: $[\hat{a}_{\eta n}, \hat{a}^\dagger_{\eta n}]_+ = 1$ and $[\hat{b}_{\eta n}, \hat{b}^\dagger_{\eta n}]_+ = 1$. It is easy to show that

$$\kappa_{\eta n1}(E_{\eta n})\kappa_{\eta n2}(E'_{\eta n}) - \kappa_{\eta n2}(E_{\eta n})\kappa_{\eta n1}(E'_{\eta n})$$
$$\approx 2\hat{a}^\dagger_{\eta n}(E_{\eta n})\hat{a}_{\eta n}(E'_{\eta n}) - 2\hat{b}^\dagger_{\eta n}(E_{\eta n})\hat{b}_{\eta n}(E'_{\eta n}). \qquad (5.37)$$

A final adjustment is required to account for the electron's spin that has been ignored so far. As per Pauli's exclusion principle, two electrons of opposite spins can be transported through each channel without any interaction. This is accounted for by multiplying the current in Eq. (5.36) by a factor of two. With this modification, the current through the port η is given by

$$\hat{I}_\eta(t) = \frac{2q_e}{h} \iint \sum_{n\in\{\eta\}} \left[\hat{a}^\dagger_{\eta n}(E_{\eta n})\hat{a}_{\eta n}(E'_{\eta n}) - \hat{b}^\dagger_{\eta n}(E_{\eta n})\hat{b}_{\eta n}(E'_{\eta n}) \right]$$
$$\times \exp\left[-\frac{i}{\hbar}(E'_{\eta n} - E_{\eta n})t \right] dE_{\eta n}\, dE'_{\eta n}. \qquad (5.38)$$

5.2.3 Average Current through the Quantum Device

The preceding section has provided us with a current operator in terms of the creation and annihilation operators associated with electrons of different energies in different channels and ports of the quantum device. In this section we use this operator to calculate the average current passing through the device. The first thing to note is that the $\hat{b}_{\eta n}$ operators are related to the $\hat{a}_{\eta n}$ operators through the scattering matrix, as indicated in Eq. (5.21). We can write this relation as $\hat{\mathbf{b}}(E) = \mathbf{S}(E)\hat{\mathbf{a}}(E)$ for any energy E. Using it, we can replace the operator product $\hat{b}^\dagger_{\eta n}(E_{\eta n})\hat{b}_{\eta n}(E'_{\eta n})$ in Eq. (5.38) with an expression involving the elements of the scattering matrix and the operator product $\hat{a}^\dagger_{\eta n}(E_{\eta n})\hat{a}_{\eta n}(E'_{\eta n})$. Thus, the current operator $\hat{I}_\eta(t)$ depends only on the operator $\hat{a}^\dagger_{\eta n}(E_{\eta n})\hat{a}_{\eta n}(E'_{\eta n})$, which is related to the number density of incoming electrons.

In most cases, we are interested in the ensemble-averaged value $\langle\hat{I}_\eta(t)\rangle$ of the current operator because that is a measurable quantity. Given that this average is a function of the ensemble-averaged number density of electrons coming from a reservoir in thermal equilibrium, we can use the results of Section 3.1 to obtain

$$\langle\hat{a}^\dagger_{\eta n}(E_{\eta n})\hat{a}_{\zeta m}(E'_{\zeta m})\rangle = \delta_{\eta\zeta}\delta_{nm}\delta(E_{\eta n} - E'_{\zeta m})f_{F\eta}(E_{\eta n}), \qquad (5.39)$$

where $f_{F\eta}(E_{\eta n})$ is the Fermi distribution for the port η. When an electrode is in thermal equilibrium at a temperature T_η with a chemical potential μ_η and an electrical potential V_η, its Fermi distribution is given by

$$f_{F\eta}(E_{\eta n} - q_e V_\eta) = \left[\exp\left(\frac{E_{\eta n} - \mu_\eta - q_e V_\eta}{kBT_\eta}\right) + 1\right]^{-1}. \tag{5.40}$$

If we know the density operator ρ of a quantum system, the average current can be calculated using $\langle \hat{I}_\eta(t)\rangle = \mathrm{Tr}(\rho \hat{I}_\eta(t))$. Using Eqs. (5.38) and (5.39) and integrating over $E'_{\eta n}$, we obtain

$$\langle \hat{I}_\eta(t)\rangle = \frac{2q_e}{h}\int \sum_{n\in\{\eta\}} T_{\zeta\eta}(E_{\eta n})f_{F\eta}(E_{\eta n} - q_e V_\eta)\,dE_{\eta n}, \tag{5.41}$$

where $T_{\zeta\eta}$ is the transmission probability (related to the elements of the scattering matrix) for an electron to go from the port η to the port ζ. Since some electrons from the port ζ may also appear at the port η through scattering, the net current between these two ports is given by

$$\langle \hat{I}_\eta(t)\rangle = \frac{2q_e}{h}\int \sum_{n\in\{\eta\}} T_{\zeta\eta}(E_{\eta n})\left[f_{F\eta}(E_{\eta n} - q_e V_\eta) - f_{F\zeta}(E_{\eta n} - q_e V_\zeta)\right]dE_{\eta n}. \tag{5.42}$$

This equation is known as the Tsu–Esaki equation and is useful in several different contexts.

The final energy integral can be carried out under some restrictive assumptions. First, we write the transmission coefficient as $\sum T_{\zeta\eta}(E_{\eta n}) = M_{\zeta\eta}(E_{\eta n})\overline{T}_{\zeta\eta}(E_{\eta n})$, where $\overline{T}_{\eta n}(E_{\eta n})$ is the average probability that an electron with energy $E_{\eta n}$ at the port η will be transmitted to the port ζ and $M_{\eta\zeta}$ is the total number of channels available to this electron. By using this form in Eq. (5.41) and assuming that the average transmission coefficient is independent of energy and applied voltages, we obtain

$$\langle \hat{I}_\eta(t)\rangle = \frac{2q_e}{h}M_{\zeta\eta}\overline{T}_{\zeta\eta}\int \left[f_{F\eta}(E_{\eta n} - q_e V_\eta) - f_{F\zeta}(E_{\eta n} - q_e V_\zeta)\right]dE_{\eta n}. \tag{5.43}$$

The Fermi factors appearing in Eq. (5.43) can be simplified by noting that the local chemical potential in the presence of a static electrostatic potential is given by $\mu_\eta - q_e V_\eta$. Therefore, if the voltage difference between the two electrodes is given by $V_{\eta\zeta} = V_\eta - V_\zeta$, we can use the approximation

$$\int \left[f_{F\eta}(E_{\eta n} - q_e V_\eta) - f_{F\zeta}(E_{\eta n} - q_e V_\zeta)\right]dE_{\eta n} \approx q_e(V_\eta - V_\zeta) = q_e V_{\eta\zeta}. \tag{5.44}$$

Under these assumptions, we obtain the following simple expression for the average current flowing from port η to ζ:

$$\langle \hat{I}_\eta\rangle \approx \frac{2q_e^2}{h}M_{\zeta\eta}\overline{T}_{\zeta\eta}V_{\eta\zeta} = G_{\zeta\eta}V_{\eta\zeta}, \tag{5.45}$$

where the conductance of the quantum device is defined using the well-known current–voltage relation $I = GV$ and is given by

$$G_{\zeta\eta} = \frac{2q_e^2}{h}M_{\zeta\eta}\overline{T}_{\zeta\eta}. \tag{5.46}$$

This is the main result of the Landauer–Büttiker method for two-terminal quantum devices. It is known to hold for temperatures as low as close to absolute zero. It can be shown that it holds approximately at high temperatures as well [302]. To get this simple result, we had to assume that the transmission coefficient is independent of energy and applied voltages. This assumption is quite restrictive and its validity ultimately determines the validity of the preceding simple expression [302]. Even though Eq. (5.45) is derived for a two-terminal device, it can be readily extended to a multiport device using

$$\langle I_\eta \rangle = \sum_{\zeta \neq \eta} G_{\zeta\eta}(V_\eta - V_\zeta), \tag{5.47}$$

where we used the relation $V_{\eta\zeta} = V_\eta - V_\zeta$ for the voltage difference between any two ports.

5.2.4 Construction of Scattering Matrix

The first step in applying the Landauer–Büttiker method is to identify the scattering states for all ports of a quantum device and build the scattering matrix of the device. In practice, several probing techniques are used to construct the scattering matrix through experimental measurements. These techniques rely on the Landauer–Büttiker method because Kirchhoff's laws are inadequate for the probing purpose. The reason is that contact resistances cannot be ignored for the conductance because of its quantum nature (see Aside 5.2). The most striking result is that we cannot arbitrarily increase the current through a quantum device by changing its dimensions and other properties. This is because, for a given voltage, there is a limit to the maximum current a single scattering event can allow. This limit corresponds to the conductance quantum G_Q given in Eq. (5.3).

Aside 5.2   Contact Resistance

When a nanosize metallic lead (called here a quantum lead) is placed between two wider (classical) electrodes, its resistance measured through these electrodes is not zero, even when the charge transport is fully ballistic (no electrons are scattered by the quantum lead). This finite resistance occurs at the interface between the electrode and the quantum lead and its origin lies in the different numbers of conduction channels within a broad electrode and a narrow quantum lead.

Consider a quantum lead connected to two broad electrodes, labeled η and ζ in Figure 5.5. Let us focus on two points η' and ζ' inside the quantum lead but adjacent to the corresponding electrode. The relation given in Eq. (5.46) provides the conductance between η and ζ because we used the electrochemical potential to track electrons. As the electrochemical potential is the total potential and includes both the chemical potential and the electric potential, it can be used to calculate the distribution of electrons in thermal equilibrium, a condition satisfied easily for a wide (classical) electrode.

How do we find the conductance between the points η' and ζ' inside the quantum lead that is not expected to be in thermal equilibrium? It turns out that we can extend the same

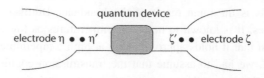

Fig. 5.5 Schematic illustration of the contact resistance. The outer black circles (η and ζ) are inside the electrodes, whereas the inner circles (η' and ζ') are within the quantum device but close to their corresponding electrodes.

reasoning to the quantum lead under steady-state conditions because none of the parameters change with time. The chemical potential has a well-defined quasi-equilibrium value in the steady state on both sides of the scattering barrier, assumed to be located in the middle of the quantum lead. However, just after electrons are transmitted through this barrier (before energy relaxation has taken place) the distribution is severely distorted from its equilibrium Fermi distribution. As a result, strictly speaking, chemical potential is not a well-defined quantity inside the quantum lead. We can define a pseudochemical potential such that when we integrate the corresponding Fermi distribution over energy, we obtain the correct number of electrons. However, this pseudochemical potential cannot be used to calculate the electron distribution within the quantum lead (even though it provides the correct number of electrons) because electrons are "hot" and their distribution deviates from the equilibrium Fermi distribution.

Our goal is to use this pseudochemical potential to establish the potential difference between η' and ζ' so that we can calculate the conductance between those points. Consider electrons moving along the path $\eta \rightarrow \eta' \rightarrow \zeta' \rightarrow \zeta$. Even though $\mu_{\eta'} = \mu_\eta$, we expect $\mu_{\zeta'} \neq \mu_\zeta$ because the number of electrons on this side of the potential barrier is smaller as some of them have been scattered back. The pseudochemical potential in this situation can be written as

$$\mu_{\zeta'} = \overline{T}_{\zeta\eta}\mu_\eta + (1 - \overline{T}_{\zeta\eta})\mu_\zeta, \qquad (5.48)$$

where $1 - \overline{T}_{\zeta\eta}$ represents the number of electrons scattered back. Therefore, the potential difference $V_{\eta'\zeta'}$ seen by the electrons between the internal points η' and ζ' is given by

$$q_e V_{\eta'\zeta'} = \mu_{\eta'} - \mu_{\zeta'} = (1 - \overline{T}_{\zeta\eta})(\mu_\eta - \mu_\zeta) = (1 - \overline{T}_{\zeta\eta})q_e V_{\eta\zeta}. \qquad (5.49)$$

We get the same result for the potential difference if we track the electrons from the contact ζ to η. As the current flowing between the internal points is equal to the current flowing between the two wide contacts, we can use this result to calculate the conductance $G_{\zeta'\eta'}$ between the internal points η' and ζ':

$$G_{\zeta'\eta'} = \frac{\langle \hat{I}_\eta(t) \rangle}{V_{\eta'\zeta'}} = \frac{2q_e^2}{h}\frac{M_{\zeta\eta}\overline{T}_{\zeta\eta}}{1 - \overline{T}_{\zeta\eta}}. \qquad (5.50)$$

The contact resistance $R_{\eta\eta'}$ can now be calculated as

$$R_{\eta\eta'} = \frac{1}{G_{\zeta\eta}} - \frac{1}{G_{\zeta'\eta'}} = \frac{h}{2q_e^2 M_{\zeta\eta}}. \qquad (5.51)$$

This result shows clearly that the contact resistance is negligible when the number of available conduction channels is very large. This is the case for all classical (wide) leads for which $M_{\zeta\eta} \to \infty$. However, a quantum lead only provides a limited number of conduction channels owing to its small transverse dimensions. This feature requires a redistribution of incoming electrons among a limited number of current-carrying modes at the interface, leading to a finite contact resistance (see also Ref. [303]).

The four-port sensing method is a highly accurate probing method used in electrical engineering for measuring impedances. This method is also known as the Kelvin sensing method after its inventor, and the four probes are known as Kelvin probes. The method relies on using current and voltage probes. The current probes are assumed to have zero impedance (thus no scattering). The voltage probes when connected to the measuring sample have no current flowing through them. Engquist and Anderson applied such a measurement technique in 1981 to the Landauer–Büttiker method [304]. They found that, in principle, the conductance given in Eq. (5.46) can be measured if ideal voltage probes are used and conductance is calculated by taking the ratio of the current through the device to the voltage difference measured.

Consider the four-probe configuration shown in Figure 5.6. For simplicity, we restrict our analysis to each lead having a single channel. The procedure will be similar in the multi-channel case except for algebraic complexity. The current flows from port 1 to port 2. The voltage probes 3 and 4 are weakly coupled to the sample; weak coupling can be realized in practice by using tunnel junctions as the interface. As no current flows through the voltage probes ($I_3 = I_4 = 0$), we have $I_1 = -I_2 \approx (2q_e^2/h)(V_1 - V_2)$, as required by charge conversation. As there is no current flow between ports 3 and 4, we could assume that associated impedance is very high, enabling us to ignore the effects of G_{34} and G_{43}. With these simplifications, we obtain the result [305]

$$V_3 - V_4 = \frac{T_{31}T_{42} - T_{32}T_{41}}{(T_{31} + T_{32})(T_{41} + T_{42})}(V_1 - V_2), \tag{5.52}$$

where we used the transmission coefficients instead of conductances.

Our task is to measure the conductance between ports 3 and 4. This can be done by accounting for all scattering events between these two ports through the transmission and

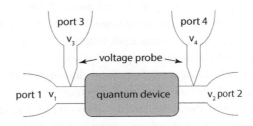

Fig. 5.6 The four-port sensing setup. Current flows from port 1 to port 2. The two voltage probes 3 and 4 are weakly coupled to the sample such that no current flows through them.

reflection coefficients such that $T + R = 1$. If the two voltage probes are identical and placed symmetrically as seen in Figure 5.6, we can assume that the transmission from port 1 to port 3 is the same as from port 4 to port 2. These transmissions include two contributions: direct tunneling with some small probability $\epsilon \ll 1$ (because electrodes 3 and 4 are weakly coupled) and tunneling after reflection from the scattering region with the probability ϵR. The application of the superposition principle leads to $T_{31} = T_{42} = \epsilon(1 + R)$. On the other hand, the probability of going from the first port to the fourth one (and from port 3 to port 2) is $T_{32} = T_{41} = \epsilon T$. Using these results in Eq. (5.52), we obtain $V_3 - V_4 = R(V_1 - V_2)$. Therefore the conductance between ports 3 and 4 is given by

$$G = \frac{I_1}{V_3 - V_4} = \frac{2q_e^2}{h}\frac{T}{R}. \tag{5.53}$$

This is identical to the expression in Eq. (5.50), confirming the validity of this method.

The main limitation of the Landauer–Büttiker method is that it breaks down as soon as charge carriers start to interact with each other or other parts of a quantum device. In many nanoscale devices, interactions of an electron with other electrons or phonons cannot be neglected. In this situation, the Landauer–Büttiker method may not describe the underlying physics even qualitatively. Over the last three decades, a method based on the nonequilibrium Green's function has emerged as the method of choice for analyzing current flow through quantum devices. As discussed in the following section, the versatility of this method stems from its ability to handle all kinds of interactions that charge particles and transport channels may endure when carrying current.

5.3 Nonequilibrium Green's Function Method

The nonequilibrium Green's function (NGF) method has been a workhorse of computational many-body studies for several decades. However, its widespread use does not imply that the underlying concepts and their implementations are easy or intuitive. Indeed, one has to spend a considerable amount of time mastering the theory and then make reasonable approximations to account for many interactions endured by the charge carriers during their transport. Being a generic technique, it has been applied to study the dynamics of plasmas, electrons, spins, and phonons in various materials. It has been found that the NGF method can provide an accurate description of a variety of nonequilibrium charge-transport scenarios under reasonable approximations.

Historically, Martin and Schwinger described many-body effects as early as 1959 using the NGF method from a unified nonperturbative point of view [306]. Further developments in this area were due to Kadanoff and Baym [307]. Later, this method was applied to superconductors and to molecular electronics. Keldysh used a graphical technique, analogous to the Feynman diagrams in field theory, to develop Green's functions for particles in a statistical system that were subjected to an external field [308]. Even though they appeared quite different, the equivalence between the technique of Kadanoff and Baym and that of Keldysh was demonstrated in 1975 by Langreth [309]. The graphical technique of Keldysh was advanced further in Refs. [310, 311, 312] using the NGF method. Meir and Wingreen showed in 1992 that the Keldysh formalism can be used to derive the Landauer formula for the current passing through a region of interacting electrons [313]. The NGF method was also shown to be a versatile tool for calculating the response of a biased, double-barrier, quantum well to a small alternating voltage [314]. Jauho et al. used this method in 1994 to calculate the time-dependent current through a mesoscopic region coupled to two leads under the influence of external time-dependent voltages [315]. These studies fueled the use of the NGF method for describing charge transport while incorporating fully the interactions of electrons with photons, phonons, or other electrons.

In spite of the preceding developments, the NGF method remained unaccessible to the engineering community owing to its complexity and its mathematical nature. Datta bridged this gap in 1989 with his work on a quantum kinetic equation [316, 317], and he made further advances in later years [318, 319, 320]. Many resources on the NGF-based transport modeling can be found at the website known as nanoHub [321]. Interested readers may also find Refs. [8, 9, 167, 302, 305, 322] useful for understanding the NGF method.

5.3.1 Evolution of Quantum Operators

The use of the NGF method for calculating currents in nanoscale devices can be illustrated by considering a simple scenario shown in Figure 5.7. The quantum device in the middle is connected to an electrode on each side through a coupling region. The two electrodes are assumed to be in thermal equilibrium with no direct coupling between them and have properties identical to those used for the Landauer–Büttiker method (see Fig. 5.2). The electrochemical potentials of the electrodes can be shifted by an external voltage to enable the flow of electrons from one electrode to the other via the quantum device. Electrons do not interact with each other inside the electrodes, but they do so inside the quantum device. To calculate the current, we treat the electrodes and the quantum device as separate entities before current begins to flow [323]. More precisely, the two electrodes are separated from each other for $t < 0$ and transport charges through the device starting at $t = 0$. As we see later, even with these simplifications, the model becomes challenging computationally if we include various interactions among electrons and phonons that lead to phenomena such as Coulomb blockade and the Kondo effect.

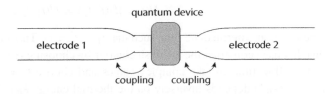

quantum device

electrode 1   electrode 2

coupling   coupling

Fig. 5.7   A quantum device coupled to two electrodes via quantum operators. The electrodes do not couple with each other directly.

To understand the advantages of the NGF method over a direct solution of the Schrödinger equation, consider the steps involved in constructing the solution for the system described in Figure 5.7. If the Hamiltonian of the whole system is given by H, one can describe its evolution by solving the Schrödinger equation, but that is not an easy task in practice as the Hamiltonian H contains multiple parts representing electrodes, coupling region, and the quantum device. The Green's function $G(E)$ provides an alternative approach to solving the same problem. It satisfies the following equation in the energy basis:

$$(E - H)G(E) = I, \tag{5.54}$$

where I is the identity matrix. The Green's function enables us to write the response of a quantum system to a constant perturbation $|\delta\Psi\rangle$ as

$$(E - H)|\Psi\rangle = -|\delta\Psi\rangle \rightarrow |\Psi\rangle = -G(E)|\delta\Psi\rangle. \tag{5.55}$$

The second equation is much easier to solve compared to the perturbed Schrödinger equation. For technical reasons, Green's function requires the use of a contour in the complex time plane known as the Keldysh contour. We discuss this aspect before describing the current flow through a quantum device.

In Section 2.2.5, we introduced the time-evolution operator $U(t, t_0)$ that describes the evolution of a quantum system from an initial time t_0 to a later time t (see Aside 2.14). As discussed there, when the system's Hamiltonian is independent of time, the time-evolution operator has the simple form: $U(t, t_0) = \exp[-(i/\hbar)(t - t_0)H]$. When the Hamiltonian depends on time, the unitary operator evolves in a more complicated manner as

$$U(t, t_0) = \mathcal{T}\left\{\exp\left(-\frac{i}{\hbar}\int_{t_0}^{t} H(\tau)\,d\tau\right)\right\}, \tag{5.56}$$

where $\mathcal{T}$ is a time-ordering operator; it rearranges the operators in chronological order such that an operator with a later time is placed to the left of operators with earlier times. However, this ordering depends on the nature of the particle (fermion or boson) whose dynamics is being studied. More specifically, the time ordering of the product of two operators, $A(t)$ and $B(t)$, is written as

$$\mathcal{T}\{A(t_1)B(t_2)\} = u(t_1 - t_2)A(t_1)B(t_2) \pm u(t_2 - t_1)B(t_2)A(t_1), \tag{5.57}$$

where $u(t)$ is the Heaviside step function and the positive or negative sign is chosen for bosons and fermions, respectively. The time-evolution operator satisfies the *group property*

$$U(t_1, t')U(t', t_2) = U(t_1, t_2), \tag{5.58}$$

where t' is an intermediate time such that $t_1 < t' < t_2$. This is an important property, and it enables us to construct the Keldysh contour by choosing t_1 to be a complex number.

We allow time to take complex values and choose $t_1 = t_0 - it_i$ and $t_2 = t_0$, where $t_i = \hbar/(k_B T)$ depends inversely on the thermal energy $k_B T$. The evolution operator for a closed system with the time-independent Hamiltonian can then be written as

$$U(t_0 - it_i, t_0) = \exp(-t_i H/\hbar) = \exp(-H/k_B T). \tag{5.59}$$

If we use a grand-canonical ensemble for a system in thermal equilibrium, we know from Section 3.1.2 that its density operator is given by $\rho_{GCE} = \exp[-(H - \mu N)/\hbar k_B T]$, where μ is the chemical potential. Using it, the ensemble-averaged value of any operator $A(t)$ is given by

$$\langle A(t) \rangle = \frac{\text{Tr}[\rho_{GCE} A(t)]}{\text{Tr}[\rho_{GCE}]} = \frac{\text{Tr}[\rho_{GCE} U^{-1}(t_0 + t, t_0) A(t_0) U(t_0 + t, t_0)]}{\text{Tr}[\rho_{GCE}]}. \tag{5.60}$$

Assuming that the operators H and N commute, ρ_{GCE} can be written in terms of the evolution operator U as

$$\rho_{GCE} = \exp(\mu N/k_B T) U(t_0 - it_i, t_0). \tag{5.61}$$

Using this form, the ensemble-averaged value can be written as

$$\langle A(t) \rangle = \frac{\text{Tr}[\exp(\mu N/k_B T) U(t_0 - it_i, t_0) U(t_0, t_0 + t) A(t_0) U(t_0 + t, t_0)]}{\text{Tr}[\exp(\mu N/k_B T) U(t_0 - it_i, t_0)]}. \tag{5.62}$$

The three U operators in the numerator of this equation show the path taken in the complex t plane and shown in Figure 5.8. This contour is called the *Kadanoff–Baym–Keldysh contour* or just the Keldysh contour [308]. The same contour is known as the *Schwinger–Keldysh contour* when the imaginary strip is neglected. The Keldysh contour has three branches: a forward branch from t_0 to $t_0 + t$, a backward branch from $t_0 + t$ to t_0, and a branch along the imaginary time axis from t_0 to $t_0 - it_i$. When using the contour, any point lying on the imaginary time axis is taken to be a later time compared to all points lying on the forward or the backward branch (along the real-time axis).

Owing to the group property of the evolution operator U, the intermediate time t' in Eq. (5.58) can take any value, including a complex one. Through the technique of analytic continuation, we extend Eq. (5.62) and replace t with a complex variable z. It is important to recognize that the operator $A(t_0)$ commutes with U, and we can lump the three U operators together and write Eq. (5.62) in the following form:

$$\langle A(z_t) \rangle = \frac{\text{Tr}\left\{\exp(\mu N/k_B T) \mathcal{T}_C \left[\exp\left(-\frac{i}{\hbar} \int_C H \, dz\right) A_{z_t}\right]\right\}}{\text{Tr}\left\{\exp(\mu N/k_B T) \mathcal{T}_C \left[\exp\left(-\frac{i}{\hbar} \int_C H \, dz\right)\right]\right\}}, \tag{5.63}$$

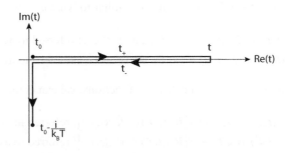

Fig. 5.8   The Kadanoff–Baym–Keldysh contour in the complex time plane. The arrows indicate time ordering.

where z_t denotes a point on the contour C shown in Figure 5.8. The contour time-ordering operator $\mathcal{T}_C$ rearranges the operators in a chronological order, after considering the commutative properties of operators involved. The operator A_{z_t} denotes the position of the operator A on the contour at any point z_t (i.e., it is an indexing variable on the contour C, consisting of three branches: a forward branch C_+, a backward branch C_-, and a vertical branch C_v). Note that the notion of earlier or later times is different on the contour from that of the real time. For example, on the branch C_-, a point earlier in real time appears later when we consider time ordering along this branch.

The behavior of the quantum system is quite different on the vertical branch C_v compared to the other two branches because $z_t = -it_i$ is purely imaginary along it. Since H is a constant, the contour integral in Eq. (5.62) along this path leads to the simple relation

$$\int_{C_v} H\, dz = H[(t_0 - it_i) - t_0] = -iHt_i = -iH\hbar/(k_B T). \tag{5.64}$$

Substituting this result in Eq. (5.62), we obtain

$$\langle A(z_t) \rangle = \frac{\mathrm{Tr}\,\{\exp[-(H - \mu N)/k_B T]A\}}{\mathrm{Tr}\,\{\exp[-(H - \mu N)/k_B T]\}} = \mathrm{Tr}[\rho_{\mathrm{GCE}} A], \tag{5.65}$$

where the cyclic property of the trace was used. We note that the resulting expression is independent of z_t and coincides with the thermal average expected before the time evolution occurred. We can draw the following conclusions from these results. The average $\langle A(z_t) \rangle$ corresponds to the statistical average of the observable A when z_t lies on the forward and backward branches. In contrast, it corresponds to the thermal average of the observable A before the system is disturbed when z_t lies on the imaginary time axis.

When using the Keldysh contour for calculating Green's function, we frequently encounter convolution-type integrals of the form

$$R(t, t') = \int_C P(t_1, \tau) Q(\tau, t')\, d\tau, \tag{5.66}$$

where P and Q are two time-dependent operators and C is the Schwinger–Keldysh contour in Figure 5.8. To evaluate such integrals, we follow the prescription provided by Langreth [309] and let $t_0 \to -\infty$. When t and t' lie on the same branch, $R(t, t')$ can be calculated using the conventional time ordering. However, when t and t' lie on different branches, we label $R(t, t')$ as follows (in a way similar to what is done for Green's functions):

- $R(t, t') = R^<(t, t')$ if $t \in C_+$ and $t' \in C_-$; referred to as "lesser" $R(t, t')$,
- $R(t, t') = R^>(t, t')$ if $t \in C_-$ and $t' \in C_+$; referred to as "greater" $R(t, t')$.

Using the "greater" and "lesser" functions, we can define two important functions as

- $R^+(t, t') = u(t - t')\left[R^>(t, t') - R^<(t, t')\right]$; referred to as "retarded" $R(t, t')$,
- $R^-(t, t') = u(t' - t)\left[R^<(t, t') - R^>(t, t')\right]$; referred to as "advanced" $R(t, t')$.

Notice the identity: $R^+(t, t') - R^-(t, t') = R^>(t, t') - R^<(t, t')$.

Let us first calculate $R^<(t,t')$. Following Langreth [309], we deform the Schwinger–Keldysh contour with another contour consisting of an upper part C_U and a lower part C_L, as shown in Figure 5.9. Then $R^<(t,t')$ can be written as:

$$R^<(t,t') = \int_{C_U} P(t,\tau)Q^<(\tau,t')\,d\tau + \int_{C_L} P^<(t,\tau)Q(\tau,t')\,d\tau. \tag{5.67}$$

In the contour integral along the path C_U, we use $Q^<(\tau,t')$ because there $\tau \in C_U$ but $t' \in C_L$. For the same reason, $P^<(t,\tau)$ is used for the contour integral along the path C_L.

The first contour integral along the path C_U consists of two parts that can be combined as follows:

$$\int_{C_U} P(t,\tau)Q^<(\tau,t')\,d\tau = \int_{-\infty}^{t} P^>(t,\tau)Q^<(\tau,t')\,d\tau + \int_{t}^{-\infty} P^<(t,\tau)Q^<(\tau,t')\,d\tau$$

$$= \int_{-\infty}^{\infty} u(t-\tau)\left[P^>(t,\tau) - P^<(t,\tau)\right]Q^<(\tau,t')\,d\tau$$

$$= \int_{-\infty}^{\infty} P^+(t,\tau)Q^<(t,t')\,d\tau. \tag{5.68}$$

The same procedure is followed for the second contour integral along the path C_L:

$$\int_{C_L} P^<(t,\tau)Q(\tau,t')\,d\tau = \int_{-\infty}^{t'} P^<(t,\tau)Q(\tau,t')\,d\tau + \int_{t'}^{-\infty} P^<(t,\tau)Q(\tau,t')\,d\tau$$

$$= \int_{-\infty}^{\infty} u(t'-\tau)P^<(t,\tau)\left[Q^<(\tau,t') - Q^>(\tau,t')\right]d\tau$$

$$= \int_{-\infty}^{\infty} P^<(t,\tau)Q^-(\tau,t')\,d\tau. \tag{5.69}$$

Combining Eqs. (5.68) and (5.69), we obtain Langreth's result

$$R^<(t,t') = \int_{-\infty}^{+\infty}\left[P^+(t,\tau)Q^<(\tau,t') + P^<(t,\tau)Q^-(\tau,t')\right]d\tau. \tag{5.70}$$

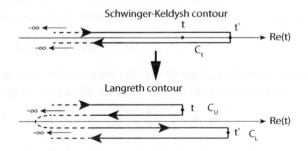

Fig. 5.9   Langreth's modification of the Schwinger–Keldysh time-contour C in the complex time plane. As a result of this modification, we have $C = C_U \cup C_L$.

Using the same procedure, we can write $R^>(t, t')$ as

$$R^>(t, t') = \int_{-\infty}^{+\infty} \left[P^+(t, \tau) Q^>(\tau, t') + P^>(t, \tau) Q^-(\tau, t') \right] d\tau. \tag{5.71}$$

These functions can be used to obtain the retarded function $R^+(t, t')$ as

$$R^+(t, t') = u(t - t') \left[R^>(t, t') - R^<(t, t') \right]$$

$$= u(t - t') \int_{-\infty}^{\infty} \left[P^+(t, \tau) Q^>(\tau, t') + P^>(t, \tau) Q^-(\tau, t') \right] d\tau$$

$$- u(t - t') \int_{-\infty}^{\infty} \left[P^+(t, \tau) Q^<(\tau, t') + P^<(t, \tau) Q^-(\tau, t') \right] d\tau. \tag{5.72}$$

Using the definitions of $P^+(t, \tau)$ and $Q^-(\tau, t')$, we can combine the two terms in the preceding equation to obtain

$$R^+(t, t') = \int_{t'}^{t} P^+(t, \tau) Q^+(\tau, t') d\tau. \tag{5.73}$$

Similarly, we can show that the advanced function $R^-(t, t')$ can be written as

$$R^-(t, t') = \int_{t'}^{t} P^-(t, \tau) Q^-(\tau, t') d\tau. \tag{5.74}$$

5.3.2 Current Flow through a Quantum Device

To apply the NGF method to the quantum system in Figure 5.7, we need to first find its Hamiltonian H_{NGF}. This Hamiltonian should include all parts of the system including the two electrodes (called contact leads), the quantum device, and the coupling regions between the contact leads and the device. Thus, the Hamiltonian has the general form

$$H_{NGF} = H_{CL1} + H_{\chi 1} + H_{QD} + H_{\chi 2} + H_{CL2}, \tag{5.75}$$

where the subscripts CL and χ are used for the coupling lead and the coupling with 1 and 2 added for the two sides of the quantum device. The device Hamiltonian H_{QD} depends on the details of charge transport and may contain terms related to interaction of an electron with excitons, phonons, and other electrons. To allow for these terms, we assume that H_{QD} is a function of multiple creation and annihilation operators,

$$H_{QD} = H_{QD} \left(\{ \hat{q}_n^\dagger, \hat{q}_n \} \right), \tag{5.76}$$

where the curly brackets denote the entire set of these operators.

The Hamiltonians of the contact leads depend on the applied voltages such that

$$H_{CL\eta} = \sum_{\varsigma} [E_{\eta\varsigma} - q_e V_\eta(t)] \hat{l}_{\eta\varsigma}^\dagger \hat{l}_{\eta\varsigma} \quad (\eta = 1, 2), \tag{5.77}$$

where the operators $\hat{l}_{\eta\varsigma}^\dagger$ and $\hat{l}_{\eta\varsigma}$ create or annihilate single electrons of energy $E_{\eta\varsigma}$ and $V_\eta(t)$ is the potential on the left or right electrode for $\eta = 1, 2$. A single index ς is used to account for the electron's momentum $\hbar k$ (discrete or continuous k), its spin, and other conserved quantities needed to represent the quantum device. This index can be conceptualized as a generalized channel vector describing the current flow, because each value corresponds to one transport channel through the device.

The coupling Hamiltonian between the leads and the quantum device contains products of the creation and annihilation operators associated with them. We use the normal ordering (or *Wick ordering*) and place all creation operators to the left of the annihilation operators. With this choice, this part of the Hamiltonian has the form

$$H_{\chi\eta} = \sum_{\varsigma,n} \left(\chi_{\eta\varsigma,n} \hat{l}^{\dagger}_{\eta\varsigma} \hat{q}_n + \chi^*_{\eta\varsigma,n} \hat{q}^{\dagger}_n \hat{l}_{\eta\varsigma} \right), \tag{5.78}$$

where the coupling coefficients $\chi_{\eta\varsigma,n}$ are found self-consistently considering changes in the distribution of electrons resulting from the current flow.

Using the total Hamiltonian given in Eq. (5.75), we can calculate the current flowing from each contact electrode toward the quantum device using $I = dQ/dt$, where $Q(t) = -q_e \langle \hat{N}(t) \rangle$ is the total charge at the electrode and $\hat{N}$ is the number operator for electrons of charge $-q_e$. The number operator at each electrode is given by

$$\hat{N}_{\eta} = \sum_{\varsigma} \hat{l}^{\dagger}_{\eta\varsigma} \hat{l}_{\eta\varsigma}, \quad \eta = 1,2. \tag{5.79}$$

In the Heisenberg picture, the evolution of $\hat{N}_{\eta}$ is governed by

$$\frac{d}{dt} \hat{N}_{\eta} = \frac{i}{\hbar} [H_{NGF}, \hat{N}_{\eta}] = \frac{i}{\hbar} [H_{\chi\eta}, \hat{N}_{\eta}], \tag{5.80}$$

where we used the fact that all parts of H_{NGF} commute with $\hat{N}_{\eta}$ except the coupling part $H_{\chi\eta}$. We can now calculate the current as

$$I_{\eta} = -q_e \left\langle \frac{d}{dt} \hat{N}_{\eta} \right\rangle = \frac{q_e}{i\hbar} \langle [H_{\chi\eta}, \hat{N}_{\eta}] \rangle. \tag{5.81}$$

This relation shows that the current depends only on a single commutator involving the coupling Hamiltonian.

We can calculate this commutator by using $H_{\chi\eta}$ from Eq. (5.78) and writing the current in the form

$$I_{\eta} = \frac{q_e}{i\hbar} \left\langle \sum_{\varsigma,n} \left(\chi_{\eta\varsigma,n} \hat{l}^{\dagger}_{\eta\varsigma} \hat{q}_n + \chi^*_{\eta\varsigma,n} \hat{q}^{\dagger}_n \hat{l}_{\eta\varsigma} \right), \sum_{\varsigma} \hat{l}^{\dagger}_{\eta\varsigma} \hat{l}_{\eta\varsigma} \right\rangle. \tag{5.82}$$

Clearly, we need to evaluate multiple commutators of the form $[\hat{l}^{\dagger}_{\eta 1\varsigma 1} \hat{q}_n, \hat{l}^{\dagger}_{\eta 2\varsigma 2} \hat{l}_{\eta 2\varsigma 2}]$. We can simplify this commutator as

$$[\hat{l}^{\dagger}_{\eta 1\varsigma 1} \hat{q}_n, \hat{l}^{\dagger}_{\eta 2\varsigma 2} \hat{l}_{\eta 2\varsigma 2}] = \left(\hat{l}^{\dagger}_{\eta 1\varsigma 1} \hat{l}^{\dagger}_{\eta 2\varsigma 2} \hat{l}_{\eta 2\varsigma 2} - \hat{l}^{\dagger}_{\eta 2\varsigma 2} \hat{l}_{\eta 2\varsigma 2} \hat{l}^{\dagger}_{\eta 1\varsigma 1} \right) \hat{q}_n$$

$$= -\delta_{\eta 1\varsigma 1, \eta 2\varsigma 2} \hat{l}^{\dagger}_{\eta 2\varsigma 2} \hat{q}_n. \tag{5.83}$$

Similarly, $[\hat{l}^{\dagger}_{\eta 1\varsigma 1} \hat{q}_n, \hat{l}^{\dagger}_{\eta 2\varsigma 2} \hat{l}_{\eta 2\varsigma 2}] = \delta_{\eta 1\varsigma 1, \eta 2\varsigma 2} \hat{q}^{\dagger}_n \hat{l}_{\eta 2\varsigma 2}$. With these results, the current in Eq. (5.81) takes the form

$$I_{\eta} = \frac{iq_e}{\hbar} \sum_{\varsigma,n} \left(\chi_{\eta\varsigma,n} \langle \hat{l}^{\dagger}_{\eta\varsigma}(t) \hat{q}_n(t) \rangle - \chi^*_{\eta\varsigma,n} \langle \hat{q}^{\dagger}_n(t) \hat{l}_{\eta\varsigma}(t) \rangle \right). \tag{5.84}$$

The preceding expression contains ensemble-averaged operator products that can be expressed as combinations of Green's functions. Several types of Green's functions are discussed in Aside 5.3. Here we use two lesser types of Green's functions defined as

$$G^<_{n,\eta\varsigma}(t-t') = -\frac{i}{\hbar}\left\langle \hat{q}_n(t)\hat{l}^\dagger_{\eta\varsigma}(t')\right\rangle = \frac{i}{\hbar}\left\langle \hat{l}^\dagger_{\eta\varsigma}(t')\hat{q}_n(t)\right\rangle \qquad (5.85)$$

$$G^<_{\eta\varsigma,n}(t-t') = -\frac{i}{\hbar}\left\langle \hat{l}_{\eta\varsigma}(t)\hat{q}^\dagger_n(t')\right\rangle = \frac{i}{\hbar}\left\langle \hat{q}^\dagger_n(t')\hat{l}_{\eta\varsigma}(t)\right\rangle, \qquad (5.86)$$

where a comma between the subscripts of a Green's function is used when the operators from two different parts of the system are involved. Such Green's functions are called mixed types.

It turns out that the energy-domain representation of Green's function is more useful for calculating the current. Taking the Fourier transform, we use the relation

$$G^<_{\eta\varsigma,n}(E) = \int G^<_{\eta\varsigma,n}(\tau)\exp\left(\frac{i}{\hbar}E\tau\right)d\tau. \qquad (5.87)$$

As $G^<_{\eta\varsigma,n}(t-t') = -\left[G^<_{n,\eta\varsigma}(t-t')\right]^*$, we can use the relation $\check{G}^<_{\eta\varsigma,n}(E) = -\left[\check{G}^<_{n,\eta\varsigma}(E)\right]^*$ and write the current in the form

$$I_\eta = \frac{q_e}{2\pi\hbar}\int \sum_{\varsigma,n}\left[\chi_{\eta\varsigma,n}G^<_{n,\eta\varsigma}(E) + \chi^*_{\eta\varsigma,n}\left[G^<_{n,\eta\varsigma}(E)\right]^*\right]dE. \qquad (5.88)$$

Using $\hbar = h/2\pi$, this equation can be written in a more compact form as

$$I_\eta = \frac{2q_e}{h}\int \sum_{\varsigma,n}\Re\left[\chi_{\eta\varsigma,n}G^<_{n,\eta\varsigma}(E)\right]dE, \qquad (5.89)$$

where $\Re$ denotes the real part.

The Green's function $G^<_{n,\eta\varsigma}(E)$, the only one needed to calculate the current, can be obtained by applying the Langreth theorem to the time-ordered Green's function encountered earlier and defined as

$$G_{n,\eta\varsigma}(t-t') = -\frac{i}{\hbar}\left\langle \mathcal{T}\left\{\hat{q}_n(t)\hat{l}^\dagger_{\eta\varsigma}(t')\right\}\right\rangle. \qquad (5.90)$$

As the leads are assumed not to interact directly with each other (e.g., via tunneling), the calculation of this Green's function does not generate the Bogoliubov hierarchy. Its equation of motion has the form (see Aside 5.4)

$$i\hbar\frac{\partial}{\partial t}G_{n,\eta\varsigma}(t-t') = \delta_{n,\eta\varsigma}\delta(t-t') + E_{\eta\varsigma}G_{n,\eta\varsigma}(t-t') + \sum_m G_{nm}(t-t')\chi^*_{\eta\varsigma,m}, \qquad (5.91)$$

showing that $G_{n,\eta\varsigma}(t-t')$ can be written in terms of the Green's function $G_{nm}(t-t')$ associated with the quantum device. Using this equation, one can show that (see Ref. [324] for details)

$$G^<_{n,\eta\varsigma}(E) = \sum_m \chi^*_{\eta\varsigma,m}\left[G^+_{nm}(E)G^<_{\eta\varsigma}(E) + G^<_{nm}(E)G^-_{\eta\varsigma}(E)\right]. \qquad (5.92)$$

Aside 5.3   Several Kinds of Green's Functions

To make the following discussion applicable to both bosons and fermions, we adopt the notation $[\alpha, \beta]_\varkappa = \alpha\beta + \varkappa\beta\alpha$, where $\varkappa = +1$ for fermions and -1 for bosons (see also Section 2.4.1). The statistical average is taken using the initial density operator at time $t = t_0$ (i.e., $\langle \hat{O} \rangle = \text{Tr}\{\rho(t_0)\hat{O}\}$). Let $\{\hat{q}_\varsigma, \hat{q}_\varsigma^\dagger\}$ represent a set of creation and annihilation operators, where the index ς spans through the available single-particle quantum states. Using these operators, several types of Green's functions can be defined as follows [322]:

Lesser and greater Green's functions
These two basic Green's functions are defined using the product of creation and annihilation operators:

$$G_{\varsigma\varrho}^<(t,t') = \frac{i}{\hbar}\left\langle \hat{q}_\varrho^\dagger(t')\hat{q}_\varsigma(t) \right\rangle, \tag{5.93}$$

$$G_{\varsigma\varrho}^>(t,t') = -\frac{i}{\hbar}\left\langle \hat{q}_\varsigma(t)\hat{q}_\varrho^\dagger(t') \right\rangle. \tag{5.94}$$

Retarded and Advanced Green's functions:
The preceding two Green's functions can be combined to construct the retarded and advanced Green's functions as

$$G_{\varsigma\varrho}^+(t,t') = u(t - t')\left(G_{\varsigma\varrho}^>(t,t') - G_{\varsigma\varrho}^<(t,t') \right), \tag{5.95}$$

$$G_{\varsigma\varrho}^-(t,t') = u(t' - t)\left(G_{\varsigma\varrho}^<(t,t') - G_{\varsigma\varrho}^>(t,t') \right). \tag{5.96}$$

The step function $u(t - t')$ ensures causality. Using the definitions of the lesser and greater Green's functions, the preceding two functions can be written as

$$G_{\varsigma\varrho}^+(t,t') = -\frac{i}{\hbar}u(t - t')\left\langle [\hat{q}_\varsigma(t), \hat{q}_\varrho^\dagger(t')]_\varkappa \right\rangle, \tag{5.97}$$

$$G_{\varsigma\varrho}^-(t,t') = \frac{i}{\hbar}u(t' - t)\left\langle [\hat{q}_\varsigma(t), \hat{q}_\varrho^\dagger(t')]_\varkappa \right\rangle. \tag{5.98}$$

The retarded Green's function is useful because it can be used to find the linear response of any quantum device. For example, using it, we can characterize deviations from equilibrium (dissipation in the case of the electrical conductivity) in terms of fluctuations, leading to the fluctuation-dissipation theorem discussed in Section 3.3.

Time-ordered Green's function
The time-ordering operator $\mathcal{T}$ rearranges a product of two time-dependent operators such that operators evaluated at a later time always appear to the left of the operators evaluated at earlier times. Such a Green's function is defined as

$$G_{\varsigma\varrho}(t,t') = -\frac{i}{\hbar}\left\langle \mathcal{T}\left\{ \hat{q}_\varsigma(t)\hat{q}_\varrho^\dagger(t') \right\} \right\rangle. \tag{5.99}$$

It can be related to the lesser and greater Green's functions as

$$G_{\varsigma\varrho}(t,t') = -\frac{i}{\hbar}u(t-t')\left\langle \hat{q}_\varsigma(t)\hat{q}_\varrho^\dagger(t')\right\rangle + \varkappa\frac{i}{\hbar}u(t'-t)\left\langle \hat{q}_\varrho^\dagger(t')\hat{q}_\varsigma(t)\right\rangle$$

$$= u(t-t')G_{\varsigma\varrho}^>(t,t') + \varkappa u(t'-t)G_{\varsigma\varrho}^<(t,t'). \tag{5.100}$$

It is possible to attach a physical meaning to this Green's function. For example, consider the case of electrons. For $t > t'$, $G_{\varsigma\varrho}(t,t')$ describes motion of an electron created at time t' in the state ϱ and detected at time t in the state ς. In contrast, for $t' > t$, $G_{\varsigma\varrho}(t,t')$ describes the motion of a hole from the state ς at time t and detected at the state ϱ at time t'.

5.3.3 Lehmann Representation of Green's Functions

Lehmann representation provides a rigorous way to describe Green's functions using eigen-energies of the device Hamiltonian. It also paves the way for the energy-basis representation of a Green's function, which is often suited for calculations in practice.

As we have seen, Green's functions require averaging over an equilibrium ensemble. This must be done by invoking a grand-canonical ensemble because the creation and annihilation operators change the number of particles. As we discussed in Section 3.1.3, ensemble averaging of any operator $\hat{O}$ in this situation is carried out using

$$\langle \hat{O} \rangle = \text{Tr}[\rho_{\text{GCE}}\hat{O}]/Z_\mu, \tag{5.101}$$

where $\rho_{\text{GCE}} = \exp(-H_{\text{eff}}/k_BT)$, $Z_\mu = \text{Tr}[\rho_{\text{GCE}}]$, $H_{\text{eff}} = H - \mu N$, H is the device Hamiltonian, μ is the chemical potential, and N is the number of particles.

We denote by $\{|n\rangle\}$ the complete set of orthonormal eigenstates of the Hamiltonian H_{eff} obtained by solving $H_{\text{eff}}|n\rangle = E_n|n\rangle$, where E_n is the energy associated with the state $|n\rangle$. It is easy to show that

$$\langle m|\hat{q}_\varsigma(t)|n\rangle = \langle m|\exp(iH_{\text{eff}}t/\hbar)\hat{q}_\varsigma \exp(-iH_{\text{eff}}t/\hbar)|n\rangle$$

$$= \exp[i(E_m - E_n)t/\hbar]\langle m|\hat{q}_\varsigma|n\rangle, \tag{5.102}$$

where we used $\exp(-iH_{\text{eff}}t/\hbar)|n\rangle = \exp(-iE_nt/\hbar)|n\rangle$. Using this result, we can calculate the average in Eq. (5.101) in the basis $\{|n\rangle\}$ to obtain

$$\langle \hat{O} \rangle = \langle n|\exp(-H_{\text{eff}}/k_BT)\hat{O}|n\rangle/Z_\mu = \exp(-E_n/k_BT)\langle n|\hat{O}|n\rangle/Z_\mu. \tag{5.103}$$

The preceding result can be used to calculate the lesser Green's function as

$$G_{\varsigma\varrho}^<(t,t') = \frac{i}{\hbar Z_\mu}\sum_n \langle n|\exp(-H_{\text{eff}}/k_BT)\hat{q}_\varrho^\dagger(t')\hat{q}_\varsigma(t)|n\rangle$$

$$= \frac{i}{\hbar Z_\mu}\sum_{n,m} \langle n|\exp(-H_{\text{eff}}/k_BT)\hat{q}_\varrho^\dagger(t')|m\rangle\langle m|\hat{q}_\varsigma(t)|n\rangle$$

$$= \frac{i}{\hbar Z_\mu}\sum_{n,m} \exp[-E_n/k_BT + \frac{i}{\hbar}(E_n - E_m)(t-t')]\langle n|\hat{q}_\varrho^\dagger|m\rangle\langle m|\hat{q}_\varsigma|n\rangle. \tag{5.104}$$

Similarly, we obtain the greater Green's function in the form

$$G^>_{\varsigma\varrho}(t,t') = -\frac{i}{\hbar Z_\mu} \sum_{n,m} \exp[-E_n/k_B T + \frac{i}{\hbar}(E_n - E_m)(t - t')] \langle n|\hat{q}_\varsigma|m\rangle \langle m|\hat{q}^\dagger_\varrho|n\rangle . \quad (5.105)$$

The retarded and advanced Green's functions can now be calculated as they are related to $G^<_{\varsigma\varrho}(t,t')$ and $G^<_{\varsigma\varrho}(t,t')$ as simple linear combinations. We note that the eigenstates $\{|n\rangle\}$ can always be written as a linear combination of the Fock-space states, if the occupation-number representation is employed.

Even though all Green's functions are indicated as functions of two time variables (t and t'), the preceding results show that, in fact, they depend on time only through the difference $t - t'$. This result is very important because it allows us to introduce one time parameter $\tau = t - t'$ and to write any Green's function in the frequency space as a Fourier transform with respect to τ. As we have already discussed, the Fourier domain provides a compact and intuitive approach to such complex problems.

We can also show that Green's functions depend only on the time difference $t - t'$ by analyzing the product $\langle \hat{q}_\varsigma(t)\hat{q}^\dagger_\varrho(t)\rangle$. Noting that the time-dependent operator has the form $\hat{q}_\varsigma(t) = \exp(iH_{\text{eff}}t/\hbar)\hat{q}_\varsigma \exp(-iH_{\text{eff}}t/\hbar)$, we obtain

$$\begin{aligned}\langle \hat{q}_\varsigma(t)\hat{q}^\dagger_\varrho(t)\rangle &= \frac{i}{\hbar Z_\mu} \text{Tr}[\rho_{\text{GCE}}\hat{q}^\dagger_\varrho(t')\hat{q}_\varsigma(t)] \\ &= \frac{i}{\hbar Z_\mu} \text{Tr}\left[\rho_{\text{GCE}} \exp\left(\frac{i}{\hbar}H_{\text{eff}}(t - t')\right)\hat{q}^\dagger_\varrho \exp\left(-\frac{i}{\hbar}H_{\text{eff}}(t - t')\right)\hat{q}_\varsigma\right], \quad (5.106)\end{aligned}$$

which is clearly a function of $t - t'$. This conclusion is a direct consequence of the time-evolution operator, $U(t) = \exp(-iH_{\text{eff}}t/\hbar)$, which commutes with ρ_{GCE} because both are functional of the same Hamiltonian H_{eff}.

Using $\tau = t - t'$, the Fourier transform $\check{G}(E)$ of $G(\tau)$ is defined as

$$\check{G}(E) = \mathcal{F}\{G(\tau)\} = \int_{-\infty}^{\infty} G(\tau)\exp\left(\frac{i}{\hbar}E\tau\right) d\tau. \quad (5.107)$$

When a Green's function contains the step function $u(t)$, we make use of the Sokhotski–Plemelj formula (see Aside 3.3):

$$\int_{-\infty}^{\infty} u(\tau)\exp\left(\frac{i}{\hbar}E\tau\right) d\tau = \frac{i}{E/\hbar + i\eta}, \quad (5.108)$$

$$\lim_{\eta \to 0^+} \frac{1}{E/\hbar + i\eta} = \text{P.V.}\left(\frac{\hbar}{E}\right) - i\hbar\pi\,\delta(E), \quad (5.109)$$

where η is an infinitely small positive quantity.

As an example, consider the time-ordered Green's function $G_{\varsigma\varrho}(t,t')$. Its Fourier transform (or the energy space representation) is given by:

$$\check{G}_{\varsigma\varrho}(E) = \mathcal{F}\left\{u(t - t')G^>_{\varsigma\varrho}(t,t')\right\} + \varkappa\mathcal{F}\left\{u(t' - t)G^<_{\varsigma\varrho}(t,t')\right\}. \quad (5.110)$$

Using Eqs. (5.104), (5.105), and (5.108), the preceding two Fourier transforms become

$$\mathcal{F}\left\{u(t-t')G^>_{\varsigma\varrho}(t,t')\right\} = -\frac{i}{\hbar Z_\mu}\sum_{n,m}\frac{\exp(-E_n/k_BT)\,\langle n|\hat{q}_\varsigma|m\rangle\,\langle m|\hat{q}^\dagger_\varrho|n\rangle}{E/\hbar + (E_n - E_m)/\hbar + i\eta}, \tag{5.111}$$

$$\mathcal{F}\left\{u(t-t')G^<_{\varsigma\varrho}(t,t')\right\} = \frac{i}{\hbar Z_\mu}\sum_{n,m}\frac{\exp(-E_n/k_BT)\,\langle n|\hat{q}^\dagger_\varrho|m\rangle\,\langle m|\hat{q}_\varsigma|n\rangle}{E/\hbar + (E_m - E_n)/\hbar + i\eta}. \tag{5.112}$$

The resulting expression for $\check{G}(E)$ is called the Lehmann representation of the time-ordered Green's function. It shows explicitly the dependence of the Green's function on the eigen-energies of the effective Hamiltonian of the system. We can simplify this result further by swapping the dummy indices m and n in the preceding equation to obtain

$$\mathcal{F}\left\{u(t-t')G^<_{\varsigma\varrho}(t,t')\right\} = \frac{i}{\hbar Z_\mu}\sum_{nm}\frac{\exp(-E_m/k_BT)\,\langle m|\hat{q}^\dagger_\varrho|n\rangle\,\langle n|\hat{q}_\varsigma|m\rangle}{E/\hbar + (E_n - E_m)/\hbar + i\eta}. \tag{5.113}$$

It is useful to define a new function, called the *spectral function*, as

$$\Gamma(E) = \frac{2\pi}{Z_\mu}\sum_{nm}\exp(-E_m/k_BT)\,\langle m|\hat{q}^\dagger_\varrho|n\rangle\,\langle n|\hat{q}_\varsigma|m\rangle\,\delta\,(E + E_n - E_m)\,. \tag{5.114}$$

Using the spectral function, we can write the lesser and greater Green's functions as

$$G^>_{\varsigma\varrho}(t) = -\frac{i}{2\pi\hbar}\int_{-\infty}^{\infty}\Gamma(E)\exp\left(\frac{E}{k_BT} - \frac{i}{\hbar}Et\right)dE \tag{5.115}$$

$$G^<_{\varsigma\varrho}(t) = \frac{i}{2\pi\hbar}\int_{-\infty}^{\infty}\Gamma(E)\exp\left(-\frac{i}{\hbar}Et\right)dE. \tag{5.116}$$

The energy-space representation of other Green's functions can also be related to the spectral function. It is easy to deduce from Eq. (5.114) that $\Gamma(E)$ is real and positive when $\varsigma = \varrho$.

Aside 5.4   Equation of Motion for Green's Functions

In this Aside, we derive an equation of motion for the retarded Green's function, $G^+_{\varsigma\varrho}$, defined in Eq. (5.95). It can be used to find such equations for other types of Green's functions. We start with Eq. (2.62)) governing temporal evolution of an operator in the Heisenberg picture. In the case of the retarded Green's function, the relevant operator is $\hat{q}_\varsigma(t)$ and its evolution is governed by

$$i\hbar\frac{d\hat{q}_\varsigma}{dt} = [\hat{q}_\varsigma(t), H_{\text{eff}}]. \tag{5.117}$$

We use this result to calculate the time derivative of the retarded Green's function given in Eq. (5.95). Recalling that any Green's function depends on its two arguments such that $G(t,t') = G(t - t')$ and setting $t' = 0$, the equation of motion is found to contain the following two terms:

$$\frac{d}{dt}G^+_{\varsigma\varrho} = -\frac{i}{\hbar}\delta(t)\left\langle[\hat{q}_\varsigma(t),\hat{q}^\dagger_\varrho]_{\varkappa}\right\rangle - \frac{i}{\hbar}u(t-t')\left\langle[d\hat{q}_\varsigma/dt,\hat{q}^\dagger_\varrho]_{\varkappa}\right\rangle$$

$$= -\frac{i}{\hbar}\delta(t)\left\langle[\hat{q}_\varsigma(t),\hat{q}^\dagger_\varrho]_{\varkappa}\right\rangle - \frac{1}{\hbar^2}u(t-t')\left\langle[[\hat{q}_\varsigma(t),H_{\text{eff}}],\hat{q}^\dagger_\varrho]_{\varkappa}\right\rangle. \qquad (5.118)$$

One may think of the second term as a higher-order Green's function and derive a new equation of motion for this function, but that equation would contain a new Green's function of still higher order, resulting in an infinite hierarchical chain of equations of motion.

This complication can be avoided in the case of noninteracting systems, because the second term in Eq. (5.118) can be expressed in terms of $G^+_{\varsigma\varrho}$. However, for an interacting many-body quantum system, a closed-form solution exists only if a conserved quantity can be found such that the commutator associated with one of the high-order Green's functions vanishes. If that is not the case, one needs to truncate the infinite hierarchy by expressing one high-order Green's function in terms of lower-order Green's functions, resulting in an approximate solution. Such an approach is not systematic, and the solution depends on the specific truncation point used to find it.

5.3.4 Current in Terms of Green's Functions

We are now ready to use Eq. (5.89) derived earlier for the current. We need to make sure that this equation leads to the Landauer–Büttiker expression obtained under the assumption that the two leads were not interacting with each other. The evaluation of the associated Green's function becomes considerably simpler under this assumption.

We start by writing the Green's functions for the leads (electrodes) in the Lehmann representation. Because the applied voltage V_η remains constant with time, we obtain

$$G^<_{\eta\varsigma}(t_1 - t_2) = \frac{i}{\hbar}f_{F\eta}(E_{\eta\varsigma})\exp\left[i\int_{t_1}^{t_2}(E_{\eta\varsigma} - q_eV_\eta)\,dt\right]$$

$$= \frac{i}{\hbar}f_{F\eta}(E_{\eta\varsigma})\exp\left[-i(E_{\eta\varsigma} - q_eV_\eta)(t_1 - t_2)\right] \qquad (5.119)$$

This result can be used to write the retarded and advanced Green's functions in the form

$$G^+_{\eta\varsigma}(t_1 - t_2) = -\frac{i}{\hbar}u(t_2 - t_1)\exp\left[-i(E_{\eta\varsigma} - q_eV_\eta)(t_1 - t_2)\right], \qquad (5.120)$$

$$G^-_{\eta\varsigma}(t_1 - t_2) = \frac{i}{\hbar}u(t_2 - t_1)\exp\left[-i(E_{\eta\varsigma} - q_eV_\eta)(t_1 - t_2)\right]. \qquad (5.121)$$

The Green's function required for the current in Eq. (5.89) is a mixed-type Green's function defined as $G^<_{n,\eta\varsigma}(t - t') = -(i/\hbar)\left\langle\hat{q}_n(t)\hat{l}^\dagger_{\eta\varsigma}(t')\right\rangle$. Using the anticommutator properties of fermion operators in the leads and the quantum device, we express this Green's function in terms of Green's functions associated with the lead and the quantum device:

$$G_{n,\eta\varsigma}^{<}(t_1, t_2) = \sum_m \int \chi_{\eta\varsigma,m}^{*} \left[G_{nm}^{+}(t_1, t_3) G_{\eta\varsigma}^{<}(t_3, t_2) + \right.$$

$$\left. G_{nm}^{<}(t_1, t_3) G_{\eta\varsigma}^{-}(t_3, t_2) \right] dt_3. \tag{5.122}$$

This expression contains time-domain convolutions. However, as convolutions turn to products in the Fourier domain, we take advantage of this property and express the current in the energy basis, as we did earlier. Using Eq. (5.89), we obtain

$$I_\eta = \frac{2q_e}{h} \int \sum_{\varsigma,n,m} \Re \left[\chi_{\eta\varsigma,n} \chi_{\eta\varsigma,m}^{*} \left[G_{nm}^{+}(E) G_{\eta\varsigma}^{<}(E) + G_{nm}^{<}(E) G_{\eta\varsigma}^{-}(E) \right] \right] dE. \tag{5.123}$$

We can simplify the preceding expression by introducing physically meaningful spectral functions for the two electrodes in a way similar to the spectral function introduced in Section 5.3.3. Using the definition

$$\{ \Gamma_\eta(E) \}_{mn} = 2\pi \sum_\varsigma \delta(E - E_{\eta\varsigma}) \chi_{\eta\varsigma,n}(E) \chi_{\eta\varsigma,m}^{*}(E), \tag{5.124}$$

where $\eta = 1$ for the left electrode and $\eta = 2$ for the right electrode, we can write the current in the form

$$I_\eta = \frac{iq_e}{2\pi h} \int \mathrm{Tr} \left\{ \Gamma_\eta(E - q_e V_\eta) \times \right.$$

$$\left. \left[G^{<}(E) + f_{F\eta}(E - q_e V_\eta)(G^{+}(E) - G^{-}(E)) \right] \right\} dE, \tag{5.125}$$

where the trace operation represents the double sum over m and n and $f_{F\eta}$ is the Fermi function in Eq. (5.40). This equation contains three Green's functions, $G^{<}(E)$, $G^{+}(E)$, and $G^{-}(E)$, all of which correspond to the quantum device coupled to the two leads. It is important to recognize that they often cannot be calculated without making a few reasonable approximations (see Aside 5.4).

Equation (5.3.4) can be used to find current I_1 flowing through the left electrode using $\eta = 1$ and the current I_2 flowing through the right electrode using $\eta = 2$. Under ideal steady-state conditions, Kirchhoff's current law requires $I_2 = -I_1$. Using this result, the total current can be written as $I = (I_1 - I_2)/2$ and is given by

$$I = \frac{iq_e}{2\pi h} \int \mathrm{Tr} \left\{ \left[\Gamma_1(E - q_e V_1) - \Gamma_2(E - q_e V_2) \right] G^{<}(E) \right.$$

$$+ \left[f_{F1}(E - q_e V_1) \Gamma_1(E - q_e V_1) - f_{F2}(E - q_e V_2) \Gamma_2(E - q_e V_2) \right]$$

$$\left. \left[G^{+}(E) - G^{-}(E) \right] \right\} dE, \tag{5.126}$$

where we included spin degeneracy by multiplying I with a factor of two.

We need to introduce further approximations to write this result in the Landauer–Büttiker form. As our focus is on the steady-state situation in the case of noninteracting leads, we calculate the lesser Green's function by using the Keldysh equation

$$G^{<}(E) = G^{+}(E) \Sigma^{<}(E) G^{-}(E), \tag{5.127}$$

where $\Sigma^<(E)$ is the lesser self-energy and can be calculated from the equilibrium relations using

$$\Sigma^<(E) = f_{F1}(E - q_e V_1)\Gamma_1(E - q_e V_1) + f_{F2}(E - q_e V_2)\Gamma_2(E - q_e V_2), \qquad (5.128)$$

where $\Gamma_1(E)$ and $\Gamma_2(E)$ are the spectral functions defined earlier. We also note that

$$G^+(E) - G^-(E) = -iG^+(E)\left[\Gamma_1(E - q_e V_1) + \Gamma_2(E - q_e V_2)\right]G^-(E). \qquad (5.129)$$

Substituting the preceding results in the expression for the current I, and simplifying, we finally obtain the Landauer–Büttiker result:

$$I = \frac{iq_e}{\pi h} \int \text{Tr}\left[\Gamma_1(E - q_e V_1)G^+(E)\Gamma_2(E - q_e V_2)G^-(E)\right]$$
$$\times \left[f_{F1}(E - q_e V_1) - f_{F2}(E - q_e V_2)\right]dE. \qquad (5.130)$$

An advantage of the NGF method is that it provides more insight than the single-particle scattering matrix method. The steady-state analysis carried out here with the NGF formalism is known as the Keldysh formalism [308]. It has been used to investigate the steady-state quantum transport in many mesoscopic systems [324].

The NGF method has also been used to calculate time-dependent currents by considering voltages varying with time [315, 325]. This work fueled the use of the NGF method to describe charge transport while incorporating the interaction of electrons with photons, phonons, and other electrons. The time dependence is included into this formalism by making the parameters appearing in the Hamiltonian a function of time. As a result, $E_n \to E_n(t)$, $E_{n\varsigma} \to E_{n\varsigma} + \Delta E_{n\varsigma}(t)$, and $\chi_{n\varsigma,n}$ also become a function of time. With these changes, the time-dependent current through each electrode becomes ($\eta = 1, 2$):

$$I_\eta(t) = -\frac{2q_e}{h} \iint \Im\left\{\text{Tr}\left[\Gamma_\eta(E, t', t)\exp\left(-\frac{i}{\hbar}(t' - t)\right) \times \right.\right.$$
$$\left.\left. \left(G^<(t, t') + f_{F\eta}(E)G^+(t, t')\right)\right]\right\} dt' \, dE, \qquad (5.131)$$

where the spectral function Γ_η is also time dependent:

$$\Gamma_\eta(E, t_1, t_2) = 2\pi \sum_\varsigma \chi_{n\varsigma,n}(t_1)\exp\left(-\frac{i}{\hbar}\int_{t_2}^{t_1}\Delta E_{n\varsigma}(\tau)\,d\tau\right)\chi_{n\varsigma,m}^*(t_2)\delta(E - E_{n\varsigma}(t)). \qquad (5.132)$$

The Keldysh equation, $G^<(E) = G^+(E)\Sigma^<(E)G^-(E)$, in the energy domain becomes a double convolution when converted to the time domain. As a result, the two Green's functions appearing in Eq. (5.3.4) are given by

$$G^<(t, t') = G_0^+(t, t') + \iint G_0^+(t_1, t_2)\Sigma^<(t_1, t_2)G_0^-(t_2, t')\,dt_1\,dt_2 \qquad (5.133)$$

$$G^+(t, t') = G_0^+(t, t') + \iint G_0^+(t_1, t_2)\Sigma^+(t_1, t_2)G_0^+(t_2, t')\,dt_1\,dt_2. \qquad (5.134)$$

Here, the subscript "0" represents the unperturbed Green's function of the system before the time-dependent voltage was applied. The self energies appearing in the preceding equations are also functions of time and have the form

$$\Sigma_{mn}^{<}(t_1, t_2) = \frac{i}{h} \sum_{\eta} \int f_{F\eta}(E) \Gamma_{\eta mn}(E, t_1, t_2) \exp\left(-\frac{i}{\hbar} E(t_1 - t_2)\right) dE, \qquad (5.135)$$

$$\Sigma_{mn}^{+}(t_1, t_2) = -\frac{i}{h} u(t_1 - t_2) \sum_{\eta} \int \Gamma_{\eta mn}(E, t_1, t_2) \exp\left(-\frac{i}{\hbar} E(t_1 - t_2)\right) dE. \qquad (5.136)$$

These equations can even be used to calculate the nonlinear response of a quantum device controlled with external voltages. The reason is that the NGF method takes into account all correlations from the beginning of a time-dependent perturbation. In contrast, the equilibrium formalism does not consider correlations at all. This is the main reason why the Keldysh formalism is widely used for describing nonequilibrium operation of quantum devices.

6 Quantum Tunneling

"All right," said the Cat; and this time it vanished quite slowly, beginning with the end of the tail, and ending with the grin, which remained some time after the rest of it had gone.

Lewis Carroll, *Alice's Adventures in Wonderland*

6.1 Physics of Tunneling

Consider a particle of energy E approaching a potential barrier, whose maximum height U_{max} is such that it exceeds the particle's energy (see Figure 6.1). According to classical mechanics, the particle cannot penetrate the potential barrier, nor does it have enough energy to get over the barrier. As a result, in a classical world, the particle must get reflected at the barrier. Indeed, we observe such a behavior when billiard balls hit the boundaries of the table used to play the game.

6.1.1 What Is Tunneling?

A new possibility opens up in a quantum world, where quantum mechanics makes possible what is impossible classically. In quantum mechanics, the motion of a particle such as an electron is governed by the wave function Ψ, which satisfies the Schrödinger equation, first encountered in Section 2.2. Even though the amplitude of Ψ must decay exponentially inside the classically forbidden energy region where $E < U_{max}$, as seen in Figure 6.1, it may still have a finite value on the other side of the potential barrier, if the barrier is not too wide. This indicates a finite probability of the particle to be found on the opposite side of the barrier and gives the appearance that the particle has penetrated the potential barrier. Such a phenomenon is referred to as quantum tunneling, or simply tunneling [326].

Tunneling is a genuine quantum effect with no counterpart in classical physics. Owing to its nonintuitive nature, this phenomenon remains an enigma to this very day. Tunneling is a direct consequence of the wave–particle duality, a core concept in quantum mechanics [40, 96]. Its existence leads to questions such as what path a tunneling particle takes through a potential barrier, or how much time it takes to tunnel through this barrier. There exists a vast amount of literature on the tunneling time and its dependence on the shape and other features of the potential barrier, leading to many controversies and paradoxes. The

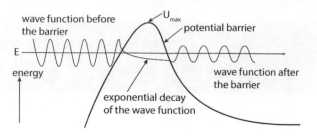

Fig. 6.1 Schematic illustration of tunneling across a potential barrier whose height U_{max} exceeds the particle's energy E. The behavior of the particle's wave function is also shown qualitatively, both inside and outside the barrier.

source of confusion lies in the observation that no general method exists for determining the Hamiltonian that is canonical self-adjoint of the time operator [327].

As an example of a controversy, some measurements of the tunneling time seem to predict superluminal speeds of the particle inside the potential barrier, leading to violation of the fundamental speed limit imposed by the theory of relativity. A well-known paradox is the Hartman effect discovered in 1962 [328]. It states that the tunneling time through a relatively thick barrier is independent of the actual thickness of the barrier. This statement implies that the tunneling speed has no upper limit, which also appears to agree with several experiments. Indeed, a 2019 experiment suggests that tunneling across a barrier occurs instantaneously [329]. Regardless of these paradoxes, tunneling is a real quantum phenomenon with a wide range of applications in areas such as scanning tunneling microscopy [330], nanoelectronics [331], and attosecond nanophysics [332]. Examples from everyday life include the flash-memory cards exploiting Fowler–Nordheim tunneling and a thermonuclear process inside the sun that fuses high-speed protons into helium nuclei.

One can identify three distinct scenarios based on the boundary conditions imposed on the tunneling process resulting from the specific shape of a potential barrier. These three generic barrier shapes are depicted in Figure 6.2. Part (a) shows a double-well potential profile with a middle hump. In this case, the particle is confined to the regions bounded at both ends, but tunneling may still occur through the middle hump. Such potential barriers occur inside devices known as superconducting quantum interference devices (SQUID) and that are used for quantum computing [333]. Part (b) of Figure 6.2 shows another common potential profile where a particle decays from a metastable state via tunneling. An example of this type of tunneling is provided by radioactive decay of polonium-212, leading to the emission of alpha particles [334]. The potential barrier in this case is formed by the combined action of the strong and weak forces inside the nucleus and the electromagnetic force. Part (c) of Figure 6.2 shows the conventional tunneling case where a particle is not confined on either side of the barrier. Tunnel junctions provide a good example of such a potential profile [335]. The three basic potential profiles serve as building blocks of more esoteric potential profiles for which tunneling can take place. For example, we study later a resonant tunneling device, where tunneling via two conventional barriers is exploited to enhance the tunneling process. In this section, we focus for simplicity on potential barriers

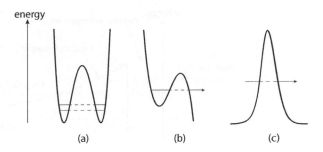

energy

(a)   (b)   (c)

Fig. 6.2   Schematic illustration of three generic shapes of one-dimensional potential barriers that serve as building blocks of more complex potential profiles.

that are one-dimensional. We should stress that the results obtained in this case may not always apply to higher-dimensional tunneling owing to the appearance of peculiarities that have no one-dimensional counterparts.

6.1.2 Gamow's Theory of Tunneling

The tunneling concept was first invoked by Gamow in 1928 to explain the emission of alpha particles during radioactive decay of certain heavy nuclei [334]. Alpha particles are helium atoms whose two electrons have been removed, resulting in ions with a net positive charge of $2q_e$. These helium ions have two protons and two neutrons, tightly bound together by the strong nuclear force. The radioactive nuclei of heavy atoms (such as polonium) were observed to emit alpha particles (helium ions) at random times through a process known as the alpha decay. This emission was explained as tunneling of alpha particles (with energies 4–9 MeV) trapped inside a potential barrier (height about 30 MeV) formed by the combination of an attractive nuclear force and a repulsive Coulomb force resulting from the remaining charge of $(Z_p - 2)q_e$, where Z_p is the atomic number of the heavy nucleus. Classically, an alpha particle of energy $E \ll U_{max}$ cannot escape the potential barrier. The tunneling theory of Gamow describes how an alpha particle may occasionally tunnel through the barrier because of its oscillating wave function. His theory not only pointed to the underlying mechanism for this type of radioactive decay but also explained large variations in the mean lifetimes of various radioactive nuclei.

Figure 6.3 shows the potential experienced by an alpha particle while it is confined inside the nucleus of a heavy atom such as polonium-212 ($Z = 84$). Gamow modeled the barrier as a spherical potential well of radius r_n representing the size of the nucleus. Outside the nucleus ($r > r_n$), alpha particles experience Coulomb repulsion arising from the remaining charge $Z_p - 2$ inside the nucleus. This potential can be written as

$$U(r) = \frac{(2q_e)(Z_p - 2)q_e}{4\pi \epsilon_0 r}, \qquad r > r_n. \tag{6.1}$$

It can be used to find the distance r_f (see Figure 6.3) at which an alpha particle escapes the Coulomb barrier. Using $U(r_f) = E_i$, where E_i is the initial energy of the alpha particle inside the nucleus, we find

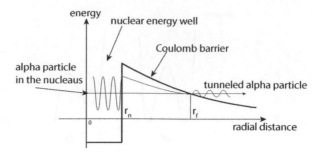

Fig. 6.3 Emission of an alpha particle from a radioactive nucleus through tunneling. The schematic shows the nuclear energy well where the alpha particle resides and the Coulomb barrier that it must tunnel through.

$$r_f = \frac{(2q_e^2)(Z_p - 2)}{4\pi\epsilon_0 E_i}. \tag{6.2}$$

Typical values of E_i are estimated to be 4 to 9 MeV, whereas the barrier height at $r = r_n$ is about 30 MeV.

Experiments indicate that the emission rate of alpha particles is proportional to the number N_R of radioactive atoms present at that time. Mathematically, we can write this relation as

$$\frac{dN_R}{dt} = -K_R N_R, \tag{6.3}$$

where K_R is called the decay constant. It is related to the transmission coefficient T_t across the potential barrier, which depends on the shape of the barrier and can be calculated with the WKB method discussed in Section 6.2.1. This method shows that K_R is of the form $K_R = C_R \exp(-2\gamma)$, where $C_R = \hbar/(2m_\alpha r_n^2)$ for an alpha particle with mass m_α, and γ is given by [see Eq. (6.36)]

$$\gamma = \frac{1}{\hbar}\int_{r_n}^{r_f} \sqrt{2m_\alpha[U(r) - E_i]}\,dr = \frac{\sqrt{2m_\alpha E_i}}{\hbar}\int_{r_n}^{r_f}\sqrt{\frac{r_f}{r} - 1}\,dr, \tag{6.4}$$

where we used $U(r)$ given in Eq. (6.1). The integration over r can be done with the variable change $r = r_f \cos^2\theta$ and leads to the following result:

$$\gamma = \frac{r_f}{\hbar}\sqrt{2m_\alpha E_i}\left[\cos^{-1}\sqrt{\eta} - \sqrt{\eta(1-\eta)}\right], \tag{6.5}$$

where $\eta = r_n/r_f$ is a dimensional ratio of two distances. For heavy radioactive atoms $(Z_p \gg 1)$, this ratio is a relatively small number proportional to $1/Z_p$. If we assume $\eta \approx 0$ and use $\cos^{-1}(0) = \pi/2$, we obtain the following result for $K_R = C_R \exp(-2\gamma)$:

$$K_R \approx \frac{\hbar}{2m_\alpha r_n^2}\exp\left[-\frac{2\pi}{\hbar}\sqrt{\frac{2m_\alpha}{E_i}}\frac{(Z_p - 2)q_e^2}{4\pi\epsilon_0}\right]. \tag{6.6}$$

As seen in Table 6.1, this expression provides values of the decay constant that are quite close to those measured experimentally.

Table 6.1 Measured and predicted values of decay constants for three heavy nuclei [336].

Nucleus	K_R (experimental)	K_R (theory)
^{148}Gd	2.2×10^{-10}	2.2×10^{-10}
^{214}Po	4.23×10^3	4.9×10^3
^{230}Th	2.09×10^{-13}	1.7×10^{-13}

6.1.3 Applications of Tunneling

In a series of three papers, Hund used the concept of tunneling to describe the observed features of molecular spectra [337, 338, 339]. He analyzed the oscillations of electrons between two atomic bound states, modeled as a double potential well. A necessary assumption in his analysis was the separation of the motion of electrons from vibrations and rotations of atoms; an approximation later made quantitative by Born and Oppenheimer. Hund studied the dynamics of a bound pair of atoms and noted the presence of reflection-symmetric potentials with classically impenetrable potential barriers. He found the two stationary states, even and odd combinations of the atomic states, for a molecule made with two atoms, which can be identical or distinct atoms. Hund noted that the superposition of these particular stationary states turned the system into a nonstationary oscillatory state, resulting from tunneling between the associated atomic quantum wells. He used the distance between the atoms to determine the width of the potential barrier and derived an expression for the beat period. The success of this theory cemented the validity of the tunneling concept as a genuine physical effect.

Another area where the tunneling concept improved our understanding is related to the emission of electrons from a metal surface when a strong electrostatic field is applied to accelerate electrons toward the metal's surface. This type of emission is known as Fowler–Nordheim tunneling, after the two scientists who showed that the effect is purely quantum mechanical and can be explained by considering tunneling of electrons through the potential barrier at the metal's surface [340]. Their simplified model described the effect as a simple one-dimensional problem where the electrons tunnel through a thin barrier subjected to a uniform electrostatic field perpendicular to the surface of the metal. The metal was modeled as an ideal Fermi gas, a method pioneered by Sommerfeld, and the thin barrier was modeled as a rectangular potential barrier; such a barrier is now used in textbooks for understanding the tunneling phenomenon [46]. The resulting solution showed that the tunneling probability has exponential dependence on the tunneling distance, which depends inversely on the applied electrostatic field. This type of field emission has found practical applications in solid-state electronic components such as tunneling diodes [341].

The operation of Schottky junctions, used for rectifying the flow of current, is also based on tunneling. Until the end of World War II, many attempts were made to relate the current flow in a metal–semiconductor junction (a Schottky junction) to the tunneling of electrons in solids. But the models were not realistic enough, and theory predicted current flow in the direction opposite of the observed one [341]. Unlike the field emission effect, where current flow is a result of the tunneling of electrons through a thin surface barrier, the electrostatic field lowers the surface barrier height in a Schottky junction, enabling electrons to get over the barrier.

The invention of the transistor in 1947 rekindled the interest in tunneling of electrons in rectifying junctions made with doped semiconductors. A significant advance occurred in 1958 when Esaki made tunnel diodes that were capable of oscillating at frequencies as high as 100 GHz [342] (see Aside 6.1). Such a tunnel diode exhibits a negative differential resistance in its voltage-current (V-I) characteristics over a wide range of frequencies. It consists of a heavily doped p–n junction (only about 10 nm wide), inside which interband tunneling takes electrons from the valence band to the conduction band. This behavior showed conclusively that electrons can tunnel between two metals, a topic pursued by many scientists for decades. Owing to a relatively high parasitic junction capacitance, tunnel diodes are not much used in modern devices.

Aside 6.1   Tunnel Junctions

A tunnel junction is essentially a capacitor with a sufficiently thin dielectric layer between two metallic plates through which an electron can tunnel from one plate to the other. Even though several methods exist for estimating the tunneling time, recent experiments indicate that tunneling is almost instantaneous [329]. Figure 6.4 shows a tunneling junction schematically. Two metal plates (or electrodes) of cross-sectional area A are separated by a short gap d (width of the junction).

When a voltage V_{12} is applied across the tunnel junction, the Fermi levels μ_1 and μ_2 of the electrodes become separated by $q_e V_{12}$ such that $\mu_2 = \mu_1 - q_e V_{12}$. If U_{TJ} is the height of the potential barrier between the two electrodes, the average transmission probability T_{TJ} for an electron of mass m_e can be obtained from Eq. (6.36). Assuming U_{TJ} is nearly constant and much larger than E, the integration can be easily done to obtain

$$T_{TJ} = \exp\left(-\frac{2d}{\hbar}\sqrt{2m_e U_{TJ}}\right). \tag{6.7}$$

Physically speaking, U_{TJ} is the difference between the work functions of the metal used for the electrodes and the insulator used for the barrier. The occupation probability of the energy states in electrodes 1 and 2 is governed by the distribution functions, $f_1(E)$ and $f_2(E)$, respectively. Using them, the current flowing between the two electrodes is given by (see Chapter 5)

$$I_{1\to 2} \propto -q_e A T_{TJ} \int_{-\infty}^{\infty} D_{os1}(E) D_{os2}(E)$$
$$\times \Big(f_1(E)[1-f_2(E)] - f_2(E)[1-f_1(E)]\Big)\, dE, \tag{6.8}$$

where the product $f_1(E)[1-f_2(E)]$ indicates that an electron in a filled energy state of electrode 1 can only tunnel to electrode 2 if there is a vacancy in the corresponding energy level to receive it. The term $f_2(E)(1-f_1(E))$ corresponds to tunneling of electrons from electrode 2 to electrode 1. These probabilities are multiplied by their respective density of

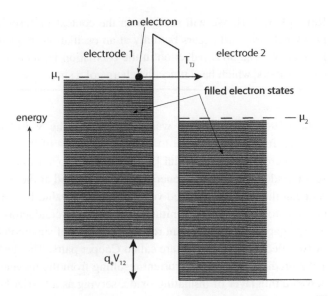

Fig. 6.4 Schematic showing operation of a tunnel junction. Two metal plates (electrodes) are separated by a short gap (width of the junction). Their filled energy levels are shown by a set of dense lines, while two dashed lines indicate their Fermi levels.

states and then integrated over all available energy states. The net current is related to the difference of the tunneling rates of electrons in the opposite directions.

If we assume that the two electrodes are in thermal equilibrium, then the distribution functions are just the Fermi–Dirac distributions (i.e., $f_j(E) \rightarrow f_{Fj}(E)$ for $j = 1, 2$). The current is then given by

$$I_{1\rightarrow 2} \propto -q_e A T_{TJ} \int_{-\infty}^{\infty} D_{os1}(E) D_{os2}(E)[f_{F1}(E) - f_{F2}(E)]\, dE. \tag{6.9}$$

The integration over energy can be carried out if we replace the density of states for each electrode with its value at the Fermi energy E_F, which is a valid assumption, as we have seen in Chapters 3 and 5. The result is then given by

$$I_{1\rightarrow 2} \propto -q_e A T_{TJ} D_{os1}(E_F) D_{os2}(E_F)(\mu_1 - \mu_2) = \frac{V_{12}}{R_{TJ}}, \tag{6.10}$$

where we used $(\mu_1 - \mu_2) = q_e V_{12}$ and defined the tunnel-junction resistance R_{TJ} such that Ohm's law can be used to describe the current flow through the tunnel junction. Our result provides an expression for R_{TJ} in terms of the physically relevant parameters of the tunnel junction. Note that it does not depend on the operating temperature of the device.

If a tunnel junction is biased with a constant current I_b, it will sustain this current flow through oscillations at the frequency $f_T = I_{21}/q_e$ because electrons need to tunnel one by one through the junction's barrier. Owing to its capacitive character, a tunnel junction continues to accumulate charges until the time it becomes favorable for an electron to tunnel through the barrier. Further details on single-charge tunneling can be found in

Refs. [343, 344]. We will see later in the context of Josephson junctions that the oscillatory behavior still occurs but only at an oscillation frequency of $f_T/2$ (called the Bloch frequency). The 50% reduction in the oscillation frequency is related to the tunneling of Cooper pairs, which have twice the charge of a single electron.

In a 1960 experiment, it was observed that the V–I characteristic of a tunnel diode changed from a straight line to a curve when one of the two metal electrodes was made of a superconducting material [345]. This was found to be related to the feature that all superconductors exhibit an energy gap E_g centered at the Fermi level. As a result, no current can flow until the applied voltage reaches a value $V = E_g/(2q_e)$. This feature made it possible to measure the magnitude of E_g for superconductors with sufficient accuracy. The energy gap plays an important role in the theory of superconductivity based on the pairing of two electrons; such pairs are called Cooper pairs. The theoretical work of Josephson in 1962 predicted that a supercurrent resulting from the movement of Cooper pairs can flow across a thin layer of insulating oxide, serving as a barrier between two superconductors. The supercurrent is in addition to the current found by Giaever [345] and results from tunneling of electrons in pairs [346] (see Aside 6.2). This effect is known as the Josephson effect and occurs in two different forms. In the case of the DC Josephson effect, a constant current flows across the junction even without applying any electric or magnetic fields (no bias voltage across the tunnel junction). In the case of the AC Josephson effect, a constant voltage is applied across the junction, and the resulting current oscillates at frequencies in the range 10 to 100 MHz.

Aside 6.2   Josephson Junction

According to the theory of Bardeen, Cooper, and Schrieffer (BCS), superconductivity results from the movement of Cooper pairs [347, 348], which form through pairing of two electrons with the same charge but opposing spins and momenta. Unlike electrons, which are fermions, Cooper pairs have a net spin of zero and thus act as bosons. As we have seen in Section 2.4, while only two fermions can occupy a quantum state, an infinite number of bosons could be in the same quantum state. As a result, all Cooper pairs inside a superconductor can be described by a single wave function and have the same phase. This *phase coherence* of Cooper pairs plays a critical role in Josephson junctions.

As shown schematically in Figure 6.5, a Josephson junction is made with two superconductors with a thin layer between them. This layer is typically an insulator but can even be made of a metal or a semiconductor, as its role is only to provide a potential barrier for the Cooper pairs inside two superconductors. If the barrier is sufficiently thin, Cooper pairs can tunnel through it, and a current would flow through the Josephson junction. The tunneling can be understood as follows. The wave function Ψ_1 of the superconductor 1 does not vanish at the barrier but decays exponentially inside the barrier. If its magnitude remains finite at the other end of the barrier, it can couple with the wave function Ψ_2 of the superconductor 2, as seen in Figure 6.5. We denote this quantum-mechanical coupling

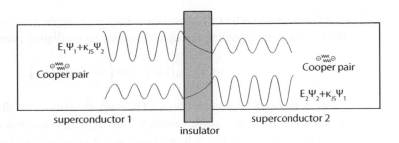

Fig. 6.5 Schematic of a Josephson junction. Two superconductors are separated by a thin insulator providing a potential barrier for tunneling of Cooper pairs. The wave functions Ψ_1 and Ψ_2 of Cooper pairs become coupled weakly through the barrier.

of the two superconductor wave functions by κ_{JS}. It can be viewed as a mechanism that maintains the phase coherence of Cooper pairs over the entire Josephson junction. This long-range phase coherence is fundamental to the Josephson effect.

Mathematically, the Josephson effect is governed by the following two equations known as *Josephson equations* [349]:

$$\frac{d\phi(t)}{dt} = \frac{2q_e}{\hbar} V_{12}(t), \qquad I_{1 \to 2}(t) = I_{cc} \sin \phi(t), \tag{6.11}$$

where $V_{12}(t)$ is the applied voltage, $I_{1 \to 2}(t)$ is the current flowing across the Josephson junction, and $\phi(t)$ is the phase difference between the wave functions in the two super-conductors. The quantity I_{cc} is known as the critical current and is proportional to the coupling constant κ_{JS}. The critical current is an important phenomenological parameter of the device; it can be controlled by varying temperature or by applying a magnetic field. The physical constant $2q_e/h$ is known as the *Josephson constant*, and its inverse, $h/(2q_e)$, is called the *magnetic flux quantum*. It is possible to combine the two Josephson equations and obtain the following expression for the tunneling current across the device:

$$I_{1 \to 2}(t) = I_{cc} \sin \left(\frac{2q_e}{\hbar} \int_0^t V_{12}(\tau) \, d\tau + \phi_0 \right), \tag{6.12}$$

where ϕ_0 is an integration constant that depends on the initial setup of the junction. This result shows that a Josephson junction can be operated in the following two regimes:

- **DC Josephson effect:** Even in the absence of any voltage ($V_{12} = 0$), a current flows through the Josephson junction because of tunneling of Cooper pairs through the thin barrier. This current is given by

$$I_{1 \to 2}(t) = I_{cc} \sin (\phi_0). \tag{6.13}$$

This result shows that the magnitude and direction of the current are set by the initial value of the phase difference ϕ_0 between the wave functions in two superconductors. If $\phi \neq 0$, a finite constant current flows through a Josephson junction even without any applied voltage. Depending on the initial phase value, this current can take any value between $-I_{cc}$ and I_{cc}.

- **AC Josephson effect:** If a constant voltage V_0 is applied across a Josephson junction, the Cooper pairs tunneling through it generate an oscillating current such that

$$I_{1 \to 2}(t) = I_{cc} \sin \left(\frac{2q_e}{\hbar} V_0 t + \phi_0 \right). \qquad (6.14)$$

Being a sinusoidal current without a DC term, the current oscillates between $-I_{cc}$ and I_{cc} such that its time average vanishes. The oscillation frequency of this AC current is given by $\omega = (2q_e/\hbar)V_0$, indicating that ω is proportional to V_0 and can be controlled through the applied voltage. For this reason, the AC Josephson effect can be used as a voltage-to-frequency converter.

The Josephson effect is affected considerably by the presence of a magnetic field. As a result, a Josephson junction can be transformed into a Giaever tunneling junction, also called a superconducting tunneling junction, by applying a small magnetic field [350]. Such tunneling junctions have found applications as sensitive detectors of electromagnetic radiation, capable of operating in a wide frequency range from the infrared to the X-ray region. Such detectors exploit the ability of a superconducting tunneling junction to detect photons with an energy approximately equal to twice the value of the gap parameter of the material of the junction. Thus, arrays of superconducting tunneling junctions can be used to construct highly accurate spectrometers. Another application employs the high sensitivity of such a junction to magnetic fields to measure extremely weak magnetic fields [351] through a device known as a superconducting quantum interference device (SQUID).

The current advances in this field are such that one can monitor the motion of individual hydrogen atoms on a metal surface by using a scanning tunneling microscope. One could invoke classical arguments to argue that the motion of atoms would be inhibited at low temperatures owing to thermal diffusion effects. However, it was found experimentally that atoms remain mobile down to temperatures as low as 9K [352]. Moreover, the tunneling rate of atoms through the metal's surface was found to increase as temperature was lowered! Quantum tunneling continues to lead to new advances as we synthesize new materials and invent novel quantum devices.

6.2 Quantum Description of Tunneling

Tunneling through a rectangular potential barrier is a common textbook problem [96, 341]. We begin this section by considering a more general scenario shown in Figure 6.6, where the shape of the barrier is not rectangular but varies in a continuous fashion in the region $z_1 < z < z_2$. This region is classically forbidden for a particle whose energy E is less than $V(z)$ over the entire region. We first discuss the Wentzel–Kramers–Brillouin (WKB) method used extensively in quantum mechanics and then use it to solve the tunneling problem.

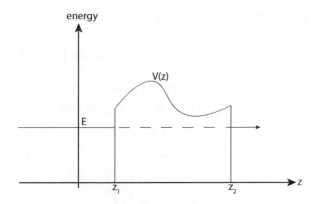

Fig. 6.6 Schematic showing a nonuniform potential barrier $V(z)$ for a particle of energy E. The classically forbidden region is confined to $z_1 < z < z_2$.

6.2.1 Wentzel–Kramers–Brillouin (WKB) Method

The WKB method is useful for finding the wave function of quantum particles whose potential energy varies slowly relative to other length scales [353, 354]. More precisely, the potential energy remains almost constant over the de Broglie wavelength of the particle. In this situation, both the amplitude and the phase of the wave function vary slowly over this length scale. The WKB method exploits this feature to find an approximate solution for the wave function. In quantum mechanics, the WKB method is often used for calculating energies of the bound states as well as tunneling rates through a potential barrier. For simplicity, we focus on the one-dimensional case here. The method can be extended to multiple dimensions, but the resulting equations are not always analytically solvable.

Consider the motion of a particle of mass m subjected to a slowly varying potential $V(z)$. As discussed in Section 2.2, the energy eigenstates of this particle are found by solving the Schrödinger equation

$$-\frac{\hbar^2}{2m}\frac{d^2\Psi}{dz^2} + V(z)\Psi(z) = E\Psi(z), \tag{6.15}$$

where E is the energy of a stationary eigenstate. This equation can be written in the form

$$\frac{d^2\Psi}{dz^2} + \frac{1}{\hbar^2}p^2(z)\Psi(z) = 0, \tag{6.16}$$

where $p(z)$ is defined as

$$p(z) = \begin{cases} \sqrt{2m[E - V(z)]}, & \text{if } E \geq V(z) \\ i\sqrt{2m[V(z) - E]}, & \text{if } V(z) > E. \end{cases} \tag{6.17}$$

Clearly $p(z)$ can be interpreted as the classical momentum of the particle when its value is real.

The WKB method solves Eq. (6.16) with the ansatz,

$$\Psi(z) = A(z) \exp\left(\frac{i}{\hbar} S(z)\right), \tag{6.18}$$

where both the amplitude $A(z)$ and phase $S(z)$ are real functions. Substituting this ansatz in Eq. (6.16) and equating the real and imaginary parts, we obtain two equations:

$$\frac{d^2 A}{dz^2} - \frac{A}{\hbar^2}\left[\left(\frac{dS}{dz}\right)^2 - p^2(z)\right] = 0, \tag{6.19}$$

$$2\frac{dA}{dz}\frac{dS}{dz} + A\frac{d^2 S}{dz^2} = 0. \tag{6.20}$$

These equations can be solved approximately if A varies slowly enough that its second derivative in Eq. (6.19) can be neglected. This equation then leads to

$$\frac{dS(z)}{dz} = \pm p(z) \rightarrow S(z) = \pm \int_{z_i}^{z} p(z)\,dz, \tag{6.21}$$

where z_i is the initial position of the particle. Using this result in the second equation, amplitude is found to be given by

$$\frac{d}{dz}\left(A^2 p\right) = 0 \rightarrow A(z) = \frac{C}{\sqrt{|p(z)|}}, \tag{6.22}$$

where C is a constant that needs to be determined. Since the phase equation has two solutions, the general solution of Eq. (6.16) is a linear combination of these two solutions.

As an example, consider the solution in the classically allowed region where $E > V(z)$. In this case $p(z)$ is real and positive and the general solution has the form

$$\Psi(z) = \frac{C_+}{\sqrt{p(z)}} \exp\left(\frac{i}{\hbar} \int_{z_i}^{z} p(z)\,dz\right) + \frac{C_-}{\sqrt{p(z)}} \exp\left(-\frac{i}{\hbar} \int_{z_i}^{z} p(z)\,dz\right). \tag{6.23}$$

In the classically forbidden region, $V(z) > E$, Eq. (6.17) shows us that $p(z)$ is purely imaginary. Using $p(z) = i|p(z)|$, the general solution takes the form

$$\Psi(z) = \frac{C}{\sqrt{|p(z)|}} \exp\left(-\frac{1}{\hbar} \int_{z_i}^{z} |p(z)|\,dz\right) + \frac{C'}{\sqrt{|p(z)|}} \exp\left(\frac{1}{\hbar} \int_{z_i}^{z} |p(z)|\,dz\right). \tag{6.24}$$

However, there is another situation not covered by the above two scenarios; it corresponds to $E \approx V(z)$ or $p(z) \approx 0$. As $p(z)$ appears in the denominator of the preceding solutions for $\Psi(z)$, we have a singular situation near $p(z) = 0$. The z values where $p(z)$ vanishes are known as the *turning points* because a classical particle stops and reverses its direction of motion at those points. In quantum mechanics, the behavior of the wave function changes from being oscillatory to decaying exponential at a turning point. To gain further insight, Figure 6.7 shows the potential $V(z)$ near a turning point, assumed to be located at $z = 0$. In the region on the left of the turning point ($z < 0$), the solution is a linear combination of the forward and backward propagating waves, whereas it must decay exponentially in the region on the right of the turning point ($z > 0$). Thus, the general

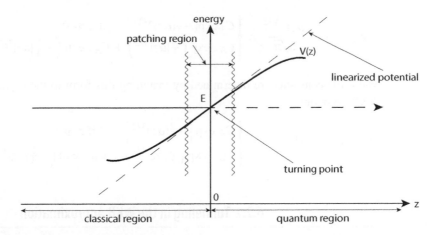

Fig. 6.7 Schematic showing the potential $V(z)$ near a turning point located at $z = 0$. At the turning point, solution of the wave function changes from being oscillatory to exponential and it should be forced to be continuous in the patching region.

solution can be written as

$$\Psi(z) = \begin{cases} \frac{C_+}{\sqrt{p(z)}} \exp\left(\frac{i}{\hbar} \int_{z_i}^{z} p(z)\,dz\right) + \frac{C_-}{\sqrt{p(z)}} \exp\left(-\frac{i}{\hbar} \int_{z_i}^{z} p(z)\,dz\right), & \text{if } z < 0 \\ \frac{C}{\sqrt{|p(z)|}} \exp\left(-\frac{1}{\hbar} \int_{z_i}^{z} |p(z)|\,dz\right), & \text{if } z > 0. \end{cases} \tag{6.25}$$

The three constants are found using the normalization condition and forcing the requirement that the wave function $\Psi(z)$ and its derivative $d\Psi/dz$ should be continuous at the turning point $z = 0$.

One way to enforce the continuity requirement is to linearize the potential $V(z)$ around this turning point at $z = 0$ using $V(z) \approx E + V_d z$, where V_d is the value of dV/dz at $z = 0$. If we introduce a new variable $x = \alpha z$ with $\alpha = (2mV_d/\hbar^2)^{1/2}$, we can write Eq. (6.16) in the form

$$\frac{d^2}{dx^2} \Psi(x) - x\Psi(x) = 0. \tag{6.26}$$

This is the *Airy equation* and its general solution can be written in terms of the Airy functions $\text{Ai}(x)$ and $\text{Bi}(x)$ as

$$\Psi(x) = C_A \, \text{Ai}(x) + C_B \, \text{Bi}(x), \tag{6.27}$$

where the constants C_A and C_B must be determined from the boundary conditions. This solution must match the WKB solution for values of z outside the patching region in Figure 6.7. However, the Airy function $\text{Bi}(x)$ grows exponentially for large values of x, whereas the WKB solution requires exponential decay. Because $\text{Ai}(x)$ decays exponentially for large values of x, matching becomes possible if we set $C_B = 0$. Using the well-known asymptotic forms of $\text{Ai}(x)$ for positive and negative values of x, the wave function has the following functional form:

$$\Psi_{\text{Airy}}(z) = \frac{|\alpha z|^{-1/4}}{2\sqrt{\pi}} \times \begin{cases} C_A \exp[-\frac{2}{3}(\alpha z)^{3/2}] & \text{if } z \gg 0 \\ C_{A+} \exp\left(\frac{2i}{3}|\alpha z|^{3/2}\right) + C_{A-} \exp\left(-\frac{2i}{3}|\alpha z|^{3/2}\right) & \text{if } z \ll 0, \end{cases}$$

(6.28)

where the constants are determined by matching this form to the asymptotic form of the WKB solution:

$$\Psi_{WKB}(z) = |\alpha^3 z|^{-1/4} \times \begin{cases} C_W \exp\left[-\frac{2}{3}(\alpha z)^{3/2}\right] & \text{if } z \gg 0 \\ C_{W+} \exp\left(\frac{2i}{3}|\alpha z|^{3/2}\right) + C_{W-} \exp\left(-\frac{2i}{3}|\alpha z|^{3/2}\right) & \text{if } z \ll 0. \end{cases}$$

(6.29)

6.2.2 Tunneling in the WKB Approximation

We now apply the WKB approximation to the tunneling problem. We assume that the potential $V(z)$ exceeds the particle's energy E in the entire region $z_1 < z < z_2$. This is not a severe restriction because the barrier can always be partitioned into several segments, each satisfying the preceding condition. Even though classically forbidden, there is a finite probability of the particle's tunneling through the barrier that we want to calculate.

As discussed in the preceding section, the Schrödinger equation can be solved analytically in the WKB approximation. The resulting wave function has different forms in three regions of Figure 6.6:

$$\Psi(z) = \begin{cases} A_+ \exp\left(\frac{i}{\hbar}\sqrt{2mE}z\right) + A_- \exp\left(-\frac{i}{\hbar}\sqrt{2mE}z\right) & \text{if } z < z_1, \\ \frac{B_+}{\sqrt{|p(z)|}} \exp\left(-\frac{1}{\hbar}\int_{z_1}^{z}|p(z)|\,dz\right) & \text{if } z_1 < z < z_2, \\ C_+ \exp\left(\frac{i}{\hbar}\sqrt{2mE}z\right) & \text{if } z > z_2, \end{cases}$$

(6.30)

where $p(z)$ in the barrier region is given in Eq. (6.17). On physical grounds, a backward propagating wave must be included in the region $z < z_1$ but not in the region $z > z_2$. The four constants, A_+, A_-, B_+, and C_+ are determined using the continuity of the wave function and its derivative at the two boundaries located at $z = z_1$ and $z = z_2$. It is important to note that we only kept the exponentially decaying solution in the barrier region because the exponentially growing solution cannot be sustained for wide barriers without violating the conversation of energy and momentum principles.

Invoking the continuity of $\Psi(z)$ and its derivative $d\Psi/dz$ at $z = z_1$, we obtain the relations

$$A_+ \exp\left(\frac{i}{\hbar}\sqrt{2mE}z_1\right) + A_- \exp\left(-\frac{i}{\hbar}\sqrt{2mE}z_1\right) = \frac{1}{\sqrt{|p(z_1)|}}B_+$$

(6.31)

$$A_+ \exp\left(\frac{i}{\hbar}\sqrt{2mE}z_1\right) - A_- \exp\left(-\frac{i}{\hbar}\sqrt{2mE}z_1\right) = i\sqrt{\frac{|p(z_1)|}{2mE}}B_+.$$

(6.32)

Similarly, the continuity of $\Psi(z)$ and $d\Psi/dz$ at the boundary $z = z_2$ provides the relations

$$\frac{B_+}{\sqrt{|p(z_2)|}} \exp\left(-\frac{1}{\hbar}\int_{z_1}^{z_2}|p(z)|\,dz\right) = C_+ \exp\left(\frac{i}{\hbar}\sqrt{2mE}z_2\right)$$

(6.33)

$$B_+ \exp\left(-\frac{1}{\hbar}\int_{z_1}^{z_2} |p(z)|\,dz\right) = -i\sqrt{\frac{2mE}{|p(z_2)|}}\,C_+ \exp\left(\frac{i}{\hbar}\sqrt{2mE}z_2\right). \tag{6.34}$$

We can solve these four equations to find the four coefficients and obtain the wave function $\Psi(z)$.

The quantity of primary interest is the tunneling probability given by the transmissivity $T_t = \left|\frac{C_+}{A_+}\right|^2$. This ratio is easy to calculate and is given by

$$T_t = \frac{4|p(z_1)|}{|p(z_2)|}\left[1 + \frac{|p(z_1)|^2}{2mE}\right]^{-1}\exp\left(-\frac{2}{\hbar}\int_{z_1}^{z_2}|p(z)|\,dz\right). \tag{6.35}$$

We can write the transmission coefficient in the form

$$T_t = C_t \exp(-2\gamma), \qquad \gamma = \frac{1}{\hbar}\int_{z_1}^{z_2}\sqrt{2m[V(z) - E]}\,dz, \tag{6.36}$$

where the prefactor C_t is ~ 1 and its value depends on particle's energy E and the potentials at the end points z_1 and z_2 of the barrier. In the case of a constant potential V over the entire barrier region, the transmission coefficient becomes

$$T_t = \frac{4E}{V}\exp\left(-\frac{2}{\hbar}(z_2 - z_1)\sqrt{2m(V - E)}\right). \tag{6.37}$$

This is the main result of the theory of quantum tunneling in the WKB approximation. It predicts an exponential reduction in the tunneling probability of a quantum particle with increasing width or height of the potential barrier.

6.3 Transfer Hamiltonian Method

An alternative approach is based on a *tunneling Hamiltonian* and is called the *transfer Hamiltonian method*. It is used to describe the tunneling phenomenon without making the WKB approximation. Historically, the tunneling Hamiltonian was introduced in 1961 by Bardeen to explain Giaever's observation of tunneling in a Josephson junction [355]. It was further developed by Harrison [356], and formulated in its second-quantized form by Cohen et al. [357]. This approach has been used extensively for describing tunnel junctions (including Josephson junctions) and the Coulomb blockade phenomenon. The tunneling Hamiltonian of the whole system is divided into three parts as

$$H_S = H_L + H_R + H_T, \tag{6.38}$$

where H_L is the Hamiltonian of the left electrode, H_R is the Hamiltonian of the right electrode, and H_T is the Hamiltonian of the tunneling barrier. The last part, H_T, is treated as a perturbation to the system Hamiltonian [358, 355, 359].

The rationale behind this method is to exploit the well-understood features of the system in the absence of the tunneling barrier, and use time-dependent perturbation theory to study the system's response to the tunneling barrier. This method also allows us to calculate the current flowing through the tunneling barrier by calculating the transfer rate of charged particles across the potential barrier in the forward and backward directions. Owing to the use of perturbation theory, the validity of this approach is limited to the cases where H_T is relatively small compared to other parts of the system Hamiltonian.

In practice, the analytical expressions resulting from the application of perturbation theory are relatively cumbersome. One way out of this complexity is to exploit the well-established diagrammatic techniques used in quantum electrodynamics and many-body theory [131]. Besides the appealing aspect of representing perturbative expressions with drawings, the diagrammatic method can also be used for reasoning and problem solving. The easily recognizable topology of diagrams makes the diagrammatic method a powerful tool for constructing equations that may hold even beyond perturbation theory. With the use of diagrammatic techniques, the transfer Hamiltonian method has been used to include the many-body effects such as quasi-particle tunneling or phonon-assisted tunneling. In this section we only present the key features of the transfer Hamiltonian method by applying time-dependent perturbation theory [305].

6.3.1 Time-Dependent Tunneling Theory

To understand the basics of the transfer Hamiltonian method, we consider a simple setup shown in Figure 6.8, where a thin rectangular-shape barrier is placed between two metallic electrodes with the Hamiltonians H_L and H_R. It is assumed that the left and right electrodes interact only through this tunneling barrier of width w occupying the region $-w/2 < z < w/2$. We also neglect Coulomb interactions among the electrons inside each electrode. The wave function $\Psi_S(t)$ of the entire system under these conditions can be found by solving

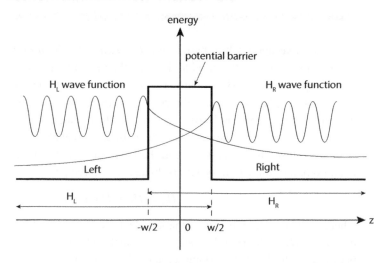

Fig. 6.8 Schematic illustration of the tunneling-Hamiltonian approach. The tunneling barrier occupies the middle region. The wave functions for the left and right regions are governed by the Hamiltonians H_L and H_R, respectively; double arrows mark their domain size.

the single-electron Schrödinger equation:

$$i\hbar \frac{\partial \Psi_S}{\partial t} = H_S \Psi_S(t) \qquad (6.39)$$

where the system Hamiltonian H_S is given by

$$H_S = \begin{cases} -\frac{\hbar^2}{2m} \frac{d^2}{dz^2} + U_L & \text{if } z < -\frac{w}{2}, \\ -\frac{\hbar^2}{2m} \frac{d^2}{dz^2} + U_0 & \text{if } -\frac{w}{2} < z < \frac{w}{2}, \\ -\frac{\hbar^2}{2m} \frac{d^2}{dz^2} + U_R & \text{if } z > \frac{w}{2}. \end{cases} \qquad (6.40)$$

Here U_0 is the potential of the barrier and U_L and U_R are the potentials on the left and right sides of the barrier region, respectively. It is difficult to solve this time-dependent problem exactly. We use the following strategy to construct an approximate solution by using time-dependent perturbation theory.

We partition the system into two tractable subsystems by considering the left and right electrodes separately. We modify the potentials on each side to include the barrier's potential. With this change, the potential for the left electrode takes the form

$$U_L(z) = \begin{cases} U_L, & \text{if } z < -w/2 \\ U_0, & \text{if } -w/2 < z < w/2, \\ 0, & \text{if } z > w/2. \end{cases} \qquad (6.41)$$

The potential for the right electrode has the same form except $U_R(z) = 0$ for $z < -w/2$ and takes the value U_R for $z > w/2$. Note that the potentials coincide with the actual potential of each electrode but vanish at the other electrode. This simplification allows one to find the stationary solutions of the left and right electrodes and use them to approximately construct the time evolution of the entire system. The stationary states of H_L and H_R are found by solving the following time-independent Schrödinger equations:

$$-\frac{\hbar^2}{2m} \frac{d^2 \Psi_{L\eta}}{dz^2} + U_L(z) \Psi_{L\eta}(z) = E_{L\eta} \Psi_{L\eta}(z) \qquad (6.42)$$

$$-\frac{\hbar^2}{2m} \frac{d^2 \Psi_{R\zeta}}{dz^2} + U_R(z) \Psi_{R\zeta}(z) = E_{R\zeta} \Psi_{R\zeta}(z), \qquad (6.43)$$

where we used the index η for the left electrode and ζ for the right electrode. With this notation, the eigen-energies of the two electrodes are $E_{L\eta}$, and $E_{R\zeta}$.

Consider an electron in the left electrode with energy $E_{L\eta}$ and assume that its interaction via the barrier is switched on at time $t = 0$. If there is no coupling, its wave function evolves as $\Psi_{L\eta} \exp(-iE_{L\eta}t/\hbar)$. In the presence of the coupling, we can express the time-dependent wave function $\Psi_S(t)$ of the total Hamiltonian H_S using the eigenfunctions $\Psi_{L\eta}(z)$ and $\Psi_{R\zeta}(z)$. Noting that H_S and the two partial Hamiltonians (H_L and H_R) are identical in the barrier region, the solutions of H_L and H_R in the overlapping regions must satisfy the relations

$$H_S \Psi_{L\eta}(z) = E_{L\eta} \Psi_{L\eta}(z) \qquad \text{if } z < w/2, \qquad (6.44)$$

$$H_S \Psi_{R\zeta}(z) = E_{R\zeta} \Psi_{R\zeta}(z) \qquad \text{if } z > -w/2. \qquad (6.45)$$

We express the time-dependent wave function $\Psi_S(t)$ using these eigenfunctions and write it in the form

$$\Psi_S(t) = \Psi_{L\eta} \exp\left(-\frac{i}{\hbar}E_{L\eta}t\right) + \sum_\zeta \kappa_{R\zeta}(t)\Psi_{R\zeta}, \tag{6.46}$$

where the sum is over all bound states of the right electrode. For each bound state, $\kappa_{R\zeta}(t)$ represents the time-dependent coupling owing to leaking of the wave function into the potential barrier. To simplify the notation, we have absorbed the time evolution of the states $\Psi_{R\zeta}$ into the coupling coefficient. As the coupling is switched on at $t = 0$, the initial condition is $\kappa_{R\zeta}(0) = 0$ for all the bound states. Moreover, the condition $|\kappa_{R\zeta}(t)| \ll 1$ must hold for all $t > 0$ in view of our assumption of weak coupling.

6.3.2 Calculation of the Coupling Coefficient

We need an expression for the coupling coefficient $\kappa_{R\zeta}(t)$. For this purpose, we substitute Eq. (6.46) on the right side of Eq. (6.39) to obtain

$$i\hbar\frac{\partial\Psi_S}{\partial t} = H_S\Psi_{L\eta}\exp\left(-\frac{i}{\hbar}E_{L\eta}t\right) + \sum_\zeta \kappa_{R\zeta}(t)H_S\Psi_{R\zeta}. \tag{6.47}$$

Replacing H_S with $H_L + (H_S - H_L)$ in the first term and with $H_R + (H_S - H_R)$ in the second term, we obtain

$$i\hbar\frac{\partial\Psi_S}{\partial t} = [E_{L\eta} + (H_S - H_L)]\Psi_{L\eta}\exp\left(-\frac{i}{\hbar}E_{L\eta}t\right)$$
$$+ \sum_\zeta \kappa_{R\zeta}(t)[E_{R\zeta} + (H_S - H_R)]\Psi_{R\zeta}. \tag{6.48}$$

But, we can also directly differentiate $\Psi_S(t)$ with respect to t to get

$$i\hbar\frac{\partial\Psi_S}{\partial t} = E_{L\eta}\Psi_{L\eta}\exp\left(-\frac{i}{\hbar}E_{L\eta}t\right) + i\hbar\sum_\zeta \frac{d\kappa_{R\zeta}(t)}{dt}\Psi_{R\zeta}. \tag{6.49}$$

Equating the preceding two equations, we obtain the relation

$$i\hbar\sum_\zeta \frac{d\kappa_{R\zeta}}{dt}\Psi_{R\zeta} = (H_S - H_L)\Psi_{L\eta}\exp\left(-\frac{i}{\hbar}E_{L\eta}t\right)$$
$$+ \sum_\zeta \kappa_{R\zeta}(t)[E_{R\zeta} + (H_S - H_R)]\Psi_{R\zeta}. \tag{6.50}$$

We discard the last sum in this equation owing to our assumption $|\kappa_{R\zeta}(t)| \ll 1$.

The next step is to multiply the preceding equation with $\Psi_{R\zeta'}^*(z)$ and integrate over z. Using the orthogonal property of the bound states $\Psi_{R\zeta}$, the result can be written as

$$i\hbar\frac{d\kappa_{R\zeta}}{dt} = V_{\eta\zeta}\exp\left[-\frac{i}{\hbar}(E_{L\eta} - E_{R\zeta})t\right], \tag{6.51}$$

where the matrix element representing coupling between the eigenstates of the left and right electrodes (responsible for tunneling) is defined as

$$V_{\eta\zeta} = \int_{-\infty}^{\infty} \Psi_{R\zeta}^*(z)(H_S - E_{L\eta})\Psi_{L\eta}(z)\,dz. \tag{6.52}$$

The preceding first-order differential equation can be easily integrated to obtain the following solution for $t \geq 0$:

$$\kappa_{R\zeta}(t) = \frac{V_{\eta\zeta}}{E_{L\eta} - E_{R\zeta}}\left[\exp\left(-\frac{i}{\hbar}(E_{L\eta} - E_{R\zeta})t\right) - 1\right]. \tag{6.53}$$

Finally, we need to calculate the matrix element $V_{\eta\zeta}$. As $H_S\Psi_{L\eta}(z) = E_{L\eta}\Psi_{L\eta}(z)$ for any $z < w/2$, the lower limit of the integral can be set to z_0 where z_0 is any value in the range $[-w/2, w/2]$. Thus,

$$V_{\eta\zeta} = \int_{z_0}^{\infty} \Psi_{R\zeta}^*(z)(H_S - E_{L\eta})\Psi_{L\eta}(z)\,dz. \tag{6.54}$$

Noting that $H_S\Psi_{R\zeta}(z) = E_{R\zeta}\Psi_{R\zeta}(z)$ for any $z > -w/2$, we have the relation

$$\int_{z_0}^{\infty} \Psi_{L\eta}(z)(H_S - E_{R\zeta})\Psi_{R\zeta}^*(z)\,dz = 0. \tag{6.55}$$

Subtracting this integral from $V_{\eta\zeta}$, we obtain

$$V_{\eta\zeta} = \int_{z_0}^{\infty} \left[\Psi_{R\zeta}^*(z)(H_S - E_{L\eta})\Psi_{L\eta}(z) - \Psi_{L\eta}(z)(H_S - E_{R\zeta})\Psi_{R\zeta}^*(z)\right]dz. \tag{6.56}$$

Using the form of H_S given in Eq. (6.40), we can write $V_{\eta\zeta}$ in the form

$$V_{\eta\zeta} = \frac{\hbar^2}{2m}\int_{z_0}^{\infty} \left[\Psi_{R\zeta}^*(z)\left(-\frac{d^2}{dz^2}\right)\Psi_{L\eta}(z) - \Psi_{L\eta}(z)\left(-\frac{d^2}{dz^2}\right)\Psi_{R\zeta}^*(z)\right]dz. \tag{6.57}$$

Consider the first integral. Integrating in parts, we obtain

$$\int_{z_0}^{\infty} \Psi_{R\zeta}^*(z)\frac{d^2\Psi_{L\eta}}{dz^2}\,dz = \Psi_{R\zeta}^*(z_0)\frac{d\Psi_{L\eta}}{dz}\Big|_{z=z_0} - \int_{z_0}^{\infty} \frac{d\Psi_{R\zeta}^*}{dz}\frac{d\Psi_{L\eta}}{dz}\,dz. \tag{6.58}$$

A similar expression is obtained for the second integral. Subtracting the two, the integrals cancel out and we obtain only the two boundary terms. The resulting expression is

$$V_{\eta\zeta} = \frac{\hbar^2}{2m}\left[\Psi_{L\eta}(z_0)\frac{d}{dz}\Psi_{R\zeta}^*(z_0) - \Psi_{R\zeta}^*(z_0)\frac{d}{dz}\Psi_{L\eta}(z_0)\right]. \tag{6.59}$$

As our lower integration limit was set within the barrier region ($-w/2 < z_0 < w/2$), this expression is valid for any value of z_0 falling within the barrier. Clearly, the magnitude of $V_{\eta\zeta}$ depends on the overlap of the left and right wave functions within the barrier region. Even though we have done our calculation for a single variable, the procedure can be readily extended to two or three dimensions but with considerably more algebra (see Ref. [355]).

6.3.3 Tunneling Current

To calculate the current resulting from tunneling of electrons across the potential barrier, we need to first calculate the probability of an electron tunneling through this barrier. This probability for an electron on the left electrode to tunnel to the bound state $\Psi_{R\zeta}^*$ in the right electrode can be calculated using

$$\int_{-\infty}^{\infty} \Psi_{R\zeta}^*(z)\Psi_S(t)\,dz = \kappa_{R\zeta}(t) + \exp\left(-\frac{i}{\hbar}E_{L\eta}t\right)\int_{-\infty}^{\infty}\Psi_{R\zeta}^*(z)\Psi_{L\eta}(z)\,dz, \quad (6.60)$$

where we used $\Psi_S(t)$ given in Eq. (6.46). Making use of *Oppenheimer perturbation theory*, we can discard the second term. This amounts to assuming that the eigenstates of the left and right electrodes are approximately orthogonal.

The rate R_{LR} at which an electron initially in the left electrode ends up on the right electrode can be calculated by assuming that both electrodes are in nearly thermal equilibrium. In this situation, we can use their Fermi–Dirac distribution for the occupancy probability of a state. Summing over all the states on the left electrode, we obtain

$$R_{LR} = \frac{d}{dt}\sum_{\zeta} F_{F\eta}(E_{L\eta})[1 - F_{F\zeta}(E_{R\zeta})]\left|\int_{-\infty}^{\infty}\Psi_{R\zeta}^*(z)\Psi_S(t)\,dz\right|^2$$

$$= \frac{d}{dt}\sum_{\zeta} F_{F\eta}(E_{L\eta})[1 - F_{F\zeta}(E_{R\zeta})]\left|\kappa_{R\zeta}(t)\right|^2, \quad (6.61)$$

where $F_{F\eta}(E_{L\eta})$ is the occupancy probability of an electron in the state η of the left electrode and the term $1 - F_{F\zeta}(E_{R\zeta})$ represents the vacancy probability of the state ζ at the right electrode to receive this electron. Both of these probabilities are needed to satisfy the Pauli exclusion principle.

We can now calculate the total current, $I_{L\to R}$, from the left electrode to the right electrode. For this, we need to consider the tunneling rate through the barrier in both directions (i.e., $I_{L\to R} = q_e(R_{LR} - R_{RL})$). The calculation of R_{RL} mirrors the preceding calculation of R_{LR}. The resulting expression is

$$I_{L\to R} = -q_e\frac{d}{dt}\sum_{\eta}\sum_{\zeta}\Big[F_{F\eta}(E_{L\eta})[1 - F_{F\zeta}(E_{R\zeta})]\left|\kappa_{R\zeta}(t)\right|^2$$

$$- F_{F\eta}(E_{R\zeta})[1 - F_{F\eta}(E_{L\eta})]\left|\kappa_{L\eta}(t)\right|^2\Big]. \quad (6.62)$$

We calculate $|\kappa_{R\zeta}(t)|^2$ using Eq. (6.53) and obtain

$$\left|\kappa_{R\zeta}(t)\right|^2 = 4\left|V_{\eta\zeta}\right|^2\frac{\sin^2[(E_{L\eta} - E_{R\zeta})t/2\hbar]}{(E_{L\eta} - E_{R\zeta})^2}. \quad (6.63)$$

The quantity $|\kappa_{L\eta}(t)|^2$ is given by a similar expression.

The final step is to replace the double sum in the preceding expression for $I_{L\to R}$ with two energy integrals using the density of states of the left and right electrodes. In the long time limit, we can also make use of the known result $\delta(x) = \lim_{t\to\infty}[\sin^2(xt)/(\pi x^2 t)]$.

This result is valid in the long-time limit, but the largest value of t should be such that $|\kappa_{R\zeta}(t)| \ll 1$ is maintained. Using it, we can write

$$\sum_{\zeta} |\kappa_{R\zeta}(t)|^2 \approx \frac{2\pi}{\hbar} |V_{\eta\zeta}|^2 D_{osR}(E_{R\zeta})\delta(E_{L\eta} - E_{R\zeta})t, \tag{6.64}$$

which could be seen as an application of the Fermi's golden rule. Here, $D_{osR}(E_{R\zeta})$ represents the density of states at the right electrode. It enters in Eq. (6.64) when we replace the sum over the quantum states with an energy integral.

Using Eq. (6.64), the total current can be written as

$$I_{L \to R} = \frac{2\pi q_e}{\hbar} \iint_{-\infty}^{\infty} \left[|V_{\eta\zeta}|^2 D_{osR}(E_{R\zeta})D_{osL}(E_{L\eta}) \right.$$

$$\left. \times F_{F\zeta}(E_{R\zeta})[1 - F_{F\eta}(E_{L\eta})]\delta(E_{L\eta} - E_{R\zeta}) \right] dE_{L\eta}\, dE_{R\zeta}. \tag{6.65}$$

One of the integrals can be readily done because of the presence of the delta function, resulting in

$$I_{L \to R} = \frac{2\pi q_e}{\hbar} \int_{-\infty}^{\infty} |V_{\eta\zeta}|^2 D_{osR}(x)D_{osL}(x)F_{F\zeta}(x)[1 - F_{F\eta}(x)]\, dx. \tag{6.66}$$

In this expression, the applied voltage V_{LR} appears through the Fermi–Dirac distributions containing the chemical potentials of two electrodes that are related by $\mu_R = \mu_L - q_e V_{LR}$. As we saw in Aside 6.1, it is possible to carry out the preceding integral through a linearization process and show that the current satisfies Ohm's law.

6.4 Sequential Tunneling

So far we have considered a simple device in which two metallic electrodes are connected to a thin barrier and where charges transfer from one electrode to the other through tunneling of electrons across this barrier. We call this type of charge transport fully coherent if the electron's transfer can be described by a single process, whose probability can be calculated using the Schrödinger equation. This is a reasonably accurate description when the average time electrons spend in the resonant state is much less than the scattering time. In other words, the lifetime of each energy eigenstate is much smaller than the scattering time.

In more complicated devices, electrons may undergo two sequential tunneling events such that an electron first tunnels into a central island, and then tunnels out of this island after losing memory of its phase. As a result of this memory loss, the two sequential tunneling events can be considered uncorrelated. Moreover, these two processes are not identical, as they face different conditions. The first tunneling process requires the availability of an empty state in the central island at the same energy level as in the left electrode (conservation of energy) and with the same lateral momentum (conservation of momentum).

The second tunneling process is less constrained because of the abundance of energy levels in the receiving electrode.

As we have already seen, each tunneling event can be described using first-order perturbation theory. However, at low temperatures, the sequential tunneling may be suppressed by a process known as the Coulomb blockade, and the conductance of the system may become exponentially small. In this scenario, where the first-order contribution vanishes or becomes insignificant, the higher-order contributions may need to be considered. An example of the second-order process is provided by an event where two electrons with different energies tunnel simultaneously (called inelastic co-tunneling), or one electron tunnels coherently twice (called elastic co-tunneling); the latter process is equivalent to simultaneous tunneling of two electrons of the same energy [8, 360]. Note that the terms elastic and inelastic are not related to the presence or absence of a specific scattering process. Rather, they refer to whether the energy is lost (inelastic) or conserved (elastic) during the tunneling process.

6.4.1 A Quantum Device Separated by Two Barriers

Figure 6.9 shows a schematic of a quantum device separated from the left and right electrodes through two potential barriers. The left electrode is at a higher potential than the right one. Any electron must first tunnel from the left electrode into the quantum device, spend sometime there, and then tunnel to the right electrode. Such sequential tunneling events can take place through elastic or inelastic co-tunneling processes, depending on whether the electron's energy is conserved or not. In both cases, change in free energy is $-q_e V$, if V is the potential difference between the electrodes. As there is no vacant state for electrons

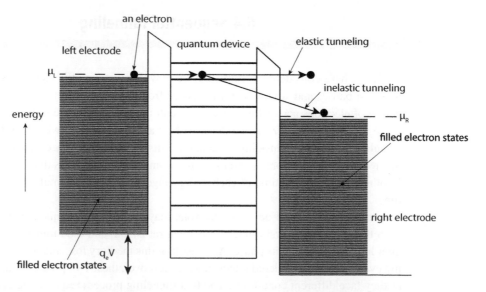

Fig. 6.9   Schematic of a quantum device, separated from electrodes through two barriers, showing differences between the elastic and inelastic co-tunneling processes.

to be received by the quantum device, the first-order tunneling is forbidden. However, electrons overcome this difficulty by tunneling via a virtual vacant state in the quantum device. Two simultaneous tunneling events via this virtual vacant state can transfer electrons from the left electrode to the right electrode, resulting in current flow.

Consider an electron at the left electrode overcoming the energy mismatch in the quantum device by violating energy conservation for a short time allowed by Heisenberg's uncertainty principle (see Aside 2.11). If a different electron from the quantum device tunnels to the right electrode during this short duration, one can view the two tunneling events as charge transfer from the left electrode to the right electrode. This process is called inelastic co-tunneling, because it produces an electron-hole excitation in the quantum device, which is eventually dissipated through carrier-carrier interactions. Inelastic co-tunneling in normal-metal tunnel junctions (as opposed to superconducting tunnel junctions) was first observed in 1990 by Geerligs *et al.* [361]. Matsuoka and Kimura observed the same phenomenon in 1995 using a silicon quantum dot [362].

In the case of elastic co-tunneling, the same electron tunnels into and out of the virtual state inside the quantum device. In this process, the phase of the electron's wave function is preserved, making elastic co-tunneling a coherent process. Elastic co-tunneling depends strongly on the internal structure of the quantum device. Usually, inelastic co-tunneling dominates in comparison to elastic co-tunneling, except at very small bias voltages and temperatures or when the density of energy states is very low in the quantum device [363]. Hanna *et al.* were the first to observe elastic co-tunneling in a silicon quantum dot [364].

The tunneling of each particle with charge q increases energy of the quantum devices by $E_C = q^2/(2C)$, where C is the capacitance of this device. As the electron's energy is conserved during tunneling, there is no source available for this energy! The need for energy conservation arises from the application of Noether's theorem to the time-translation invariance of the Schrödinger equation. However, we should recall Heisenberg's uncertainty principle requiring $\Delta t \Delta E \geq \hbar/2$. It implies that energy conservation can be violated as long as this violation lasts for a short time Δt set by the uncertainty principle (see Aside 2.11). Thus, an electron may enter and stay in the quantum device for a duration $\Delta t \approx \hbar/E_C$, where we use $\Delta E = E_C$. There exists a small but finite chance for co-tunneling to occur within this time window (i.e., the same electron or another electron should leave the quantum device and tunnel to the right electrode within a duration of Δt). The overall effect is that the system complies with the energy conservation principle, even though a co-tunneling event has taken place.

It is useful to estimate the tunneling rates for the two co-tunneling processes. Suppose Γ_L is the tunneling rate of an electron from the left electrode to the quantum device. If the tunneling rate to the right electrode is Γ_R, we can write the elastic co-tunneling rate using $\Delta t \approx \hbar/E_C$, as

$$\Gamma_{CT-el} = \Gamma_L \Gamma_R (\Delta t) \approx \Gamma_L \Gamma_R (\hbar/E_C). \tag{6.67}$$

The tunneling rates Γ_L and Γ_R can be estimated using the theory in Ref. [365]. For an electrostatic energy difference ΔE, they are given by

$$\Gamma_L = \frac{G_L \Delta E}{q_e^2 \left[1 - \exp(-\Delta E/k_B T)\right]}, \qquad \Gamma_R = \frac{G_R \Delta E}{q_e^2 \left[1 - \exp(-\Delta E/k_B T)\right]}, \tag{6.68}$$

where G_L and G_R are the conductance of two tunneling junctions and T is the absolute temperature. At a temperature near 0K, only transitions that decrease the electrostatic energy are possible. Using $\Delta E = E_C \gg k_B T$, we obtain the simple relations

$$\Gamma_L = \frac{G_L \Delta E}{q_e^2} \approx \frac{G_L E_C}{q_e^2} \approx \frac{G_L}{G_Q} \frac{2 E_C}{h}, \tag{6.69}$$

$$\Gamma_R = \frac{G_R \Delta E}{q_e^2} \approx \frac{G_R E_C}{q_e^2} \approx \frac{G_R}{G_Q} \frac{2 E_C}{h}, \tag{6.70}$$

where the conductance quantum is defined as $G_Q = 2q_e^2/h$.

To estimate the inelastic co-tunneling rate Γ_{CT-in}, we consider two different electrons, the first tunneling from the left electrode to the quantum device, and the second from the quantum device to the right electrode. This rate can be approximately written with the help of Eqs. (6.67), (6.69), and (6.70) as

$$\Gamma_{CT-in} \approx \sqrt{\Gamma_L \Gamma_R} \frac{\sqrt{G_L G_R}}{G_Q}. \tag{6.71}$$

This expression shows that the inelastic co-tunneling rate is low compared to the elastic one if both G_L and G_R are much smaller than G_Q. The same condition holds even in the presence of the Coulomb blockade. When both of them are comparable to G_Q, the inelastic co-tunneling rate becomes comparable to the single-charge tunneling through either junction. Under such conditions, the Coulomb blockade disappears.

The preceding estimate for the inelastic co-tunneling rate can be improved with the following argument [8]. Consider an event involving two different electrons to complete the transfer of a single charge from the left electrode to the right electrode. During such an event, an electron-hole pair is created in the quantum device. This means, as a result of each two-electron transfer, four excitations are created in the system: a hole in the left electrode, an electron and a hole in the quantum device, and an electron in the right electrode. We label these four events using the index $i = \{1, 2, 3, 4\}$ and denote the corresponding excitation energies by $\epsilon_i > 0$. If there are no energy restrictions imposed on these excitations, all ϵ_i are of the order of the charging energy E_C introduced earlier. Suppose the electrodes are kept at a potential difference V that is much smaller than the Coulomb-blockade threshold $(q_e V \ll E_C)$. As all excitations must receive their energy from this external biasing, we can conclude that the sum $\sum \epsilon_i$ equals $q_e V$. Given that each $\epsilon_i > 0$, it follows that $\epsilon_i < q_e V$ for $i = 1, 2, 3, 4$.

Consider the number of available electronic states in a certain energy range. As this number is proportional to the magnitude of the energy range, the number of states available for the excitations is reduced by a factor $\approx (q_e V/E_C)^3$ compared to those available in the range 0 to E_C. It is important to note that this factor is applicable only if the temperature is low enough that energy provided by the bias $q_e V$ is much higher than the thermal energy $k_B T$. Thus, when $k_B T \ll q_e V$, the rate of inelastic co-tunneling can be written as

$$\Gamma_{CT-inV} \approx \frac{G_L}{G_Q} \frac{G_R}{G_Q} \left(\frac{q_e V}{E_C} \right)^3 E_C. \tag{6.72}$$

When $k_B T \gg q_e V$, the excitations can appear only in the energy range of 0 to $k_B T$, and this rate is replaced with

$$\Gamma_{CT-inT} \approx \frac{G_L}{G_Q} \frac{G_R}{G_Q} \left(\frac{k_B T}{E_C} \right)^2 k_B T. \tag{6.73}$$

In the case of thermally activated co-tunneling, there is no preferred direction in which tunneling is dominant. Owing to the presence of electrical bias, the electron co-tunneling rates from left to right and from right to left can differ only by a small factor ($\approx q_e V / k_B T$) when $k_B T \gg q_e V$. Therefore, thermally activated co-tunneling current is estimated to be

$$I_{CT} \approx V \left(\frac{G_L}{G_Q} \frac{G_R}{G_Q} \right) \left(\frac{k_B T}{E_C} \right)^2. \tag{6.74}$$

It is clear from this discussion that thermally activated co-tunneling is an inelastic process, and the associated I–V curve not only is nonlinear but also depends on temperature.

Apart from the inelastic co-tunneling process involving two different electrons, there is a finite probability that the same electron that enters the quantum device from the left tunnels out to the right. Co-tunneling in this case is referred to as elastic because this single electron keeps its energy constant and does not create any excitations in the quantum device during its transit via a virtual state. Apart from a smaller probability of the occurrence of this event, there is no other difference in the process, and we can adopt the same methodology to estimate its tunneling rate that we used earlier for the inelastic co-tunneling rate.

So far, we have considered charging energy E_C as an energy uncertainty and related it to the tunneling rate Γ_L given in Eq. (6.69). However, in the case of elastic co-tunneling, we can improve our estimate of Γ_L by noting that a quantum device has discrete energy levels. If an electron transits through the device elastically, the energy uncertainty of such a transfer is given by $\Delta E \approx \delta E_{QD}$, where δE_{QD} is the average spacing of energy levels in the quantum device. With this choice, the tunneling rate from the left electrode to the quantum device can be written as

$$\Gamma_L \approx \frac{G_L}{G_Q} \frac{\delta_{QD}}{\hbar}. \tag{6.75}$$

The tunneling rate Γ_R from the quantum device to the right electrode is then obtained by just replacing G_L with G_R in this equation (see Eq. (6.70)). This reasoning shows that the probability of the same electron participating in both tunneling events is approximately given by $\frac{\delta_{QD}}{E_C}$. As no excitations are left in the quantum device and the electron retains its original energy at the conclusion of the event, we should remove the extra factor of $(q_e V / E_C)^2$, which was introduced in the inelastic case. It follows from Eq. (6.72) that the co-tunneling rate at a small bias voltage V is given by

$$\Gamma_{CT-el} \approx \frac{G_L}{G_Q} \frac{G_R}{G_Q} \left(\frac{q_e V}{E_C} \right) E_C \frac{\delta_{QD}}{E_C}. \tag{6.76}$$

Comparing it with Eq. (6.72), we conclude that the elastic co-tunneling dominates at low energies such that $\Delta E \leq \sqrt{\delta_{QD} E_C}$.

6.4.2 Scanning Tunneling Microscopy

As an important application of co-tunneling, we consider scanning tunneling spectroscopy. This technique provides a direct way of measuring spatial variations in the density of electronic states at a metallic surface within the energy window set by the bias voltage between the surface and a sharp metal tip. The measurement is carried out by scanning the metal tip over the conducting surface, while maintaining a small gap between the two. This gap (typically <1 nm wide) is chosen such that electrons can tunnel from the tip to the surface, resulting in a current flow. Binnig and Rohrer invented in 1983 the instrument known as the scanning tunneling microscope (STM) and used it to reconstruct an image of the atomic structure of a silicon surface [366]. For this work, they were awarded the 1986 Nobel Prize in Physics. A detailed review of the STM-based technique can be found in Ref. [367].

Figure 6.10(a) shows schematically the major parts of an STM. A scanning tip is mounted on a piezo tube whose mechanical deformation is controlled by an external voltage. As a result, the tip can be precisely positioned on the sample in both the lateral and vertical directions. The current through the vacuum gap is set to a specific value at the start of the measurement process. As the tip is raster-scanned over the surface, the tunneling current is kept constant by varying the tiny gap between the tip and the sample. Changes in the vertical position of the tip are used for creating an image of the surface. These images contain information about the geometry and the electronic structure of the surface. The spatial resolution of an STM is remarkable owing to the extremely high sensitivity of the tunneling current to the tip-sample distance. Recent advances in scanning tunneling microscopy are such that spatial features shorter than 0.1 nm can be resolved using a commercial STM equipment [368].

We can understand the operation of an STM by considering the tunneling current through the narrow gap between the metal tip and the surface at a fixed voltage V. If this voltage

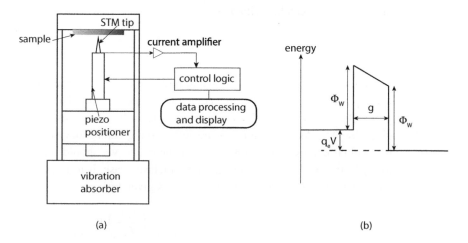

(a)   (b)

Fig. 6.10   (a) Schematic of an STM showing its major parts; (b) energy diagram of the tunneling gap between the sample and the STM tip.

is small compared to the work function Φ_W of the tip surface or the sample surface, the tunneling barrier has a shape similar to that shown in Figure 6.10(b). The work functions of the two surfaces set the barrier height, as seen there. In the WKB approximation, the wave function $\Psi_n(z)$ of an electron of energy E_n inside the potential barrier $\Phi_b(z)$ can be written as (see Section 6.2.1):

$$\Psi_n(z) = \Psi_n(0) \exp\left(-\frac{1}{\hbar}\int_{z_i}^{z} \sqrt{2m_e[\Phi_b(z) - E_n]}\,dz\right), \tag{6.77}$$

where z is the distance along the potential barrier ($z = 0$ at the start of the barrier). The integral can be evaluated for an arbitrary potential barrier using the saddle-point method (see Ref. [369]).

In scanning tunneling microscopy, a small bias voltage V_b is applied to initiate a tunneling current. Noting that the work function of a metal represents the additional energy required for an electron to leave the metal relative to its Fermi energy, the height of the potential barrier can be approximated by the average of the work functions of the sample and the tip as

$$\Phi_b(z) \approx \frac{1}{2}(W_{\text{tip}} + W_{\text{sample}}) \approx \Phi_W, \tag{6.78}$$

where Φ_W has a constant value. With this approximation, the potential barrier becomes rectangular in shape, and the integration can be done without needing the saddle-point method. If we also assume $E_n \ll \Phi_W$, the wave function takes the form

$$\Psi_n(z) \approx \Psi_n(0) \exp\left(-\frac{z}{\hbar}\sqrt{2m_e\Phi_W}\right). \tag{6.79}$$

The tunneling current $I_T(z)$ is proportional to the probability of electrons tunneling through the barrier. Summing over all possible energies, it can be written as

$$I_T(z) \propto -q_e \sum_{En}^{E_F} |\Psi_n(0)|^2 \exp\left(-\frac{2z}{\hbar}\sqrt{2m_e\Phi_W}\right), \tag{6.80}$$

where E_F is the Fermi level and the lower limit E_n depends on the applied bias voltage as $E_n = E_F - q_e V_b$. We can carry out the summation using the definition of the density of states:

$$D_{os}(E, 0) = \lim_{\epsilon \to 0} \frac{1}{\epsilon} \sum_{E_n = E - \epsilon}^{E} |\Psi_n(0)|^2 . \tag{6.81}$$

As $q_e V_b \ll E_F$, we approximate the sum using the Taylor expansion of the function $D_{os}(E_F - q_e V_b, 0)$ around the Fermi energy E_F and retain only the first two terms. The result is given by

$$\sum_{E_n}^{E_F} |\Psi_n(0)|^2 \approx -q_e V_b D_{os}(E_F, 0). \tag{6.82}$$

Substituting this expression back in Eq. (6.80) we obtain

$$I_T(g) \propto q_e^2 V_b D_{os}(E_F, 0) \exp\left(-\frac{2g}{\hbar}\sqrt{2m_e\Phi_W}\right). \tag{6.83}$$

This equation shows that the tunneling current is proportional to the product of the local density of states of the sample at the Fermi energy and the applied voltage. Note also that the tunneling current obeys Ohm's law (i.e., the current is proportional to the voltage). It is evident from this result that an STM image does not represent the position of surface atoms. Rather, it is a map of spatial variations in the density of electronic states at the surface, within the energy window set by the bias voltage. The exponential dependence of this current on the gap distance means that the tunneling current is dominated by the shortest path from a single atom at the end of the scanning probe. This is the reason why the spatial resolution of an STM can be better than 0.1 nm.

Depending on the polarity of the applied bias, electrons can flow from the metal tip to the sample surface, or vice versa. It is possible to exploit this feature to gain more information about the density of states of in the sample. For example, when the tip is negatively biased with respect to the conductive sample, electrons can only tunnel from occupied states of the tip into empty states of the sample, within the energy window $q_e V_b$. On the other hand, when the tip is positively biased, tunneling of electrons occurs from the sample surface. The measured tunneling current in this case originates from the occupied valence-band states in the sample. Therefore, depending on the bias direction, occupied or empty states of the surface can be probed [370].

6.5 Resonant Tunneling

The resonant tunneling differs from the generic tunneling process by the presence of quasi-bound states within the potential barrier that are classically forbidden for the tunneling particle. Tunneling is facilitated if energy of the tunneling particle matches approximately the energy associated with one of these quasi-bound states. As a result, the tunneling transmission coefficient peaks sharply and becomes close to unity when the electron's energy is close to the energy of a quasi-bound state. This resonance phenomenon is similar to the resonances seen in a Fabry–Perot resonator.

An excellent example of the resonant tunneling is provided by a device called the resonant tunneling diode (RTD) and shown schematically in Figure 6.11(a). The RTD Device consists of a quantum-dot island (made often with GaAs) that is separated from the left and right electrodes by two thin barriers (made of AlAs). Electrons inside the quantum dot can only belong to discrete energy levels that contribute to the resonant tunneling effect. A voltage difference between the two electrodes can be used to modulate the location of these resonances, as shown in part (b) of Figure 6.11. Owing to the deformation of the potential barrier, one of the discrete energy levels in the quantum dot may coincide with the energy of the incident electron, facilitating resonance tunneling.

In an RTD, one of the electrodes (called emitter) supplies electrons, and the quantum dot acts as a bandpass filter that probes the emitter through its discrete energy levels. This electrical engineering analogy explains why the tunneling current increases when a discrete energy level gets close to the electron's energy in the emitter and drops when the increase in applied voltage moves it beyond the electron's energy. This current drop acts

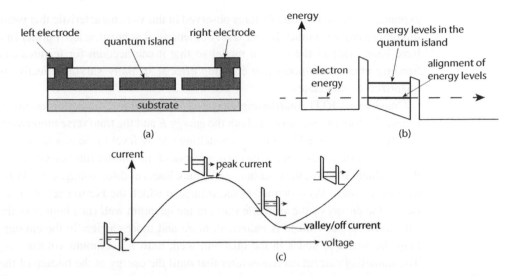

Fig. 6.11 (a) Schematic of a resonant tunneling diode; (b) energy diagram showing distortion of the potential energy curve when one of the states in the quantum island is resonant with the energy of the incoming electron; (c) the V–I curve of the device with the alignment of energy levels at low/high current points.

like a negative differential resistance, a property that is of immense value for high-speed electronics applications. The speed of such a device is limited only by its RC time constant and other parasitics. As a result, a RTD can operate at terahertz frequencies and act as a high-speed switch in nanoscale circuits.

The first experimental demonstration of resonant tunneling was carried out in 1974 using a double-barrier semiconductor heterostructure [371]. Since then, several types of RTDs have been fabricated using different material technologies, including the III-V and II-VI semiconductors. There are many variants of the original structure, such as the use of heavily doped p–n junctions in Esaki diodes or the use of quantum wells or (quantum wires). The RTD structure based on the use of silicon for the quantum island and SiGe for the barriers is suitable for integration with modern CMOS technology. The RTD is different from other switching technologies because its operation cannot be explained using classical transport models and requires the use of quantum mechanics. Because of this, the RTDs are often used as a conceptual playground for exploring the physics of quantum devices. Moreover, an extension of the double-barrier resonant tunneling problem to the more general case of a periodic rectangular potential offers a simple model, known as the Kronig–Penney model, for the behavior of electrons in a crystal lattice [372]. As is well known, the allowed energies of electrons experiencing a periodic potential form continuous bands separated by forbidden energy gaps.

Tsu and Esaki developed in 1973 a model capable of explaining the operation of a RTD [373]. Their formulation assumed equal carrier masses in both the emitter and quantum-well regions and derived an expression for the tunneling current through the device. This model was extended in 1998 by Schulman to include the effects of different in-plane masses in the emitter and quantum-well regions [374]. This enhancement provided an explanation of the intricate features observed in the V–I characteristic that were previously thought to require a much higher level of theoretical sophistication. Further improvements have been made to the original model so that it can account for features such as band-bending effects, self-consistent charging effects, spatially varying effective masses, and quantized-emitter states.

Consider an RTD transporting electrons via the elastic tunneling events as shown in Figure 6.12(a). In this situation, both the energy E and the transverse momentum $k_\parallel$ of each electron are conserved because of a matching energy level in the quantum well. Part (b) of Figure 6.12 shows the energy-dispersion curve of the emitter (dashed curve) together with the available states in the quantum well (thick lines) at three voltages. In elastic tunneling, the onset voltage V_O is defined as the voltage at which the Fermi energy E_F in the emitter equals the energy of the available states in the quantum well (i.e., bottom of the resonance subband). As the voltage is increased, more and more carriers in the emitter are able to map the available states in the quantum well, until a maximum voltage V_P is reached. The tunneling current decreases after that until the energy at the bottom of the resonance

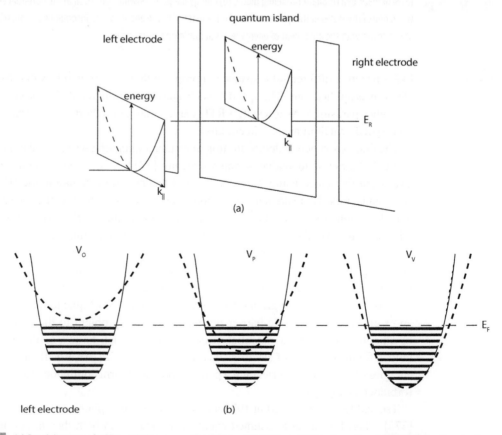

Fig. 6.12   (a) Band diagram of a RTD with the energy-dispersion curves in the emitter and quantum island showing E versus $k_\parallel$. (b) Relative position of the parabolic energy-dispersion curve (dashed) in the emitter region at various voltages: V_0 (onset), V_P (peak), and V_V (valley). Solid lines show the available states in the quantum well.

subband equals the band-edge energy of the emitter. This voltage is referred to as the valley voltage V_V.

In the following discussion, m_E and m_{QW} are effective masses of electrons in the emitter and the quantum well, respectively. The effective masses determine the intersection points of the two relevant energy dispersion curves, and hence affect directly the tunneling current. To simplify the analysis, we ignore many second-order effects such as band bending. Assuming that exactly half of the voltage drop occurs at the center of the quantum well, the three voltages (V_O, V_P, and V_V) are given by [373, 374, 375]

$$V_O = -\frac{2}{q_e}(E_R - E_F), \quad V_P = -\frac{2}{q_e\alpha}(\alpha E_R - E_F), \quad V_V = -\frac{2}{q_e}E_R, \qquad (6.84)$$

where $\alpha = m_{QW}/(m_{QW} - m_E)$ and E_R is the energy of the resonance with respect to the emitter's conduction-band edge at zero bias. If $T(E, V)$ is the tunneling transmission coefficient at energy E and voltage V, the current density $J(V)$ can be written as

$$J(V) = -\frac{2q_e}{\hbar}\frac{1}{(2\pi)^3}\int \frac{\partial E}{\partial k_z}T(E, V)\left[f_F(E) - f_F(E - q_eV)\right]d^3k, \qquad (6.85)$$

where $f_F(E)$ is the Fermi–Dirac distribution and the integration is over the entire k-space in the emitter.

As $m_E < m_{QW}$ is generally true, we introduce a new energy variable U as

$$U = E + \frac{\hbar^2 k_\parallel^2}{2m_E} - \frac{\hbar^2 k_\parallel^2}{2m_{QW}}. \qquad (6.86)$$

With this definition, we can replace the variable E with $U + \hbar^2 k_\parallel^2/(2m_{QW})$ and assume that $k_\parallel$ varies between 0 and $\frac{1}{\hbar}\sqrt{2\alpha U m_E}$. The integration over $k_\parallel$ then leads to

$$J(V) = -\frac{q_e m_{QW} k_B T}{2\pi^2\hbar^3}\left\{\int_0^\infty T(U, V)\ln\left[\frac{1 + \exp\left(E_F - U\right)/k_B T}{1 + \exp\left(E_F - U + q_eV\right)/k_B T}\right]dU \right.$$
$$\left. + \int_0^\infty T(U, V)\ln\left[\frac{1 + \exp\left(E_F - \alpha U + q_eV\right)/k_B T}{1 + \exp\left(E_F - \alpha U\right)/k_B T}\right]dU\right\}, \qquad (6.87)$$

where the integration variable k_z has been replaced with U. As we have seen before, several techniques can be used to calculate $T(U, V)$ and use it in Eq. (6.87) to get the resonant tunneling current numerically. Useful analytical results can also be obtained with appropriate approximations [374].

Quantum Noise

The noise is the signal.

Rolf Landauer

7.1 Sources of Noise and Basic Concepts

Even though "noise" seems something that we could live without, engineers and physicists have learned not only to deal with it but also to exploit it. Quantum mechanics enables us to see the underlying physical structure of noise so that we do not have to view it as random glitches disrupting the operation of our devices. Instead, as we see later, the origin of quantum noise can be traced back to physical processes taking place inside a quantum device.

Classical noise arises from fluctuations in the number of particles and in their movement within a given volume. Examples include thermal noise and shot noise. Quantum noise arises from uncertainty concerning the position and momentum of quantum particles. It can also occur when particles are photons. An example is provided by the unavoidable noise of optical amplifiers. There is a fundamental difference between the classical and quantum noises. The former is attributed to the "lack of knowledge" of the system and is modeled as a random stochastic process. This is because the deterministic classical view requires that if all the information about a system is known at some time, its trajectory in the relevant phase space can be predicted for all future times without any uncertainty. Contrary to this, quantum noise is fundamental in the sense that even a complete knowledge of the system would leave us with some uncertainty. What is more intriguing is that quantum noise sets a fundamental limit on the operation of quantum devices. We can see this clearly by investigating why it is not possible to build a noiseless optical amplifier.

7.1.1 Noise Introduced by Optical Amplifiers

Following the invention of the laser and optical amplifiers, two groups considered noise in linear optical amplifiers. Haus and Mullen [376] analyzed amplifier noise in the two field quadratures, without restricting assumptions on the noise statistics. They found that the minimum uncertainty in the output of a high-gain amplifier corresponds to $\frac{1}{2}\hbar\omega$ for photons of energy $\hbar\omega$. A similar conclusion was reached by Heffner [377] using a different approach. Heffner's method clearly shows that an ideal, noise-free, optical amplifier violates Heisenberg's uncertainty principle. It also shows that quantum noise is fundamentally different from our intuitive understanding of classical noise.

Suppose it is possible to construct an ideal optical amplifier with no noise at its output when a monochromatic noisy signal of frequency ω is incident on it. Let us assume that this amplifier receives a stream of input photons whose number is given by $n_I + \Delta n_I$, where n_I is the average number and Δn_I is the uncertainty in this count. If the gain G of this ideal amplifier is high, the resulting output photon number will be $n_O + \delta n_O = G(n_I + \Delta n_I)$. This result indicates that the input signal is linearly amplified with gain G, without any added noise. As this ideal optical amplifier works as a photon-number multiplier, it does not change phase uncertainty of the incoming signal given by $\Delta\phi_I$. However, quantum mechanics demands that the uncertainty relation, $\Delta E \Delta t \geq \hbar/2$, be satisfied at both the input and output ends of the amplifier.

We can calculate this uncertainty at the input end using $\Delta E = \hbar\omega\Delta n_I$ and $\Delta t = \Delta\phi_I/\omega$. The result is

$$\Delta n_I \Delta\phi_I \geq \frac{1}{2}. \tag{7.1}$$

This relation is known as the *number-phase uncertainty relation*. It implies that the photon number and the phase of an optical field are canonically conjugate variables (like position and momentum). As a result, it is not possible to know exactly both the phase and the number of photons simultaneously. It is easy to show that our ideal noiseless optical amplifier violates the preceding relation. Because the phase remains unchanged, Eq. (7.1) requires $\Delta n_O \Delta\phi_I \geq \frac{1}{2}$ at the output end of the amplifier. Given that $\Delta n_O = G\Delta n_I$, we get the relation $\Delta n_I \Delta\phi_I \geq \frac{1}{2G}$, which clearly violates the lower bound in Eq. (7.1) for any amplifier with $G > 1$. This contradiction implies that an optical amplifier must add noise to its amplified signal.

We can calculate the added noise as follows. Assume that the optical amplifier is followed by an ideal detector capable of counting photons. Even an ideal detector should obey quantum rules during its detection process. We assume that the detector has photon number and phase uncertainties given by Δn_D and $\Delta\phi_D$, respectively. Noting that the detector and amplifier operate independently of each other, we can add their variances as

$$\Delta n^2 = \Delta n_O^2 + \Delta n_D^2, \tag{7.2}$$

$$\Delta\phi^2 = \Delta\phi_O^2 + \Delta\phi_D^2, \tag{7.3}$$

where Δn^2 and $\Delta\phi^2$ are the photon number and phase uncertainties of the overall process. They can be directly related to the input signal uncertainties using $\Delta n^2 = G^2 \Delta n_I^2$ and $\Delta\phi^2 = \Delta\phi_I^2$. Using these relations, we can write the number-phase uncertainty relation at the input end in the form

$$\Delta n_I^2 \Delta\phi_I^2 = \frac{1}{G^2}\left[\Delta n_O^2 \Delta\phi_O^2 + \Delta n_D^2 \Delta\phi_D^2 + \Delta n_O^2 \Delta\phi_O^2\left(\frac{\Delta\phi_D^2}{\Delta\phi_O^2} + \frac{\Delta n_D^2}{\Delta n_O^2}\right)\right]. \tag{7.4}$$

The number-phase uncertainty for an ideal detector takes its minimum value, $\Delta n_D \Delta \phi_D = \frac{1}{2}$. We also impose the additional requirement that the detector's uncertainty ratio $\frac{\Delta \phi_D}{\Delta n_D}$ is such that it minimizes the output number-phase uncertainty. The resulting condition is found to be

$$\frac{\Delta n_O}{\Delta \phi_O} = \frac{\Delta n_D}{\Delta \phi_D}. \tag{7.5}$$

With these choices, Eq. (7.4) takes the form

$$\Delta n_I^2 \Delta \phi_I^2 = \frac{1}{G^2} \left[\Delta n_O^2 \Delta \phi_O^2 + \frac{1}{4} + \Delta n_O \Delta \phi_O \right]. \tag{7.6}$$

Finally, we assume that the input signal is the best-quality signal allowed by quantum mechanics. That is, the number-phase uncertainty relation for this signal is minimum with $\Delta n_I \Delta \phi_I = \frac{1}{2}$. With this choice, the product $\Delta n_O \Delta \phi_O$ satisfies the quadratic equation

$$(\Delta n_O \Delta \phi_O)^2 + \Delta n_O \Delta \phi_O - (G^2 - 1)/4 = 0. \tag{7.7}$$

Solving this quadratic equation and picking the positive root, we find that the output uncertainty product is given by

$$\Delta n_O \Delta \phi_O = \frac{1}{2}(G - 1). \tag{7.8}$$

It is clear that this product exceeds the input value of $1/2$ for $G > 2$.

To calculate the noise power P_n, we assume that the noise added by an amplifier is additive white noise with Gaussian statistics. If the signal is large compared to noise, the statistics of the output signal will also be nearly Gaussian. In this situation, both its power and phase fluctuate with variances given by $\Delta P_O^2 = 2 P_O P_n$ and $\Delta \phi_O^2 = P_n/(2 P_O)$. Here P_n is the noise power and P_O is the average output power. It can be related to the output photon number as $P_O = \hbar \omega n_O B_W$, where B_W is the bandwidth of the amplifier. Using these relations in Eq. (7.8), we obtain the following expression for the noise power:

$$P_n = \frac{1}{2} \hbar \omega B_W (G - 1). \tag{7.9}$$

Further details about this fundamental result can be found in the original references [376, 377]. The enhanced noise at the amplifier's output degrades the signal-to-noise ratio (SNR) of the input signal. The important concepts of SNR and the noise figure of an amplifier are discussed in Aside 7.1.

The preceding result shows that the noise power of an optical amplifier depends on its gain G and bandwidth B_W. The factor $\frac{1}{2} \hbar \omega$ can be viewed as the *zero-point energy* of vacuum. Quantum mechanics shows us that the ground-state energy of a harmonic oscillator oscillating at the frequency ω is exactly equal to $\frac{1}{2} \hbar \omega$. The concept of zero-point energy or zero-point fluctuations is nonclassical and has its origin in the quantization of an electromagnetic field. The minimum detectable noise power is often referred to as *quantum noise* in view of its quantum origin (see discussion in Ref. [378]). At short wavelengths, this quantum noise can be greater than thermal noise by several orders of magnitude.

Aside 7.1   Signal-to-Noise Ratio and Noise Figure

The phrase *signal-to-noise ratio* (SNR) has its origin in telecommunication engineering, but the concept is valid in many other areas including optics, biology, and financial modeling. The SNR compares the strength of a desired signal to the level of background noise, and it can be considered a measure of the signal's quality. In optics, the SNR is determined through the signal and noise levels in the current of the photodetector used to measure the optical power. If the measured current is $I_s(t)$ and the variance of current noise is $\langle \Delta^2 I_n(t) \rangle$, the SNR is defined as

$$\text{SNR} = \frac{\langle I_s(t) \rangle^2}{\langle \Delta I_n^2 \rangle}, \tag{7.10}$$

where the noise variance $\langle \Delta I_n^2 \rangle$ is found by integrating the spectral density of noise over the bandwidth of interest. It is important to note that the SNR is a ratio of electrical power levels rather than a ratio of optical power levels.

Just as the SNR serves as a figure of merit to characterize the quality of a signal, the quality of optical components including amplifiers is characterized using a measure called the *noise figure* (NF). The reason is that an active component such an optical amplifier may use an excited state to transfer energy to the amplified signals via stimulated emission or analogous mechanisms. However, spontaneous emission cannot be avoided, and the signal gets contaminated by noise as a result. The noise figure enables us to evaluate the impact of an active component on the SNR of a signal passing through that component. It does this by comparing the input and output values of the SNR.

Using the nomenclature of the International Electrotechnical Commission (IEC 61291-1:2018), the noise factor *NF* is defined as

$$\text{NF}(f_{op}, f) = \frac{\text{SNR}_{\text{in}}}{\text{SNR}_{\text{out}}}, \tag{7.11}$$

where SNR_{in} is the SNR at the amplifier's input end and SNR_{out} is the SNR at its output end. In general, noise factor is a function of both the optical frequency f_{op} and the baseband frequency f [379]. The noise figure (F) is the noise factor expressed in decibel (dB) units:

$$\text{F (dB)} = 10 \log_{10}[\text{NF}(f_{op}, f)]. \tag{7.12}$$

For a telecommunication link consisting of lossy fiber spans with optical amplifiers at periodic intervals, the noise figure for the entire chain can be computed from the noise figures of individual amplifiers. If the amplifier in the nth section provides the gain G_n to exactly compensate its span loss L_n and SNR_n denotes the SNR after the nth amplifier, the total noise factor is given by

$$\mathrm{NF}_{sys} = \frac{\mathrm{SNR}_0}{\mathrm{SNR}_N} = \left(\frac{SNR_0}{SNR_1}\right)\left(\frac{SNR_1}{SNR_2}\right)\cdots\left(\frac{SNR_{N-1}}{SNR_N}\right), \tag{7.13}$$

where SNR_0 is the SNR at the input end of the amplifier chain. The total noise figure of the amplifier chain is then found to satisfy the relation

$$\mathrm{F}_{sys} = \frac{\mathrm{F}_1}{L_1} + \frac{\mathrm{F}_2}{L_1 G_1 L_2} + \cdots + \frac{\mathrm{F}_N}{L_1 G_1 L_2 G_2 \cdots L_N}. \tag{7.14}$$

Consider the special case with $L_n = 1$ for all $n \in N$ representing a cascaded chain of N amplifiers with no span losses. Its noise figure is found to be

$$\mathrm{F}_{sys} = \mathrm{F}_1 + \frac{\mathrm{F}_2}{G_1} + \cdots + \frac{\mathrm{F}_N}{G_1 G_2 \cdots G_{N-1}}. \tag{7.15}$$

This relation shows that the noise figure of a multistage amplifier is dominated by the noise of the first stage.

7.1.2 Classical Spectral Density

The correspondence principle, formulated in 1923 by Niels Bohr, is a distillation of his thoughts that led him to the development of atomic theory, an early form of quantum mechanics [380]. The conceptual foundation of this principle relies on the idea that quantum mechanics contains classical mechanics as a limiting case. However, the concept is generic in the sense that any new theory of a physical phenomenon should encompass the associated old theories. Examples include the special theory of relativity, which reduces to the Newtonian mechanics at speeds low compared with the speed of light, and statistical mechanics, which reproduces thermodynamics when the number of particles is large.

The concept of classical noise is well developed and widely used. Thus, a theory of quantum noise should encompass all known results for classical noise. As we shall see, the measures of quantum noise are indeed defined such that they reduce to the classical measures, thus complying with the correspondence principle. What is intriguing is that, even though we extend the classical concepts to the quantum regime, the results and interpretations of quantum noise processes go beyond the conventional knowledge and open up new possibilities unique to quantum mechanics.

It is worthwhile to review first the most fundamental properties of classical noise that we eventually extend to the quantum domain. Noise gets introduced to a classical device via its interaction with its environment. This is the reason why a model for the environment is first constructed for any discussion of noise of a classical device. Although it is not easy to find exact models for the device-environment interaction, is possible to model noise in real physical systems by closely studying the observed properties of a system and making sure that the model agrees with them.

Suppose $\mathcal{N}(t)$ describes in such a model the noise process experienced by a classical device. A fundamental assumption in all such models is that the underlying noise process obeys the Gaussian statistics. This assumption is not severely restrictive because of the *central limit theorem*, which states that the sum of many independent random variables is itself a random variable whose distribution converges to a Gaussian distribution as the number of independent variables increases. Therefore, physical quantities that are expected to be the sum of many independent processes (such as thermal noise discussed later) often have distributions that are nearly Gaussian. It is sufficient to know the first- and second-order moments to fully describe a Gaussian random process. The first moment, the mean or the average, can always be reduced to zero (i.e., we can assume $\langle \mathcal{N}(t) \rangle = 0$). Thus, the second-order moment can fully characterize a Gaussian random process. The adopted measure is the autocorrelation function defined as

$$\Gamma_{\mathcal{N}\mathcal{N}}(t, \tau) = \langle \mathcal{N}(t)\mathcal{N}(t + \tau) \rangle. \tag{7.16}$$

As this definition employs two different times, the time difference τ can be considered a measure of the memory of the underlying noise process. The maximum duration over which a signal retains its memory is called the *correlation time*. If we denote it by τ_c, we have the condition $\Gamma_{\mathcal{N}\mathcal{N}}(t, \tau_c) = 0$. A signal is called white noise when τ_c is close to zero.

Noise in the real world has a small but finite correlation time [381]. When τ_c is much smaller than other characteristic time scales of a noisy device, it is common to employ the white-noise assumption. Such an assumption was adopted earlier for the Langevin noise. We also take the noise process to be stationary (i.e., its statistical features do not change with a shift of time origin). Mathematically, the statistical behavior remains unchanged for $\mathcal{N}(t)$ and $\mathcal{N}(t + t_0)$ for any value of t_0. A consequence of this assumption is that the autocorrelation function $\Gamma_{\mathcal{N}\mathcal{N}}(t, \tau)$ becomes the function of a single variable τ.

It is difficult to compare two time-domain traces of a stationary random process because they appear quite similar. The situation is different if we compare their spectra in the frequency domain. For this reason, it is useful to analyze noise in the frequency domain. The Wiener–Khinchin theorem states that the Fourier transform of the autocorrelation function,

$$S_{\mathcal{N}\mathcal{N}}(\omega) = \mathcal{F}\{\Gamma_{\mathcal{N}\mathcal{N}}(\tau)\} = \int_{-\infty}^{\infty} \Gamma_{\mathcal{N}\mathcal{N}}(\tau)e^{i\omega\tau} \, d\tau, \tag{7.17}$$

provides the frequency-domain representation of a noise process. This quantity is known as the spectral density. As the autocorrelation function is real valued for the noise process $\mathcal{N}(t)$, its spectral density is symmetric around $\omega = 0$ and satisfies the relation $S_{\mathcal{N}\mathcal{N}}(-\omega) = S_{\mathcal{N}\mathcal{N}}(\omega)$. This symmetry is used in electrical engineering where only positive frequencies are considered. It is common to refer to $S_{\mathcal{N}\mathcal{N}}(\omega)$ as the two-sided spectral density. When only positive frequencies are included, the resulting spectral density is known as the single-sided spectral density.

7.1.3 Thermal Noise

An example of a classical noise process is provided by thermal noise, also known as the Johnson–Nyquist noise because it was first detected in 1928 by Johnson [180] and explained by Nyquist [181]. The origin of thermal noise lies in thermal agitation of electrons inside an electrical conductor. As the thermal agitation increases with temperature, thermal noise power also increases with temperature. In contrast, thermal noise power does not depend on the applied voltage or electrical parameters of the device such as its resistance.

It is instructive to derive an expression for the spectral density of thermal noise using a microscopic model [382, 383]. Even though such a derivation can be done for more exotic structures such as a p–n junction [384], we focus on a resistor of length L with a rectangular cross section of area A. Suppose the density of electrons inside the resistor is N_e, and these electrons have a collision time τ_e (see Section 1.5 for details on these microscopic parameters). Using conduction theory in Ref. [382], we can find the conductance σ and resistance $R = L/(A\sigma)$ in terms of the microscopic parameters as

$$\sigma = \frac{N_e q_e^2 \tau_e}{m_e}, \qquad R = \frac{L m_e}{A N_e q_e^2 \tau_e}. \tag{7.18}$$

The average drift speed of electrons inside the resistor can be written as

$$\langle u \rangle = \frac{1}{N_e A L} \sum_j u_j, \tag{7.19}$$

where $N_e A L$ is the total number of electrons within the resistor and u_j is the speed of the jth electron. The current flowing along the resistor can be written as $I = q_e N_e A \langle u \rangle$. Invoking Ohm's law, the voltage V across the resistor is $V = RI = R q_e N_e A \langle u \rangle$. Substituting the value of $\langle u \rangle$ from the preceding equation, we get

$$V = \sum_j V_j, \qquad V_j = \frac{R q_e}{L} u_j, \tag{7.20}$$

where V_j is the voltage induced by the jth electron. It is a random quantity that changes as this electron suffers collisions inside the resistor. Using the definition of the collision time τ_e, we can write its correlation function in the form

$$\Gamma_{V_j V_j}(\tau) = \langle V_j(t) V_j(t + \tau) \rangle = \langle V_j^2 \rangle \exp\left(-|\tau|/\tau_e\right). \tag{7.21}$$

The two-sided spectral density of V_j is found using this form of $\Gamma_{V_j V_j}(\tau)$ in Eq. (7.17). The integral can be easily done to obtain the result [385]

$$S_{V_j V_j}(\omega) = \frac{2 \langle V_j^2 \rangle \tau_e}{1 + \omega^2 \tau_e^2}. \tag{7.22}$$

At this point we note the τ_e is relatively small (< 1 ps) in a typical resistor. As a result $\omega^2 \tau_e^2 \ll 1$ at frequencies as high as 10 GHz. Neglecting it and using $V_j = (R q_e/L) u_j$ we obtain

$$S_{V_j V_j}(\omega) \approx 2(R q_e/L)^2 \langle u_j^2 \rangle \tau_e. \tag{7.23}$$

We now invoke the law of equipartition of energy, which states that each electron contributes $\frac{1}{2} k_B T$ to the average energy [386]. Using $\frac{1}{2} m \langle u_j^2 \rangle = \frac{1}{2} k_B T$, we obtain $\langle u_j^2 \rangle = \frac{k_B T}{m_e}$. Substituting this value in the preceding expression, we finally get

$$S_{V_j V_j}(\omega) = \frac{2 R^2 q_e^2 \tau_e}{L^2} \left(\frac{k_B T}{m_e} \right). \tag{7.24}$$

Our objective is to obtain the power spectral density of $V(t)$, where $V(t) = \sum_j V_j(t)$. To calculate this quantity, we invoke Campbell's theorem; a detailed proof of this theorem can be found in Ref. [159]. Campbell's theorem states that $S_{VV} = N_p S_{V_j V_j}$, where N_p is the total number of electrons contributing to the sum in $V(t)$. In our case $N_p = N_e AL$, resulting in

$$S_{VV}(\omega) = 2k_B TR^2 \left(\frac{N_e A q_e^2 \tau_e}{L m_e} \right) = 2k_B TR. \qquad (7.25)$$

The spectral density of thermal noise depends only on the resistance R, in addition to temperature. It is independent of frequency (white noise) up to frequencies as high as 10 GHz because of a short collision time of electrons inside the resistor. The single-sided power spectral density of thermal noise is two times that of the preceding result:

$$S_{VV}(\omega) = 4k_B TR. \qquad (7.26)$$

This result is a special case of the general connection existing between fluctuations and dissipations in any physical system through the fluctuation-dissipation theorem (see Section 3.3). If we know the bandwidth B_W over which thermal noise contributes, we can calculate the noise power associated with voltage fluctuations using $P_n = S_{VV} B_W / R = (4k_B T) B_W$. This power is independent of any electrical parameters of the resistor.

7.1.4 Shot Noise

As a second example of noise, we consider the *shot noise*, arising from current fluctuations that occur because of the discrete nature of electrical charges. Shot noise can be readily observed in devices such as tunnel junctions, Schottky diodes, and p–n junctions. The discovery of shot noise can be traced back to Schottky, who observed that vacuum tubes produced two types of noise, described by him as the Warmeeffekt and the Schroteffekt [387]. The first one is the thermal noise. The second is the shot noise observed as current fluctuations. The power of this noise is proportional to the average current, instead of the square of the current as would be expected from Ohm's law. This feature is related to the random arrival times of different electrons. Even though the current is constant on average inside a vacuum tube, electrons arrive at the anode at random times, resulting in current fluctuations (shot noise). In recent years, experiments on nanoscale conductors have provided a better understanding of shot noise [388].

The spectral density of shot noise can be calculated using a simplified model where a stream of electrons, each having the charge q_e, arrives at a detector at random time intervals. Let N be the number of electrons received at the detector in a fixed time interval T. The resulting current will vary with time and can be written as

$$I(t) = \sum_{n=1}^{N} q_e \delta(t - t_n), \qquad (7.27)$$

where t_n is the arrival time of the nth electron. In the limit of large T, this current would approach a constant average value given by

$$\langle I \rangle = \lim_{T \to \infty} \frac{q_e N}{T}. \qquad (7.28)$$

Even though the current $I(t)$ fluctuates in time because t_n is a random variable, we expect this random process to be stationary in the sense that the average current $\langle I \rangle$ will be independent of the starting time t_1.

The autocorrelation function, $\Gamma_{II}(\tau) = \langle I(t)I(t+\tau)\rangle$, depends only on the time difference τ and can be calculated using

$$\Gamma_{II}(\tau) = q_e^2 \sum_{i=1}^{N}\sum_{j=1}^{N} \langle \delta(t+\tau-t_i)\delta(t-t_j)\rangle . \tag{7.29}$$

Assuming that this stationary random process is ergodic, we can replace the ensemble average with time average taken in the limit $T \to \infty$:

$$\Gamma_{II}(\tau) = \lim_{T\to\infty} \frac{q_e^2}{T}\sum_{i=1}^{N}\sum_{j=1}^{N}\int_{-T/2}^{T/2}\delta(t+\tau-t_i)\delta(t-t_j)\,dt$$

$$= \lim_{T\to\infty}\frac{q_e^2}{T}\sum_{i=1}^{N}\sum_{j=1}^{N}\delta(t_j+\tau-t_i). \tag{7.30}$$

Application of the Wiener–Khinchin theorem to this autocorrelation function provides us with the spectral density through the Fourier transform:

$$S_{II}(\omega) = \int_{-\infty}^{\infty}\Gamma_{II}(\tau)e^{i\omega\tau}\,d\tau. \tag{7.31}$$

By substituting $\Gamma_{II}(\tau)$, the spectral density can be written as the sum of two parts

$$S_{II}(\omega) = \lim_{T\to\infty}\frac{q_e^2}{T}\left[\sum_{i=j}^{N}\int_{-\infty}^{\infty}\delta(\tau)e^{i\omega\tau}\,d\tau + \sum_{i=1}^{N}\sum_{j\neq i}^{N}\int_{-\infty}^{\infty}\delta(t_j+\tau-t_i)e^{i\omega\tau}\,d\tau\right]$$

$$= \lim_{T\to\infty}\frac{q_e^2}{T}\left(N + \sum_{i=1}^{N}\sum_{j\neq i}^{N}\exp[i\omega(t_i-t_j)]\right), \tag{7.32}$$

where both integrals over τ could be done easily. The second term is a sum of a large number of random complex numbers that vanishes as $T \to \infty$. The first term is related to the average current in the same limit. As a result, the two-sided spectral density of shot noise reduces to the simple expression:

$$S_{II}(\omega) = q_e \langle I \rangle . \tag{7.33}$$

The single-sided spectral density is obtained by multiplying it with a factor of two.

7.1.5 Brownian Motion

The erratic movement of microscopic particles immersed in a fluid is called Brownian motion, after Robert Brown who first studied such fluctuations in 1827. Brownian motion provides an excellent example of classical noise, and its theory provides useful mathematical tools used to study the dynamics of nonequilibrium systems. The stochastic differential equation for describing the Brownian motion is known as the Langevin equation. It contains both the frictional force and a random force that fluctuates with time. The fluctuation-dissipation theorem, discussed in Section 3.3, relates these forces to each other and provides a firm theoretical footing for describing similar phenomena such as the motion of ions in solutions or reorientation of dipolar molecules. Also, the theory has been extended to situations where the Brownian particle is not a real particle at all but represents some collective property of a macroscopic system, which could be a quantum device coupled to a large reservoir in equilibrium.

To simplify the following analysis, we consider the motion of a spherically symmetric particle of mass m_p inside a large isotropic reservoir of fluid. In this situation, it suffices to consider one-dimensional motion of the particle in a given direction. The velocity $V_p(t)$ of the Brownian particle is affected by two forces, a friction force and a random force. The friction force can be described by the Stoke's law and has the form

$$F_f(t) = -\gamma_d V_p(t), \tag{7.34}$$

where γ_d is a constant. The second force results from collisions of the Brownian particle with particles of the fluid surrounding it. Owing to the random nature of such collisions, this force is modeled as a stochastic process $F_s(t)$. The most commonly used assumptions about this force are [389]:

1. $F_s(t)$ is independent of the velocity of the Brownian particle and has a zero mean value such that $\langle F_s(t) \rangle = 0$, where the average is taken over an ensemble containing the Brownian particle and the particles of the fluid. A consequence of the zero mean value is that the stochastic force has no effect on the average motion of the Brownian particle.
2. The stochastic process $F_s(t)$ obeys the Gaussian statistics and is thus fully characterized by its correlation function. In addition, it is a Markovian process with no past memory such that its correlation function has the form

$$\langle F_s(t)F_s(t') \rangle = \mathcal{G}\delta(t - t'). \tag{7.35}$$

The parameter $\mathcal{G}$ can be calculated using the fluctuation-dissipation theorem. For a Gaussian process, the second-order correlation function can be used to find all higher-order correlation functions of $F_s(t)$. In particular, all odd-order correlation functions vanish for a Gaussian random process. The preceding form of the correlation function implies the causality relation [390]

$$\langle V_p(t)F_s(t') \rangle = 0 \qquad \text{for} \quad t' > t, \tag{7.36}$$

(i.e., future values of the stochastic force do not influence particle's dynamics).
3. Any ensemble average over the stationary random process can be replaced with the corresponding time average defined as

$$\langle f[F_s(t)] \rangle = \lim_{T \to \infty} \frac{1}{T} \int_0^T f[F_s(\tau)] d\tau, \tag{7.37}$$

where f denotes an arbitrary function of $F_s(t)$. The two averages are related to each other through the concept of ergodicity. If the time interval T is long enough, the phase-space trajectory describing evolution of the system comes arbitrarily close to any specified point in the phase space accessible to the system. As a result, the phase-space average (or ensemble average) can be replaced with the time average of the same quantity.

With the specification of all forces, the motion of the Brownian particle can be described using Newton's second law:

$$m_p \frac{dV_p}{dt} = F_f(t) + F_s(t) = -\gamma_d V_p(t) + F_s(t). \tag{7.38}$$

This linear equation can be easily integrated to obtain the solution

$$V_p(t) = V_0 \exp(-\gamma_p t) + \frac{1}{m_p} \int_0^t \exp\left[-\gamma_p(t - \tau)\right] F_s(\tau) \, d\tau, \tag{7.39}$$

where $\gamma_p = \gamma_d/m_p$ and V_0 is the initial velocity of the particle at $t = 0$. Owing to the zero mean of the stochastic force $F_s(t)$, the average velocity of the particle is not affected by the random force and is given by

$$\langle V_p(t) \rangle = V_0 \exp\left(-\gamma_p t\right). \tag{7.40}$$

The two-time correlation function of the velocity can also be calculated as

$$\Gamma_{VV}(t, t') = \langle V_p(t) V_p(t') \rangle = V_{p0}^2 \exp[-\gamma_p(t + t')]$$

$$+ \frac{1}{m_p^2} \int_0^t \int_0^{t'} \exp\left[-\gamma_p(t + t' - \tau - \tau')\right] \mathcal{G}\delta(\tau - \tau') \, d\tau \, d\tau', \tag{7.41}$$

where we have used the relation in Eq. (7.35). The evaluation of the double integral requires some thought because the upper limits are different for τ and τ'. Owing to the presence of the delta function, this integral is not zero only when $\tau = \tau'$. If we first integrate over τ', the remaining integral over τ must stop at t if $t < t'$ or at t' if $t' < t$. Thus,

$$\langle V_p(t) V_p(t') \rangle = V_{p0}^2 \exp[-\gamma_p(t + t')] + \frac{\mathcal{G}}{m_p^2} \int_0^{\min(t,t')} \exp[-\gamma_p(t + t' - 2\tau)] \, d\tau. \tag{7.42}$$

The integration over τ can now be carried out to obtain

$$\langle V_p(t) V_p(t') \rangle = V_{p0}^2 e^{-\gamma_p(t+t')} + \frac{\mathcal{G}}{2m_p^2 \gamma_p} \left[\exp(-\gamma_p|t - t'|) - \exp(-\gamma_p|t + t'|)\right]. \tag{7.43}$$

It is clear from this expression that the two exponential terms containing $t + t'$ will become negligible for values of t and t' larger than $1/\gamma_p$. In this long time limit ($\gamma_p t \gg 1$ and $\gamma_p t' \gg 1$), the correlation function becomes independent of the initial velocity and depends only on the time difference $|t - t'|$ as

$$\langle V_p(t) V_p(t') \rangle = \frac{\mathcal{G}}{2m_p \gamma_d} \exp\left[-\frac{\gamma_d}{m_p} |t - t'|\right], \tag{7.44}$$

where we used the relation $\gamma_p = \gamma_d/m_p$.

We can calculate the average energy $\langle E \rangle$ of the Brownian particle as

$$\langle E \rangle = \frac{1}{2} m_p \langle V_p^2(t) \rangle = \frac{\mathcal{G}}{4\gamma_d}. \tag{7.45}$$

However, from the equipartition law of classical statistical mechanics $\langle E \rangle$ should be related in thermal equilibrium to the temperature T of the reservoir (fluid) as $\langle E \rangle_t = \frac{1}{2}k_B T$. Equating these two energy expressions, the noise-strength parameter $\mathcal{G}$ is found to be

$$\mathcal{G} = 2\gamma_d k_B T. \tag{7.46}$$

This result can be viewed as a consequence of the fluctuation-dissipation theorem discussed in Section 3.3. Thermal equilibrium demands that a balance must exist between friction (related to γ_d), which tends to attenuate the particle's speed, and fluctuations (related to $\mathcal{G}$), which tends to keep the particle moving.

7.1.6 Generalized Einstein Relation

The Langevin equation (7.38) obtained for the Brownian motion is not in a standard form of this equation. The standard form for a set of Langevin equations is

$$\frac{d}{dt}A_\mu(t) = D_\mu(t) + F_\mu(t), \tag{7.47}$$

where $D_\mu(t)$ is referred to as the drift term and $F_\mu(t)$ is the Langevin force. The index μ acts as a level when multiple Langevin equations are needed for a system requiring several physical variables for its description. As before, the Langevin force F_μ has zero mean and is modeled as a Markovian stochastic process with the correlation function

$$\langle F_\eta(t)F_\mu(t') \rangle = 2D_{\eta\mu}\delta(t - t'). \tag{7.48}$$

The constant $D_{\eta\mu}$ is called the diffusion coefficient of the Langevin force. The generalized Einstein relation allows one to express $D_{\eta\mu}$ in terms of four quantities: $D_\mu(t)$, $D_\eta(t)$, $A_\mu(t)$, and $A_\eta(t)$.

We start the derivation by using an identity relating A_μ at two different times through the integral

$$A_\mu(t) = A_\mu(t - \Delta t) + \int_{t-\Delta t}^{t} \frac{d}{dt'}A_\mu(t')\, dt'. \tag{7.49}$$

We multiply this equation with $F_\eta(t)$ and perform ensemble average to obtain

$$\langle F_\eta(t)A_\mu(t) \rangle = \langle F_\eta(t)A_\mu(t - \Delta t) \rangle + \int_{t-\Delta t}^{t} \langle F_\eta(t)\frac{d}{dt'}A_\mu(t') \rangle\, dt'. \tag{7.50}$$

Using the Langevin equation (7.47) for the time derivative, we get

$$\langle F_\eta(t)A_\mu(t) \rangle = \langle F_\eta(t)A_\mu(t - \Delta t) \rangle + \int_{t-\Delta t}^{t} \langle F_\eta(t)[D_\mu(t') + F_\mu(t')] \rangle\, dt'. \tag{7.51}$$

The first term vanishes, $\langle F_\eta(t)A_\mu(t - \Delta t) \rangle = 0$, owing to the causality requirement that $A_\mu(t - \Delta t)$ cannot be affected by the stochastic force $F_\eta(t)$ at a later time. Similarly, $\langle F_\eta(t)D_\mu(t') \rangle$ is zero except when $t' = t$; its integral evaluates to zero because the nonzero value occurs over a set of measure zero. The remaining integral can be simplified using $\tau = t' - t$ to obtain

$$\int_{t-\Delta t}^{t} \langle F_\eta(t)F_\mu(t') \rangle\, dt' = \int_{-\Delta t}^{0} \langle F_\eta(t)F_\mu(t + \tau) \rangle\, d\tau = \int_{-\infty}^{0} \langle F_\eta(0)F_\mu(\tau) \rangle\, d\tau, \tag{7.52}$$

where the lower limit was extended to infinity in view of Eq. (7.48). We also used the stationarity of $F_\eta(t)$, which ensures that any correlation function depends on the time difference of the two times involved. Collecting these results, we obtain

$$\langle F_\eta(t)A_\mu(t)\rangle = \frac{1}{2}\int_{-\infty}^{\infty} \langle F_\eta(t)F_\mu(t')\rangle \, dt'. \tag{7.53}$$

Using the correlation function in Eq. (7.48), we obtain the result

$$\langle F_\eta(t)A_\mu(t)\rangle = D_{\eta\mu}. \tag{7.54}$$

Using the same procedure, we can also show that

$$\langle A_\eta(t)F_\mu(t)\rangle = D_{\eta\mu}. \tag{7.55}$$

We use these two results to calculate the derivative of the correlation function $\langle A_\mu(t)A_\eta(t)\rangle$:

$$\begin{aligned}
\frac{d}{dt}\langle A_\eta(t)A_\mu(t)\rangle &= \langle \frac{dA_\eta}{dt}A_\mu(t)\rangle + \langle A_\eta(t)\frac{dA_\mu}{dt}\rangle \\
&= \langle [D_\eta(t)+F_\eta(t)]A_\mu(t)\rangle + \langle A_\eta(t)[D_\mu(t)+F_\mu(t)]\rangle \\
&= \langle D_\eta(t)A_\mu(t)\rangle + \langle A_\eta(t)D_\mu(t)\rangle + 2D_{\eta\mu}.
\end{aligned} \tag{7.56}$$

This result is called the generalized Einstein relation when written in the form

$$2D_{\eta\mu} = \frac{d}{dt}\langle A_\eta(t)A_\mu(t)\rangle - \langle A_\eta(t)D_\mu(t)\rangle - \langle D_\eta(t)A_\mu(t)\rangle. \tag{7.57}$$

It enables us to express the diffusion coefficient $D_{\eta\mu}$ of the random force in terms of the drift parameters D_η and D_μ and the time derivative of the correlation function $\langle A_\eta(t)A_\mu(t)\rangle$. It can be viewed as a manifestation of the fluctuation-dissipation theorem discussed in Section 3.3.

The classical analysis of this section can be easily extended to the quantum domain. In the quantum version of the Langevin equation given in (7.47), the variables A_μ, D_μ, and F_μ become Heisenberg operators. As a result, an ensemble average denotes quantum averaging over the initial state of the underlying quantum system. The generalized Einstein relation in Eq. (7.57) remains valid with this interpretation of the averages appearing in this relation. Also, the quantum Langevin equation provides an approximate method for solving the quantum master equations [391] of this system. We present in Aside 7.2 the quantum regression theorem that is useful for calculating the two-time correlation function of A_μ.

Aside 7.2   Quantum Regression Theorem

In the derivation of the generalized Einstein relation in Section 7.1.6, we calculated the correlation function $\langle A_\eta(t)A_\mu(t)\rangle$ involving a single time. The two-time correlation function, defined as $\langle A_\eta(t)A_\mu(t')\rangle$ where $t' < t$, provides more information, and its knowledge is desirable. However, knowledge of the density matrix is not sufficient to calculate such correlation functions because we also need to know the transition probabilities among various quantum states. The quantum regression theorem shows that, under certain conditions, it is possible to calculate the two-time correlation function if we know how $\langle A_\eta(t) \rangle$ evolves with time.

To derive this theorem, we consider the derivative of $\langle A_\eta(t) A_\mu(t') \rangle$ with respect to t. It is easy to see that

$$\frac{d}{dt} \langle A_\eta(t) A_\mu(t') \rangle = \langle \frac{dA_\eta}{dt} A_\mu(t') \rangle = \langle [D_\eta(t) + F_\eta(t)] A_\mu(t') \rangle = \langle D_\eta(t) A_\mu(t') \rangle, \quad (7.58)$$

If we use the fact that $\langle F_\eta(t) A_\mu(t') \rangle = 0$ if $t' < t$ because $A_\mu(t')$ cannot depend on the future values of the random force $F_\eta(t)$ owing to the causality requirement, we obtain the *quantum regression theorem* in the form

$$\frac{d}{dt} \langle A_\eta(t) A_\mu(t') \rangle = \langle D_\eta(t) A_\mu(t') \rangle. \quad (7.59)$$

It states that the two-time correlation function has the same time evolution as the average $\langle A_\eta(t) \rangle$ [391]. This theorem is attributed to Melvin Lax, who made many fundamental contributions in a series of papers on quantum noise [392, 393].

7.2 Quantum Spectral Density

The quantum version of classical noise replaces $\mathcal{N}(t)$ with a Hermitian operator, $\hat{\mathcal{N}}(t)$, in the Heisenberg picture. The quantum form of the autocorrelation function is then defined as

$$\Gamma_{\mathcal{N}\mathcal{N}}(\tau) = \langle \hat{\mathcal{N}}(t) \hat{\mathcal{N}}(t + \tau) \rangle = \langle \hat{\mathcal{N}}(0) \hat{\mathcal{N}}(\tau) \rangle, \quad (7.60)$$

where we assumed that the noise process is stationary. As before, the spectral density of noise is just the Fourier transform given by

$$S_{\mathcal{N}\mathcal{N}}(\omega) = \int_{-\infty}^{\infty} \Gamma_{\mathcal{N}\mathcal{N}}(\tau) \exp(i\omega\tau) \, d\tau. \quad (7.61)$$

It is remarkable that the quantum spectral density may not be a symmetric function of frequency. The reason is that it is not possible to guarantee that the quantum autocorrelation function $\Gamma_{\mathcal{N}\mathcal{N}}(\tau)$ is always real. As a result, its Fourier transform can have different values for positive and negative values of the same frequency.

To build some insight into the quantum spectral density, let us consider a quantum device with energy eigenstates $|\alpha\rangle$ that form a complete set such that $\sum_\alpha |\alpha\rangle \langle \alpha| = 1$. The density operator ρ of this device is diagonal in this basis with energy eigenvalues E_α. We can thus expand the autocorrelation function as [394]

$$\Gamma_{\mathcal{N}\mathcal{N}}(\tau) = \sum_\alpha \sum_\gamma \rho_{\alpha\alpha} \langle \alpha|\hat{\mathcal{N}}(\tau)|\gamma\rangle \langle \gamma|\hat{\mathcal{N}}(0)|\alpha\rangle$$

$$= \sum_\alpha \sum_\gamma \rho_{\alpha\alpha} \exp\left[\frac{i}{\hbar}(E_\alpha - E_\gamma)\tau\right] |\langle \alpha|\hat{\mathcal{N}}(0)|\gamma\rangle|^2, \quad (7.62)$$

where we used the time evolution operator $U(t, t_0) = \exp[-\frac{i}{\hbar}H(t - t_0)]$ given in Section 2.2.5 to replace $\hat{\mathcal{N}}(t)$ with $\hat{\mathcal{N}}(0)$. Using this form of $\Gamma_{\mathcal{N}\mathcal{N}}(\tau)$ in Eq. (7.61) and integrating over τ, the quantum spectral density can be written as

$$\hat{S}_{\mathcal{N}\mathcal{N}}(\omega) = \sum_{\alpha} \sum_{\gamma} \rho_{\alpha\alpha} 2\pi \delta \left(\omega - \frac{1}{\hbar}(E_\gamma - E_\alpha) \right) |\langle \alpha | \hat{\mathcal{N}}(0) | \gamma \rangle|^2$$

$$= \sum_{\alpha} \sum_{\gamma} \rho_{\alpha\alpha} W_{\alpha\gamma}, \qquad (7.63)$$

where $W_{\alpha\gamma}$ is the noise-induced transition rate from the state $|\alpha\rangle$ to $|\gamma\rangle$ (in the form of Fermi's golden rule).

We can now interpret the quantum spectral density as a physical quantity describing how a quantum device exchanges energy with its reservoir at a given frequency. This should be compared with the classical spectral density, which describes how much noise power the device has at that frequency. Even though we used analogous definitions for classical and quantum spectral densities, there are drastic differences in the way they provide device-specific information to us. Clearly, we should explore whether we can recover all classical features from the quantum spectral density, as one would expect from the correspondence principle.

For this purpose it is useful to partition the spectral density into its even and odd parts as $S_{\mathcal{N}\mathcal{N}}(\omega) = S_{\mathcal{N}\mathcal{N}}^{(e)}(\omega) + S_{\mathcal{N}\mathcal{N}}^{(o)}(\omega)$. This partitioning is possible for an arbitrary function when the even and odd parts are defined as

$$S_{\mathcal{N}\mathcal{N}}^{(e)}(\omega) = \frac{1}{2}\left[\hat{S}_{\mathcal{N}\mathcal{N}}(\omega) + \hat{S}_{\mathcal{N}\mathcal{N}}(-\omega)\right], \quad S_{\mathcal{N}\mathcal{N}}^{(o)}(\omega) = \frac{1}{2}\left[\hat{S}_{\mathcal{N}\mathcal{N}}(\omega) - \hat{S}_{\mathcal{N}\mathcal{N}}(-\omega)\right].$$
$$(7.64)$$

Clearly, the even and odd parts are the symmetric and antisymmetric parts of the quantum spectral density, respectively. In what follows, we apply the partitioned form of quantum spectral density to two well-known quantum systems, a two-level system and a harmonic oscillator. We find that the even part of $S_{\mathcal{N}\mathcal{N}}(\omega)$ contains the same information as the classical spectral density, whereas its odd part is related to damping of the quantum device [395, 394]. Indeed, the origin of the odd part can be traced back to the noncommutative nature of the quantum-noise operator: $[\mathcal{N}(t), \mathcal{N}(t')] \neq 0$.

7.2.1 A Two-Level System Coupled to a Noise Source

We consider an atom with two energy states such that its ground state $|g\rangle$ and excited state $|e\rangle$ have an energy difference of $E_e - E_g = \hbar\omega_{eg}$, where ω_{eg} is the transition frequency. Interaction of the atom with a noise source is included through the total Hamiltonian

$$H_{TL} = H_{eg} + H_I, \qquad (7.65)$$

where H_{eg} is the Hamiltonian of the two-level atom and H_I accounts for its interaction with the noise source. As in Section 4.1.4, we use the Pauli spin matrices to write H_{eg} in the form [251]

$$H_{eg} = \frac{1}{2}\hbar\omega_{eg}\left(|e\rangle\langle e| - |g\rangle\langle g| \right) = \frac{1}{2}\hbar\omega_{eg}\sigma_3, \qquad (7.66)$$

where energy is taken to be zero in the middle of two energy levels. Transitions between the two energy states are included through the raising and lowering operators introduced in Aside 4.3 and defined as

$$\sigma_+ = \sigma_1 + i\sigma_2, \qquad \sigma_- = \sigma_1 - i\sigma_2. \tag{7.67}$$

These operators change the state of the atom as

$$\sigma_+ |e\rangle = 0, \qquad \sigma_+ |g\rangle = |e\rangle, \qquad \sigma_- |g\rangle = 0, \qquad \sigma_- |e\rangle = |g\rangle. \tag{7.68}$$

Furthermore, they satisfy the relations [252]:

$$[\sigma_3, \sigma_\pm] = 2\sigma_\pm, \qquad [\sigma_+, \sigma_-] = \sigma_3, \qquad \sigma_+^2 = \sigma_-^2 = 0, \qquad \sigma_+^\dagger = \sigma_-. \tag{7.69}$$

The noise source is included through the operator $\hat{F}(t)$. We work in the interaction picture and write the interaction Hamiltonian as [395, 394]

$$\tilde{H}_I(t) = \eta_n \hat{F}(t)\sigma_1, \tag{7.70}$$

where a tilde denotes an operator in the interaction picture. This Hamiltonian is responsible for energy exchange between the two-level system and the noise source. We assume that the atom interacts weakly with the noise source, and first-order perturbation theory is adequate to treat the interaction. In this situation, the wave function evolves with time as

$$|\Psi_I(t)\rangle = |\Psi_I(0)\rangle - \frac{i}{\hbar} \int_0^t \tilde{H}_I(\tau) |\Psi_I(0)\rangle \, d\tau, \tag{7.71}$$

where $|\Psi_I(0)\rangle$ is the wave function of the system at $t = 0$.

Consider first the case of a two-level atom initially in its ground state. The probability amplitude for finding the system in its excited state at time t is given by

$$\alpha_e(t) = \langle e | \Psi_I(t) \rangle = -\frac{i}{\hbar} \int_0^t \langle e | \tilde{H}_I(\tau) | g \rangle \, d\tau, \tag{7.72}$$

where we used $\langle e | g \rangle = 0$ for the two orthogonal states. Using $\tilde{H}_I(t) = \eta_n \hat{F}(t)\sigma_1$ in the preceding equation together with $\langle e | \sigma_1(\tau) | g \rangle = \exp(i\omega_{eg}\tau)$, we obtain

$$\alpha_e(t) = -\frac{i\eta_n}{\hbar} \int_0^t \hat{F}(\tau) e^{i\omega_{eg}\tau} \, d\tau. \tag{7.73}$$

The probability of finding the two-level system in the excited state is thus given by

$$P_e(t) = \langle |\alpha_e|^2 \rangle = \frac{\eta_n^2}{\hbar^2} \int_0^t \int_0^t \langle \hat{F}(\tau_1)\hat{F}(\tau_2) \rangle \exp[i\omega_{eg}(\tau_1 - \tau_2)] \, d\tau_1 \, d\tau_2. \tag{7.74}$$

The preceding expression can be simplified for a stationary noise process. Using $\tau_1 = \tau_2 + \tau$, it can be written as

$$P_e(t) = \frac{\eta_n^2 t}{\hbar^2} \int_0^t \Gamma_{FF}(\tau) e^{i\omega_{eg}\tau} d\tau = \frac{\eta_n^2 t}{\hbar^2} S_{FF}(\omega_{eg}), \tag{7.75}$$

where we used definition of the spectral density after extending the limits of integration to infinity. This result is valid after a sufficiently long duration of time in the limit of weak coupling between the two-level atom and the noise source. When the system is initially in the excited state, a similar analysis can be carried out to show that the probability of finding the system in the ground state is given by

$$P_g(t) = \langle |\alpha_g|^2 \rangle = \frac{\eta_n^2 t}{\hbar^2} S_{FF}(-\omega_{eg}). \tag{7.76}$$

The time derivative of these probabilities gives us the transition rates between the ground and the excited states. Using $\uparrow$ for the upward transitions ($|g\rangle \to |e\rangle$) and $\downarrow$ for the downward transition ($|e\rangle \to |g\rangle$), these transition rates are given by

$$\Gamma_\uparrow = \frac{\eta_n^2}{\hbar^2} S_{FF}(\omega_{eg}), \qquad \Gamma_\downarrow = \frac{\eta_n^2}{\hbar^2} S_{FF}(-\omega_{eg}). \tag{7.77}$$

This result shows that positive frequencies correspond to absorption of energy, while negative frequencies correspond to emission of energy. If the two-level system is in thermal equilibrium with the noise source at some temperature T, the transition rates of the two-level system must satisfy the detailed-balance condition: $\Gamma_\uparrow / \Gamma_\downarrow = \exp(-\hbar\omega_{eg}/k_B T)$. This in turn implies that the quantum spectral density must satisfy the relation

$$S_{FF}(\omega_{eg}) = \exp(-\hbar\omega_{eg}/k_B T) S_{FF}(-\omega_{eg}). \tag{7.78}$$

7.2.2 A Harmonic Oscillator Coupled to a Noise Source

Another useful example of quantum noise is provided by a harmonic oscillator coupled to a noise source [396, 397]. As before, we write the Hamiltonian in the form

$$H_{ho} = H_0 + H_I, \tag{7.79}$$

where $H_0 = \hbar\omega_o(\hat{a}^\dagger\hat{a} + \frac{1}{2})$ is the free part (see Section 4.3.2) that has the eigenstates, $H_0|n\rangle = E_n|n\rangle$, with energies $E_n = \hbar\omega_o(n + \frac{1}{2})$; here ω_o is the oscillation frequency. The interaction part of the Hamiltonian can be written as [395]

$$\tilde{H}_I(t) = \eta_n[\hat{x}\hat{F}(t)] = \eta_n\left[x_0(\hat{a}^\dagger + \hat{a})\right]\hat{F}(t), \tag{7.80}$$

where η_n is the coupling constant, $x_0 = \sqrt{\hbar/(2m_{ho}\omega_o)}$ is the uncertainty in the position of the harmonic oscillator of mass m_{ho} in its ground state, and $\hat{F}$ is the operator responsible for the noise.

Coupling of the harmonic oscillator to the noise source causes transitions among its energy levels such that the state $|n\rangle$ changes to $|n+1\rangle$ or $|n-1\rangle$. As all transitions are between two neighboring energy levels separated by $\hbar\omega_o$, we can use the transition rates found in Section 7.2.1 even for a harmonic oscillator. Thus, the rate for increasing the number of quanta in the oscillator by one, taking the state $|n\rangle$ to $|n+1\rangle$, is given by

$$\Gamma_{n\to n+1} = \frac{\eta_n^2}{\hbar^2}\left[(n+1)x_0^2\right]S_{FF}(\omega_o) = (n+1)\Gamma_\uparrow. \tag{7.81}$$

Similarly, the rate for taking the state $|n\rangle$ to $|n-1\rangle$ is given by

$$\Gamma_{n\to n-1} = \frac{\eta_n^2}{\hbar^2}(nx_0^2)S_{FF}(-\omega_o) = n\Gamma_\downarrow. \tag{7.82}$$

Using the preceding two rates, we can construct the following master equation for the probability of finding the harmonic oscillator in the state $|n\rangle$:

$$\frac{d}{dt}P_n(t) = [n\Gamma_\uparrow P_{n-1}(t) + (n+1)\Gamma_\downarrow P_{n+1}(t)] - [n\Gamma_\downarrow P_n(t) + (n+1)\Gamma_\uparrow P_n(t)]. \tag{7.83}$$

The first two terms describe the transitions into the state $|n\rangle$ from the states $|n-1\rangle$ and $|n+1\rangle$ that increase $P_n(t)$. The last two terms describe transitions out of the state $|n\rangle$ to the states $|n-1\rangle$, and $|n+1\rangle$ that decrease $P_n(t)$.

The average energy of the oscillator at time t can be written as $\langle E(t)\rangle = \sum_{n=0}^{\infty}\hbar\omega_o$ $(n+\frac{1}{2})P_n(t)$. By differentiating this expression one can show that the average energy changes with time as [395]

$$\frac{d\langle E\rangle}{dt} = P_s - \gamma\langle E(t)\rangle, \tag{7.84}$$

where P_s and γ are defined as

$$P_s = \frac{\eta_n^2}{2m_{ho}}\left\{\frac{1}{2}\left[S_{FF}(\omega_{eg}) + S_{FF}(-\omega_{eg})\right]\right\}, \tag{7.85}$$

$$\gamma = \frac{\eta_n^2}{\hbar^2}\left[S_{FF}(-\omega_{eg}) - S_{FF}(\omega_{eg})\right]. \tag{7.86}$$

Physically, P_s denotes the power supplied to the harmonic oscillator by the noise source and γ is the rate at which energy of the harmonic oscillator is dissipated. Notice that P_s depends on the symmetrized part of the spectral density (the even part), whereas γ depends on its odd part. Clearly, the decay term has its origin in the asymmetric nature of the quantum spectral density with respect to the positive and negative frequencies. Positive frequencies represent absorption of energy by the oscillator, while negative frequencies denote loss of energy through emission. The difference between these two is the net flow of energy from the oscillator to the noise source.

Another feature that differentiates quantum spectral density from its classical counterpart is its finite value at zero absolute temperature (0 K). This feature has its origin in the zero-point energy in the ground state of the oscillator. It can be viewed as the energy that remains in the harmonic oscillator even when all motion has ceased. As we have seen, an harmonic oscillator contains $\frac{1}{2}\hbar\omega_o$ energy in this state. The origin of zero-point energy is attributed to Heisenberg's uncertainty principle associated with two conjugate variables (position and momentum in this case) (see Aside 2.11). If its position remains uncertain in the ground state, a harmonic oscillator should be moving and must have a finite energy.

The concept of zero-point energy can be extended to electromagnetic waves after recalling from Section 2.3 that each electromagnetic mode is equivalent to a harmonic oscillator and is thus subject to the same uncertainty principle. Thus, an electromagnetic mode oscillating at frequency ω_m must have a minimum energy of $\frac{1}{2}\hbar\omega_m$ on average. Even though this is a tiny amount of energy for that mode, the total zero-point energy can be considerable when an enormous number of modes oscillating at different frequencies occupy the vacuum. Casimir showed in 1948 that one consequence of the zero-point energy is the presence of an attractive force between two uncharged, conducting parallel plates [398]. Callen and Welton [175] found in 1951 the quantum version of the fluctuation-dissipation theorem, whose classical version (see Section 3.3) has been known since 1928 [175]. This theorem shows that when a device dissipates energy in an irreversible manner, fluctuations in the coupled reservoir are unavoidable. As a consequence, it is not possible to separate fluctuations from dissipations. This theorem is applicable to a resistor, which exhibits current fluctuations as it dissipates energy as heat. We discuss the quantum spectral density of a resistor in Aside 7.3 using a transmission-line model.

Aside 7.3   Quantum Noise of a Resistor The impedance of a semi-infinite, lossless, transmission line is a purely real quantity that is also frequency independent. Energy supplied by a source at one end of such a line is transmitted through the line without being dissipated along the line. However, because any launched energy does not return, it can be considered to be lost. For this reason, a lossless transmission line can be used to model an ideal resistor. If the transmission line has an inductance L per unit length and capacitance C per unit length, then the resistance is given by $R = \sqrt{L/C}$. We calculate the quantum spectral density of a resistor using this approach [263, 394]. The main reason behind our choice is that a transmission line is also equivalent to a large collection of harmonic oscillators that can be readily quantized.

As seen in Figure 7.1, a transmission line can be divided into small sections of length Δz such that each section contains an LC circuit. As $\Delta z \to 0$, the lumped model turns into a distributed model. Consider the nth section at a distance $z = z_n$ from the origin. Its LC circuit has lumped capacitance $C\Delta z$ and lumped inductance $L\Delta z$. Suppose this section has a loop current $I_n(t, z_n)$. Associated with this current is the loop charge, $Q_n(t, z_n) = \int_0^t I(t', z_n)\, dt'$. We use this charge as the generalized coordinate and write the Lagrangian of the system as a sum over an infinite number of LC sections. Noting that the Lagrangian for an electrical system represents the difference of energies stored in inductors and capacitors, we obtain

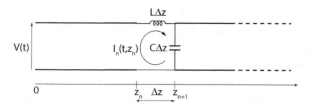

Fig. 7.1   Transmission line model of a resistor containing a large number of cascaded LC sections. The nth infinitesimal LC section of length Δz is shown with a loop current $I_n(t, z_n)$.

$$\mathcal{L} = \sum_n \left[\frac{L\Delta z}{2} \left(\frac{\partial}{\partial t} Q_n(t, z_n) \right)^2 - \frac{1}{2C\Delta z} [Q_{n+1}(t, z_{n+1}) - Q_n(t, z_n)]^2 \right]$$

$$= \sum_n \left[\frac{L}{2} \left(\frac{\partial}{\partial t} Q_n(t, z_n) \right)^2 - \frac{1}{2C} \left(\frac{Q_{n+1}(t, z_{n+1}) - Q_n(t, z_n)}{\Delta z} \right)^2 \right] \Delta z. \qquad (7.87)$$

Taking the limit $\Delta z \to 0$ and writing the sum as an integral, we obtain

$$\mathcal{L} = \int_0^\infty \left[\frac{L}{2} \left(\frac{\partial Q}{\partial t} \right)^2 - \frac{1}{2C} \left(\frac{\partial Q}{\partial z} \right)^2 \right] dz, \qquad (7.88)$$

where Q stands for $Q_n(t, z)$ in the limit $\Delta z \to 0$.

We can now use the standard variational method. Applying the Euler–Lagrange equation to the preceding Lagrangian, we obtain the wave equation

$$\frac{\partial^2 Q}{\partial z^2} - \frac{1}{v_m^2} \frac{\partial^2 Q}{\partial t^2} = 0, \qquad (7.89)$$

where the velocity is defined as $v_0 = \frac{1}{\sqrt{LC}}$. To find the modes supported by the transmission line, we first assume that its length l is finite, and let $l \to \infty$ later to mimic a semi-infinite transmission line. For our finite-length transmission line, the boundary conditions at the two ends are $Q(t, 0) = 0$, and $Q(t, l) = 0$ because no currents can flow beyond these points. The general solution of the wave equation satisfying these boundary conditions is found to be

$$Q(t, z) = \sum_{n=1}^\infty \varphi_n(t) u_n(z), \qquad u_n(z) = \sqrt{\frac{2}{l}} \sin(n\pi z/l), \qquad (7.90)$$

where $u_n(z)$ is the nth mode of the transmission line assumed to be normalized such that $\int_0^l u_m(z) u_n(z) \, dz = \delta_{mn}$.

When we substitute the preceding normal-mode expansion into the original Lagrangian and integrate over the length of the transmission line, we obtain the reduced Lagrangian

$$\mathcal{L}_r = \sum_{n=1}^\infty \left[\frac{L}{2} \left(\frac{d\varphi_n}{dt} \right)^2 - \frac{1}{2C} K_n^2 \varphi_n^2 \right] = \frac{L}{2} \sum_{n=1}^\infty \left[\left(\frac{d\varphi_n}{dt} \right)^2 - \omega_n^2 \varphi_n^2 \right], \qquad (7.91)$$

where $K_n = n\pi/l$ and $\omega_n = K_n/v_0$. This form is identical to the Lagrangian of a set of harmonic oscillators with oscillation frequencies ω_n. Each of these harmonic oscillators can be quantized using the appropriate creation ($\hat{a}_n^\dagger$) and annihilation ($\hat{a}_n$) operators (see Section 2.3).

One can use the preceding analysis to calculate the noise voltage $V(t)$ at $z = 0$. Once the correlation function $\Gamma_{VV}(\tau)$ is found, its Fourier transform provides the quantum spectral density (two-sided) in the form

$$S_{VV}(\omega) = \frac{2\hbar\omega R}{1 - \exp\left(-\frac{\hbar\omega}{k_B T} \right)}. \qquad (7.92)$$

For frequencies such that $\hbar\omega \ll k_B T$, we can use the approximation $e^x \approx 1 - x$ to find $S_{VV}(\omega) = 2k_B TR$, the same result we obtained earlier for the two-sided spectral density of thermal noise. The case corresponds to the classical limit in which electronic devices operate. In the opposite limit, $\hbar\omega \gg k_B T$, known as the quantum limit, we can neglect the exponential in the denominator to obtain $S_{VV}(\omega) = 2\hbar\omega R$.

7.3 Quantum Langevin Equations

As discussed in Section 2.2.5, time evolution of a quantum system can be studied using three different approaches known as the Schrödinger picture, the Heisenberg picture, and the interaction picture. In the Schrödinger picture, the state of a system evolves with time but operators remain unchanged. In the Heisenberg picture, operators evolve with time but the quantum state remains unchanged. The interaction picture is an intermediate situation where both states and operators are allowed to vary with time. The Heisenberg picture is most closely related to classical mechanics, where the dynamical variables (which become operators in quantum mechanics) evolve in time. For dissipative systems, the evolution of an operator is governed by a quantum Langevin equation, which contains random force terms representing the interaction of a quantum system with a surrounding reservoir. These force terms are necessary to preserve the commutation relations involved in any quantum description.

As an example, consider the position operator q and the momentum operator p. These operators become time dependent in the Heisenberg picture: $q(t)$ and $p(t)$, respectively. Their commutator initially at time $t = 0$ satisfies the relation $[q(0), p(0)] = i\hbar$. As this relation does not change with time, it is necessary that the condition $[q(t), p(t)] = i\hbar$ is satisfied at all times. In the Heisenberg picture, any operator evolves with time as (see Section 2.2.5)

$$i\hbar \frac{d}{dt} \mathcal{O}(t) = \frac{\partial}{\partial t} \mathcal{O}(t) + [\mathcal{O}(t), H], \qquad (7.93)$$

where the Hamiltonian H may also vary with time. This equation often bears a close resemblance to the corresponding classical equation, which can be of benefit in a theoretical treatment. However, the Heisenberg equation of motion is generally nonlinear and may be harder to solve in practice.

7.3.1 Quantum Theory of a Laser

As an example of quantum Langevin equations, we consider a simple model of a laser (see Fig. 7.2) in which two-level atoms interact with the radiation field of a single excited mode of a resonant cavity [399, 400, 401, 402]. We use the Jaynes–Cummings Hamiltonian discussed in Section 4.1.4 for the two-level atoms. Each atom has a ground state $|g\rangle$ with energy E_g and an excited state $|e\rangle$ with energy E_e such that $E_e - E_g = \hbar\omega_{eg}$. To simplify the analysis, we assume that the frequency ω_k of a specific electromagnetic mode of the

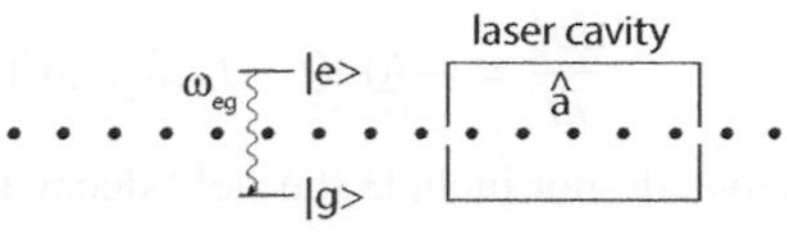

Fig. 7.2 Schematic of a laser cavity where entering atoms (represented by dots) are modeled as a two-level system. These atoms are coupled to an electromagnetic mode of the laser cavity.

laser cavity coincides with ω_{eg}. This ensures that only one mode couples strongly to the two-level atoms.

The laser model is based on the injection of atoms in their excited states into the laser's cavity at random times. These atoms interact with the radiation mode and amplify it by emitting photons through stimulated emission. We assume that the probability distribution of the arrival time of atoms does not depend on time (a constant pumping rate). We also assume that cavity losses are sufficiently small that we can adopt the mean-field approximation for the radiation field and discard its spatial variations. This field is assumed to be in the form of a plane wave with frequency ω_k and the wave number $k = \omega_k/c$. As shown in Ref. [400], such a model has a single statistical parameter in the range $0 \leq p \leq 1$, the case of Poissonian statistics corresponding to $p = 0$.

The Hamiltonian of the system can be written in the rotating-wave approximation as (see Section 4.1.4)

$$H = \hbar\omega_k \hat{a}_k^\dagger \hat{a}_k + \sum_j \left(E_e \sigma_e^j + E_g \sigma_g^j \right) + \hbar\mathcal{G} \sum_j u(t - t_j) \left(\hat{a}_k^\dagger \sigma^j + \sigma^{j\dagger} \hat{a}_k \right), \qquad (7.94)$$

where the index j accounts for a specific excited atom entering the cavity at time t_j. The step function $u(t - t_j)$ is used to represent the initiation of coupling between the field and atom at time t_j. The operators $\hat{a}_k^\dagger$ and $\hat{a}_k$ are the creation and annihilation operators of the single lasing mode. The operators, $\sigma_e^j = (|e\rangle \langle e|)^j$ and $\sigma_g^j = (|g\rangle \langle g|)^j$, are the projection operators for the kth atom. The operator $\sigma^j = (|g\rangle \langle e|)^j$ is the spin-flip operator with the properties $\sigma^j |e\rangle^j = |g\rangle^j$ and $\sigma^{\dagger j} |g\rangle^j = |e\rangle^j$. The coupling constant $\mathcal{G}$ represents how strongly the atoms interact with the cavity mode and can be written as

$$\mathcal{G} = (2\hbar\epsilon_0 \omega_k V)^{-1/2} \omega_{eg} \mu_{eg}, \qquad (7.95)$$

where μ_{eg} is the magnitude of the atomic dipole moment and V is the cavity volume.

Using the preceding Hamiltonian in the Heisenberg equation of motion given in Eq. (7.93), we obtain the following four equations for the atomic and field operators:

$$\frac{d\hat{a}_k}{dt} = -i\omega_k \hat{a}_k(t) - i\mathcal{G} \sum_j u(t - t_j)\sigma^j(t), \qquad (7.96)$$

$$\frac{d\sigma^j}{dt} = -i\omega_k \sigma^j(t) + i\mathcal{G}u(t - t_j)[\sigma_e^j(t) - \sigma_g^j(t)]\hat{a}_k(t), \qquad (7.97)$$

$$\frac{d\sigma_e^j}{dt} = i\mathcal{G}u(t - t_j)[\hat{a}_k^\dagger(t)\sigma^j(t) - \sigma^{\dagger j}(t)\hat{a}_k(t)], \qquad (7.98)$$

$$\frac{d\sigma_g^j}{dt} = -i\mathcal{G}u(t - t_j)[\hat{a}_k^\dagger(t)\sigma^j(t) - \sigma^{\dagger j}(t)\hat{a}_k(t)]. \tag{7.99}$$

These equations do not include the field-decay term resulting from loss of photons at the output mirror of the cavity and the atomic decay terms related to spontaneous emission and the interaction of atoms with an external reservoir.

The reasoning behind introducing the decay terms in the preceding set of equations is analogous to the Weisskopf–Wigner theory discussed in Section 4.1.3. As we have seen before, a Langevin force must also be included with each decay term. The resulting equations are known as the quantum Langevin equations:

$$\frac{d\hat{a}_k}{dt} = -\left(i\omega_k + \frac{\gamma}{2}\right)\hat{a}_k(t) - i\mathcal{G}\sum_j u(t - t_j)\sigma^j(t) + F_f(t), \tag{7.100}$$

$$\frac{d\sigma^j}{dt} = -(i\omega_k + \gamma_{eg})\sigma^j(t) + i\mathcal{G}u(t - t_j)[\sigma_e^j(t) - \sigma_g^j(t)]\hat{a}_k(t) + f_\sigma^j(t), \tag{7.101}$$

$$\frac{d\sigma_e^j}{dt} = -(\gamma_e + \gamma_e')\sigma_e^j(t) + i\mathcal{G}u(t - t_j)[\hat{a}_k^\dagger(t)\sigma^j(t) - \sigma^{\dagger j}(t)\hat{a}_k(t)] + f_e^j(t), \tag{7.102}$$

$$\frac{d\sigma_g^j}{dt} = -\gamma_g\sigma_g^j(t) + \gamma_e'\sigma_e^j(t) - i\mathcal{G}u(t - t_j)[\hat{a}_k^\dagger(t)\sigma^j(t) - \sigma^{\dagger j}(t)\hat{a}_k(t)] + f_g^j(t). \tag{7.103}$$

Here γ is the cavity's damping rate, γ_e and γ_g are the decay rates for the excited and ground states, respectively, γ_e' is the spontaneous decay rate, and γ_{eg} is the decay rate of the atomic polarization.

As this laser model considers only a single cavity mode, it imposes a restriction on γ: the bandwidth of the cavity mode (related to γ) must be much smaller than the cavity's mode spacing given by $\delta\omega = (2\pi c)/L$ for a cavity of length L. For purely radiative decay, the four atomic decay rates are related to each other as $2\gamma_{eg} = (\gamma_e + \gamma_g + \gamma_e')$. This relationship can be established by considering the Liouville equation of the density operator (see Section 3.1.3). However, this relation holds only for purely radiative decay and does not account for the plethora of collisions that affect the relative phase between the excited and ground states, without significantly influencing their populations. When such collisions are included, we obtain a more realistic relation: $2\gamma_{eg} \geq \gamma_e + \gamma_e' + \gamma_g$.

Equations (7.100) through (7.103) contain four Langevin forces denoted as $F_f(t), f_\sigma^j(t)$, $f_e^j(t)$, and $f_g^j(t)$. They vanish on average and thus do not contribute to the averaged equations of motion. However, their correlation functions are finite. Assuming all four random forces are Gaussian processes (see Section 7.1.5), a correlation function is all we need to fully specify statistical properties of each Langevin-noise operator. However, we need to be careful because noise operators are noncommutative, and the order of the operators in an expression matters. In practice, we need to adopt some meaningful ordering to make the resulting expressions unique. The conventional approach is to use the normal ordering through Wick's theorem (see Aside 2.14).

The correlation functions of the Langevin force $F_f(t)$ have the form

$$\langle F_f^\dagger(t)F_f(t')\rangle = \gamma\langle n\rangle_{th}\delta(t - t'), \tag{7.104}$$

$$\langle F_f(t)F_f^\dagger(t')\rangle = \gamma[\langle n\rangle_{th} + 1]\delta(t - t'), \tag{7.105}$$

$$\langle F_f(t)F_f(t')\rangle = 0, \quad \langle F_f^\dagger(t)F_f^\dagger(t')\rangle = 0, \tag{7.106}$$

where $\langle n\rangle_{th}$ is the average number of thermal photons in the laser cavity at frequency ω_k. The presence of delta functions in the preceding expression indicates that the underlying random process is Markovian with no past memory.

To simplify the form of correlation function for the atomic Langevin forces, we assume that the reservoir is at zero absolute temperature ($T = 0$ K) so that the average number of thermal photons in the cavity is zero ($\langle n\rangle_{th} = 0$). With this assumption, the nonvanishing correlation functions of the atomic Langevin forces are given by

$$\langle f_e^j(t)f_e^j(t')\rangle = (\gamma_e + \gamma_e')\,\langle \sigma_e^j(t)\rangle\,\delta(t - t'), \tag{7.107}$$

$$\langle f_g^j(t)f_g^j(t')\rangle = \gamma_g\,\langle \sigma_g^j(t)\rangle\,\delta(t - t') + \gamma_e'\,\langle \sigma_e^j(t)\rangle\,\delta(t - t'), \tag{7.108}$$

$$\langle f_e^j(t)f_g^j(t')\rangle = -\gamma_e'\,\langle \sigma_e^j(t)\rangle\,\delta(t - t'), \tag{7.109}$$

$$\langle f_\sigma^j(t)f_g^j(t')\rangle = -\gamma_e'\,\langle \sigma^j(t)\rangle\,\delta(t - t'), \tag{7.110}$$

$$\langle f_\sigma^{\dagger j}(t)f_g^j(t')\rangle = \gamma_g\,\langle \sigma^{\dagger j}(t)\rangle\,\delta(t - t'), \tag{7.111}$$

$$\langle f_\sigma^j(t)f_e^j(t')\rangle = \gamma_e\,\langle \sigma^j(t)\rangle\,\delta(t - t' + \gamma_e'\,\langle \sigma^j(t)\rangle\,\delta(t - t'), \tag{7.112}$$

$$\langle f_\sigma^j(t)f_\sigma^{\dagger j}(t')\rangle = (2\gamma_{eg} - \gamma_g)\,\langle \sigma_g^j(t)\rangle\,\delta(t - t') + \gamma_e'\,\langle \sigma_e^j(t)\rangle\,\delta(t - t'), \tag{7.113}$$

$$\langle f_\sigma^{\dagger j}(t)f_\sigma^j(t')\rangle = (2\gamma_{eg} - \gamma_e - \gamma_e')\,\langle \sigma_e^j(t)\rangle\,\delta(t - t'). \tag{7.114}$$

As discussed in Section 7.1.6, the correlation functions of the Langevin forces are generally written in the form $\langle F_\eta(t)F_\mu(t')\rangle = 2D_{\eta\mu}\delta(t - t')$, where $D_{\eta\mu}$ is a diffusion coefficient. These coefficients can be calculated using the generalized Einstein relation derived in Section 7.1.6. As an example, let us calculate $D_{ee}(t)$ using Eq. (7.57):

$$2D_{ee}(t) = -\langle \sigma_e^j(t)\left(\frac{d}{dt}\sigma_e^j(t)\right)\rangle - \langle \left(\frac{d}{dt}\sigma_e^j(t)\right)\sigma_e^j(t)\rangle + \frac{d}{dt}\langle (\sigma_e^j(t))^2\rangle. \tag{7.115}$$

Recalling that $[\sigma_e^j(t)]^2 = \sigma_e^j(t)$ and using the relation $\langle d\sigma_e^j/dt\rangle = -(\gamma_e + \gamma_e')\sigma_e^j(t)$ [obtained from Eq. (7.102)], we obtain

$$2D_{ee}(t) = (\gamma_e + \gamma_e')\,\langle \sigma_e^j(t)\rangle, \tag{7.116}$$

which agrees with the result in Eq. (7.107).

7.3.2 Macroscopic Atomic Variables

The Langevin equations (7.100) and (7.101) contain rapidly oscillating terms at the cavity mode frequency ω_k. It is useful to introduce slowly varying operators $A_k(t)$ and $\sigma_s^j(t)$ as

$$\hat{a}_k(t) = A_k(t)\exp(-i\omega_k t), \qquad \sigma^j(t) = \sigma_s^j(t)\exp(-i\omega_k t). \tag{7.117}$$

The Langevin equations for the two slowly varying operators remain the same as those for $\hat{a}_k(t)$ and $\sigma^j(t)$, with the only difference that the terms containing ω_k disappear. Let us point out that the correlation functions do not change in the rotating frame because fast variations cancel out during the averaging procedure.

At this point, we introduce three macroscopic atomic operators as [400],

$$M(t) = -i \sum_j u(t - t_j)\sigma_s^j(t), \tag{7.118}$$

$$N_e(t) = \sum_j u(t - t_j)\sigma_e^j(t), \tag{7.119}$$

$$N_g(t) = \sum_j u(t - t_j)\sigma_g^j(t). \tag{7.120}$$

The operator $M(t)$ corresponds to the atomic polarization, while the operators $N_e(t)$ and $N_g(t)$ represent the population of atoms in the excited and ground states, respectively. It is important to note that when calculating the average values or correlation functions associated with these macroscopic operators, one has to perform not only the quantum mechanical average but also the classical average over the random arrival time t_j of the atoms injected into the cavity.

We can now write Eq. (7.100) for the field as

$$\frac{d}{dt}A_k(t) = -\frac{\gamma}{2}A_k(t) + \mathcal{G}M(t) + F_f(t). \tag{7.121}$$

To find an equation for the operator $N_e(t)$, we differentiate Eq. (7.119) to obtain

$$\frac{d}{dt}N_e(t) = \sum_j \delta(t - t_j)\sigma_e^j(t_j) + \sum_j u(t - t_j)\frac{d}{dt}\sigma_e^j(t), \tag{7.122}$$

where we use the relation $d\theta/dt = \delta(t)$ and replace $\sigma_e^j(t)$ with its value at the time t_j because of the presence of the delta function. Substituting the time derivative from Eq. (7.101), we obtain

$$\frac{d}{dt}N_e(t) = \sum_j \delta(t - t_j)\sigma_e^j(t_j) + \sum_j u(t - t_j)f_e^j(t)$$
$$- (\gamma_e + \gamma_e')N_e(t) - \mathcal{G}A_k^\dagger(t)M(t) - \mathcal{G}M^\dagger(t)A_k(t). \tag{7.123}$$

To proceed further, we relate the sum in the first term to the mean pumping rate R. As all injected atoms into the cavity are initially in the excited state, we have the relation $\langle\sigma_e^j(t_j)\rangle = 1$. Thus,

$$\left\langle \sum_j \delta(t - t_j)\sigma_e^j(t_j) \right\rangle = \left\langle \sum_j \delta(t - t_j) \right\rangle = \sum_j \frac{1}{T}\int_0^T \delta(t - t_j)dt = R, \tag{7.124}$$

where $R = N_t/T$ is the mean pumping rate when N_t atoms enter the cavity over a sufficiently long duration T. In terms of R, we write the first term in Eq. (7.123) as

$$\sum_j \delta(t - t_j)\sigma_e^j(t_j) \equiv R + \left[\sum_j \delta(t - t_j)\sigma_e^j(t_j) - R\right], \tag{7.125}$$

where the term inside the square brackets acts as a pumping-noise term that vanishes on average. Substituting this expression in Eq. (7.123), we obtain

$$\frac{d}{dt}N_e(t) = R - (\gamma_e + \gamma_e')N_e(t) - \mathcal{G}A_k^\dagger(t)M(t) - \mathcal{G}M^\dagger(t)A_k(t) + F_e(t), \tag{7.126}$$

where the new Langevin stochastic force is defined as

$$F_e(t) = \sum_j u(t - t_j)f_e^j(t) + \sum_j \delta(t - t_j)\sigma_e^j(t_j) - R. \tag{7.127}$$

The first term in $F_e(t)$ accounts for population fluctuations, while the second term accounts for pump fluctuations. It is easy to show that $\langle F_e(t)\rangle = 0$. We need to calculate the correlation function of this force. Using the relation in Eq. (7.124) and noting that population fluctuations and pump fluctuations are uncorrelated, one can write this correlation function as

$$\langle F_e(t)F_e(t')\rangle = C_1(t, t') + C_2(t, t') - R^2, \tag{7.128}$$

where

$$C_1(t, t') = \left\langle \left[\sum_{j_1} u(t - t_{j_1})f_e^{j_1}(t)\right]\left[\sum_{j_2} u(t' - t_{j_2})f_e^{j_2}(t')\right]\right\rangle \tag{7.129}$$

$$C_2(t, t') = \left\langle \left[\sum_{j_1} \delta(t - t_{j_1})\sigma_e^{j_1}(t_{j_1})\right]\left[\sum_{j_2} \delta(t' - t_{j_2})\sigma_e^{j_2}(t_{j_2})\right]\right\rangle. \tag{7.130}$$

To evaluate $C_1(t, t')$, we note that only the terms with $t_{j_1} = t_{j_2}$ contribute the double sum. As a result,

$$C_1(t, t') = \left\langle \sum_j u(t - t_j)\,\langle f_e^j(t)f_e^j(t')\rangle\right\rangle. \tag{7.131}$$

Substituting for $\langle f_e^j(t)f_e^j(t')\rangle$ and simplifying, we obtain

$$C_1(t, t') = (\gamma_e + \gamma_e')\,\langle N_e(t)\rangle\,\delta(t - t'), \tag{7.132}$$

where we use Eqs. (7.107) and (7.119). To evaluate $C_2(t, t')$, we need to consider the injection statistics of atoms following Ref. [400]. The result is found to be

$$C_2(t, t') = R^2 + (1 - p)R\delta(t - t'), \tag{7.133}$$

where the parameter p lies in the range $0 < p < 1$ depending on the injection statistics.

We use the same method for $N_g(t)$. Equations for $N_g(t)$ and $M(t)$ are found to be

$$\frac{d}{dt}N_g(t) = -\gamma_g N_g(t) + \gamma_e'N_e(t) + \mathcal{G}A_k^\dagger(t)M(t) + \mathcal{G}M^\dagger(t)A_k(t) + F_g(t), \tag{7.134}$$

$$\frac{d}{dt}M(t) = -\gamma_{eg}M(t) + \mathcal{G}[N_e(t) - N_g(t)]A_k(t) + F_M(t), \tag{7.135}$$

where the new Langevin forces are given by

$$F_g(t) = \sum_j u(t - t_j)f_g^j(t) + \sum_j \delta(t - t_j)\sigma_g^j(t_j), \tag{7.136}$$

$$F_M(t) = -i\sum_j u(t - t_j)f_\sigma^j(t) - i\sum_j \delta(t - t_j)\sigma^j(t_j). \tag{7.137}$$

These forces vanish on average and their correlation functions can be calculated using the procedure described for calculating $\langle F_e(t)F_e(t')\rangle$.

The following list summarizes the correlation functions for the three Langevin forces associated with three macroscopic atomic parameters:

$$\langle F_e(t)F_e(t')\rangle = (\gamma_e + \gamma_e')\langle N_e(t)\rangle\,\delta(t-t') + R(1-p)\delta(t-t'), \qquad (7.138)$$

$$\langle F_g(t)F_g(t')\rangle = \gamma_g\langle N_g(t)\rangle\,\delta(t-t') + \gamma_e'\langle N_e(t)\rangle\,\delta(t-t'), \qquad (7.139)$$

$$\langle F_e(t)F_g(t')\rangle = -\gamma_e'\langle N_e(t)\rangle\,\delta(t-t'), \qquad (7.140)$$

$$\langle F_g(t)F_M(t')\rangle = \gamma_g\langle M(t)\rangle\,\delta(t-t'), \qquad (7.141)$$

$$\langle F_M(t)F_e(t')\rangle = (\gamma_e + \gamma_e')\langle M(t)\rangle\,\delta(t-t'), \qquad (7.142)$$

$$\langle F_M(t)F_g(t')\rangle = -\gamma_e'\langle M(t)\rangle\,\delta(t-t'), \qquad (7.143)$$

$$\langle F_M^\dagger(t)F_M(t')\rangle = (2\gamma_{eg} - \gamma_e - \gamma_e')\langle N_e(t)\rangle\,\delta(t-t') + R\delta(t-t'), \qquad (7.144)$$

$$\langle F_M(t)F_M^\dagger(t')\rangle = (2\gamma_{eg} - \gamma_g)\langle N_g(t)\rangle\,\delta(t-t') + \gamma_e'\langle N_e(t)\rangle\,\delta(t-t'). \qquad (7.145)$$

Aside 7.4 describes an intuitive way to check the accuracy of these correlation functions.

We have obtained in this section the quantum Langevin equations for the field operator A_k and three macroscopic atomic operators $N_e(t)$, $N_g(t)$, and $M(t)$. These four equations describe fully the laser dynamics and fluctuations associated with them. Their derivation is instructive to the extent it shows underlying details that can be used to construct Langevin force terms for other quantum devices. The next task involves solving these operator equations using a computer and is the topic of the following section.

Aside 7.4   A Fast Way to Evaluate Correlation Strengths of Langevin Forces

The Langevin force terms appearing in a quantum-noise analysis can often be calculated using simplified shot noise models [403, 404, 405, 392, 393, 406]. Even though such models are not always valid, they provide a valuable tool for checking the accuracy of correlation functions of the Langevin forces. We discuss such a shot-noise model for the Langevin forces associated with the quantum theory of lasers.

As we saw in Section 7.1.4, shot noise results from the discrete nature of particles (photons, electrons, etc.) flowing in and out of their reservoirs. The associated spectral density of shot noise is constant (frequency independent) and is proportional to the average rate of particle flow. In the Langevin formalism, each flow of particles from a reservoir contributes shot noise to the noise associated with that reservoir.

Consider for example two reservoirs labeled r and s, and let $F_r(t)$ and $F_s(t)$ be the Langevin forces associated with them. To calculate the strength of the correlation function $\langle F_r(t)F_r(t)\rangle$, we simply add the flow rates of particles arriving at and leaving the reservoir r. However, in the case of the correlation function $\langle F_r(t)F_s(t)\rangle$ representing the exchange of particles between the reservoirs r and s, we not only add the flow rates between the reservoirs but also multiply the final result by -1. As most of the flow rates can be found by simple inspection, it is possible to quickly estimate the correlation strengths of the Langevin forces using this method.

For example, consider the correlation function $\langle F_g(t)F_g(t')\rangle$ given in Eq. (7.139). As there are two particle flows associated with the ground state, namely the decay rate $\gamma_g\langle N_g(t)\rangle$ and the spontaneous emission rate from the excited state, $\gamma_e'\langle N_e(t)\rangle$, the correlation strength should be the sum of these values as given in Eq. (7.139). However, in the case of the correlation function $\langle F_e(t)F_g(t')\rangle$ given in Eq. (7.140), the only rate that corresponds to an exchange of particles between the two states is the spontaneous-emission rate $\gamma'_e \langle N_e(t)\rangle$. Thus the correlation strength in this case is the negative of the above rate, as seen in Eq. (7.140).

7.3.3 c-Number Langevin Equations

The Langevin operator equations often cannot be solved analytically, thus requiring the use of numerical methods. However, numerical methods cannot deal with operators. It is thus necessary to convert the operator equations into their equivalent c-number form. It is not obvious how to do that when two operators do not commute. The solution is to define the order of operators uniquely so that a direct mapping between an operator product and its equivalent c-number representation can be established. It is common to employ normal ordering in which the Hermitian conjugate of an operator precedes it. Following this rule, we adopt the following order for the atomic and field operators appearing in the quantum Langevin equations: $A_k^\dagger(t)$, $M^\dagger(t)$, $N_e(t)$, $N_g(t)$, $M(t)$, $A_k(t)$. However, fixing of the order of operators may require redefinition of some correlation relations. The resulting c-number equations are only accurate up to second-order moments, which is not a real restriction because Langevin forces are specified fully through their first two moments.

To obtain the c-number equations, we follow these steps. (1) Put the operator equations in normal order and rearrange the terms to match the operator order indicated above. (2) Replace the operators with the corresponding c numbers using the mapping

$$A_k(t) \to \mathcal{A}(t), \quad M(t) \to \mathcal{M}(t), N_e(t) \to \mathcal{N}_e(t), \quad N_g(t) \to \mathcal{N}_g(t). \tag{7.146}$$

(3) Establish the correct correlation functions by matching the first- and second-order moments with those obtained earlier for the corresponding operators.

The first two steps produce the following c-number Langevin equations for the four operators:

$$\frac{d}{dt}\mathcal{A}(t) = -\frac{\gamma}{2}\mathcal{A}(t) + \mathcal{G}\mathcal{M}(t) + \mathcal{F}_f(t), \tag{7.147}$$

$$\frac{d}{dt}\mathcal{M}(t) = -\gamma_{eg}\mathcal{M}(t) + \mathcal{G}[\mathcal{N}_e(t) - \mathcal{N}_g(t)]\mathcal{A}(t) + \mathcal{F}_M(t), \tag{7.148}$$

$$\frac{d}{dt}\mathcal{N}_e(t) = R - (\gamma_e + \gamma'_e)\mathcal{N}_e(t) - \mathcal{G}\mathcal{A}^*(t)\mathcal{M}(t) - \mathcal{G}\mathcal{M}^*(t)\mathcal{A}(t) + \mathcal{F}_e(t), \tag{7.149}$$

$$\frac{d}{dt}\mathcal{N}_g(t) = -\gamma_g\mathcal{N}_g(t) + \gamma'_e\mathcal{N}_e(t) + \mathcal{G}\mathcal{A}^*(t)\mathcal{M}(t) + \mathcal{G}\mathcal{M}^*(t)\mathcal{A}(t) + \mathcal{F}_g(t). \tag{7.150}$$

Here the random processes $\mathcal{F}_k(t)$ with $k = f, M, e,$ or g are the c-number Langevin forces with the properties

$$\langle \mathcal{F}_k(t)\rangle = 0, \qquad \langle \mathcal{F}_k(t)\mathcal{F}_l(t')\rangle = 2\mathcal{D}_{kl}\delta(t - t'), \tag{7.151}$$

where $\mathcal{D}_{kl}$ are the diffusion coefficients.

For the third step, we need to match the second-order moments for each operator and its c-number and find the correct diffusion coefficient. We illustrate the steps involved by matching the correlation functions for the operator $M(t)$ and $\mathcal{M}(t)$. In the case of the operator M, the correlation function $\langle M^\dagger(t)M(t)\rangle$ is related to dM/dt as

$$\frac{d}{dt}\langle M^\dagger(t)M(t)\rangle = \langle \frac{dM^\dagger(t)}{dt}M(t)\rangle + \langle M^\dagger(t)\frac{dM(t)}{dt}\rangle. \tag{7.152}$$

Substituting the expression for dM/dt, we obtain

$$\frac{d}{dt}\langle M^\dagger(t)M(t)\rangle = -2\gamma_{eg}\langle M^\dagger(t)M(t)\rangle + \langle M^\dagger(t)F_M(t)\rangle + \langle F_M^\dagger(t)M(t)\rangle$$
$$\mathcal{G}\langle M^\dagger(t)\left[N_e(t)-N_g(t)\right]A_k(t)\rangle + \mathcal{G}\langle A_k^\dagger(t)\left[N_e(t)-N_g(t)\right]M(t)\rangle. \tag{7.153}$$

Using the generalized Einstein relation (see Section 7.1.6), $\langle M(t)F_M(t)\rangle + \langle F_M(t)M(t)\rangle = 2D_{MM}$, and a similar relation for $M^\dagger(t)$, we can write the preceding equation in the form

$$\frac{d}{dt}\langle M^\dagger(t)M(t)\rangle = -2\gamma_{eg}\langle M^\dagger(t)M(t)\rangle + \mathcal{G}\langle M^\dagger(t)\left[N_e(t)-N_g(t)\right]A_k(t)\rangle +$$
$$+ \mathcal{G}\langle A_k^\dagger(t)\left[N_e(t)-N_g(t)\right]M(t)\rangle + 2D_{M^\dagger M}, \tag{7.154}$$

where we used $[M(t), N_e(t)-N_g(t)] = 2M(t)$. The same approach is used for the correlation function $\langle M(t)M(t)\rangle$ to obtain

$$\frac{d}{dt}\langle M(t)M(t)\rangle = -2\gamma_{eg}\langle M(t)M(t)\rangle + 2\mathcal{G}\langle\left[N_e(t)-N_g(t)\right]M(t)A_k(t)\rangle +$$
$$+2\mathcal{G}\langle M(t)A_k(t)\rangle + 2D_{MM}. \tag{7.155}$$

The corresponding c-number equations for these two correlation functions are

$$\frac{d}{dt}\langle \mathcal{M}^*(t)\mathcal{M}(t)\rangle = -2\gamma_{eg}\langle \mathcal{M}^*(t)\mathcal{M}(t)\rangle + \mathcal{G}\langle \mathcal{M}^*(t)\left[N_e(t)-N_g(t)\right]\mathcal{A}(t)\rangle$$
$$+ \mathcal{G}\langle \mathcal{A}^*(t)\left[N_e(t)-N_g(t)\right]\mathcal{M}(t)\rangle + 2\mathcal{D}_{M^*M}, \tag{7.156}$$

$$\frac{d}{dt}\langle \mathcal{M}(t)\mathcal{M}(t)\rangle = -2\gamma_{eg}\langle \mathcal{M}(t)\mathcal{M}(t)\rangle + 2\mathcal{G}\langle\left[N_e(t)-N_g(t)\right]\mathcal{M}(t)\mathcal{A}(t)\rangle + 2\mathcal{D}_{MM}. \tag{7.157}$$

By matching the equations for each correlation function, we find the relations

$$\mathcal{D}_{M^*M} = D_{M^\dagger M}, \qquad \mathcal{D}_{MM} = D_{MM} + \mathcal{G}\langle \mathcal{M}(t)\mathcal{A}(t)\rangle. \tag{7.158}$$

Carrying out this procedure for all operators, diffusion coefficients for all c-number Langevin equations are found to be

$$2\mathcal{D}_{ee} = (\gamma_e + \gamma_e')\langle N_e(t)\rangle + R(1-p) - \mathcal{G}\langle \mathcal{M}^*(t)\mathcal{A}(t)+\mathcal{A}^*(t)\mathcal{M}(t)\rangle, \tag{7.159}$$

$$2\mathcal{D}_{gg} = \gamma_g\langle N_g(t)\rangle + \gamma_e'\langle N_e(t)\rangle - \mathcal{G}\langle \mathcal{M}^*(t)\mathcal{A}(t)+\mathcal{A}^*(t)\mathcal{M}(t)\rangle, \tag{7.160}$$

$$2\mathcal{D}_{eg} = -\gamma_e'\langle N_e(t)\rangle + \mathcal{G}\langle \mathcal{M}^*(t)\mathcal{A}(t)+\mathcal{A}^*(t)\mathcal{M}(t)\rangle, \tag{7.161}$$

$$2\mathcal{D}_{M^*M} = (2\gamma_{eg}-\gamma_e-\gamma_g')\langle N_e(t)\rangle + R, \tag{7.162}$$

$$2\mathcal{D}_{MM} = 2g\langle \mathcal{M}(t)\mathcal{A}(t)\rangle, \qquad 2\mathcal{D}_{gM} = \gamma_g\langle \mathcal{M}(t)\rangle. \tag{7.163}$$

At this point, we can use a numerical technique for solving the c-number Langevin equations on a computer and use the results for characterizing the noise dynamics of the laser model.

7.4 Noise Spectra

The spectral density of a noise process can be viewed as its "fingerprint." In principle, the spectral density contains the same information as temporal variations of the noise or its correlation function, but it is much more useful in practice. This is because different noise sources produce easily identifiable signatures in their noise spectral density even when their noise time traces appear indistinguishable on inspection. Thus, noise spectra play an important role in quantifying the performance of a nanoscale quantum device.

7.4.1 Langevin Formalism

In most quantum devices, the SNR is high enough that noise can be viewed as a perturbation (see Aside 7.1 for a discussion of SNR). As a result, the drift term D_μ appearing in the Langevin formalism can be considered a linear function of the system operators. This allows us to write the drift term in the Langevin equation appearing in Section 7.1.6 as $D_\mu = -L_{\mu\eta}A_\eta$ and write this equation in the form

$$\frac{d}{dt}A_\mu(t) = -\sum_\eta L_{\mu\eta}A_\eta(t) + F_\mu(t), \tag{7.164}$$

where $L_{\mu\eta}$ can be thought of as scalar elements of the matrix L. It is useful to recast this equation in a matrix form by introducing a column matrix $A(t)$ such that:

$$A(t) = \begin{bmatrix} A_1(t) \\ \vdots \\ A_\mu(t) \\ \vdots \end{bmatrix}, \qquad A^T = [A_1(t), \ldots, A_\mu(t), \ldots], \tag{7.165}$$

where the superscript T denotes the transpose operation. The resulting matrix equation is

$$\frac{d}{dt}A(t) = -LA(t) + F(t). \tag{7.166}$$

Taking the ensemble average and recalling that $\langle F(t) \rangle = 0$, we find

$$\frac{d}{dt}\langle A(t) \rangle = -L\langle A(t) \rangle. \tag{7.167}$$

It follows that the average values of A_μ will decay to zero as time increases, for any μ, if all eigenvalues of L are positive.

We proceed to calculate the spectral density, defined as the Fourier transform of the correlation function $\langle A(t)A^T(0)\rangle$. For this purpose, we invoke the quantum regression theorem (see Aside 7.2) and obtain

$$\frac{d}{dt}\langle A(t)A^T(0)\rangle = -L\langle A(t)A^T(0)\rangle, \tag{7.168}$$

$$\frac{d}{dt}\langle A(0)A^T(t)\rangle = -L\langle A(0)A^T(t)\rangle, \tag{7.169}$$

where we have assumed all noise processes to be stationary. Being first-order linear differential equations, they can be easily solved to get

$$\langle A(t)A^T(0)\rangle = \exp(-Lt)\langle A(0)A^T(0)\rangle, \tag{7.170}$$

$$\langle A(0)A^T(t)\rangle = \langle A(0)A^T(0)\rangle \exp(-Lt). \tag{7.171}$$

Invoking the Wiener–Khinchin theorem [385], the spectral-density matrix $S_A(f)$ is given by

$$S_A(\omega) = \int_{-\infty}^{\infty}\langle A(t)A^T(0)\rangle\, e^{i\omega t}\, dt. \tag{7.172}$$

The time origin ($t = 0$) is chosen such that the system has reached its stationary state long before that time. The location time origin does not matter in that situation.

In the stationary state of a system, physical parameters such as mean and the variance do not change over time. We exploit this feature and break up the Fourier integral into two parts ranging between $(-\infty, 0]$ and $[0, +\infty)$. Changing t to $-t$ in the first part, we can write the spectral matrix in the form

$$S_A(\omega) = \int_0^{\infty}\langle A(0)A^T(t)\rangle\, e^{-i\omega t}\, dt + \int_0^{\infty}\langle A(t)A^T(0)\rangle\, e^{i\omega t}\, dt, \tag{7.173}$$

where we have used the relation $A(t)A^T(0) = A(0)A^T(-t)$. Substituting the results from Eqs. (7.170) and (7.171), the integrals can be performed formally to obtain

$$S_A(\omega) = \langle A(0)A^T(0)\rangle\,(L^T - i\omega)^{-1} + (L + i\omega)^{-1}\langle A(0)A^T(0)\rangle. \tag{7.174}$$

This result shows that once we find the matrix $\langle A(0)A^T(0)\rangle$, it possible to calculate the spectral density $S_A(f)$ using simple matrix operations.

Let us apply the preceding results to the generalized Einstein relation [see Section 7.1.6 and Eq. (7.57)]:

$$2D_{\eta\mu} = \frac{d}{dt}\langle A_\eta(t)A_\mu(t)\rangle - \langle A_\eta(t)D_\mu(t)\rangle - \langle D_\eta(t)A_\mu(t)\rangle. \tag{7.175}$$

Recalling that $D_\mu = -L_{\mu\eta}A_\eta$ and that the time-derivative term vanishes in the steady state reached at $t = 0$, we obtain

$$D_{\eta\mu} = \langle A_\eta^\dagger(0)L_\mu A_\mu(0)\rangle + \langle L_\eta^\dagger A_\eta^\dagger(0)A_\mu(0)\rangle. \tag{7.176}$$

When quantum averages are carried out using the density operator, we can use the cyclic property of the trace operation to write the preceding equation in a matrix form as

$$2D = L\langle A(0)A^T(0)\rangle + \langle A(0)A^T(0)\rangle L^T, \tag{7.177}$$

where we pulled the matrix L outside the average, as its elements are c-numbers. By adding and subtracting the quantity $i\omega \langle A(0)A^T(0)\rangle$, we obtain

$$2D = (L + i\omega) \langle A(0)A^T(0)\rangle + \langle A(0)A^T(0)\rangle (L^T - i\omega), \qquad (7.178)$$

This equation suggests that the matrix D is related to the spectral density given in Eq. (7.174). If we multiply the preceding equation on the left by $(L + i\omega)^{-1}$ and on the right by $(L^T - i\omega)^{-1}$, we obtain the spectral density in the following simple form:

$$S_A(\omega) = (L + i\omega)^{-1}2D(L^T - i\omega)^{-1}. \qquad (7.179)$$

This is remarkable result. It shows that the quantum spectral density of noise can be calculated if we know the diffusion matrix and the drift-coefficient matrix for the Langevin equations governing the dynamics of a quantum device. As a useful example, we apply this method to calculate the intensity-noise spectrum and the phase-noise spectrum of a single-mode laser. However, before calculating the noise spectra, we first discuss rate equations used to model a laser.

7.4.2 Rate Equations for Lasers

Rate equations provide a simple way for describing the operation of most lasers. All lasers employ a cavity, where the pumping of a suitable gain medium is used to increase the number of photons inside the cavity through stimulated emission. For many lasers, we need just two rate equations for the variables N_p and N_c representing, respectively, the densities of photons and gain carriers inside the cavity. In the case of an atomic gain medium, N_c is the inversion density representing the population difference $N_e - N_g$. In the case of semiconductor lasers, N_c is the density of electrons in the conduction band. The applicability of rate equations is limited to problems where the phase of laser radiation plays a minor role. They also ignore all spatial inhomogeneities, assuming that N_p and N_c represent quantities spatially averaged over the cavity volume. In spite of these limitations, rate equations are very useful for gaining insight into the operation of a laser. As they capture the underlying physics by ensuring conservation of energy, their predictions are often in reasonable agreement with experiments. The most complete set of single and multimode rate equations applicable to lasers can be found in Ref. [407].

The rate equations for N_c and N_p can be written as [408]:

$$\frac{dN_c}{dt} = N_{in} - R_{nr} - R_{sp} - v_g G(N_c, N_p)N_p, \qquad (7.180)$$

$$\frac{dN_p}{dt} = \left(\Gamma v_g G(N_c, N_p) - \frac{1}{\tau_p}\right)N_p + \Gamma\beta_{sp}R_{sp}, \qquad (7.181)$$

where $\Gamma = V_p/V_c$ is the ratio of the volume V_p used by the photons and the active volume V_c to which the carriers are confined. Various terms on the right side of each equation account for the sources through which N_c and N_p increase or decrease. Thus, N_{in} is the rate at which carriers are injected into the laser (by pumping them) while R_{nr} is the non-radiative recombination at which carriers disappear without producing a photon. The next term R_{sp} represents the loss of carriers through spontaneous emission. The carriers can also vanish through stimulated emission at a rate $R_{eg}N_p$ and reappear through absorption at a rate $R_{ge}N_p$, where the order of the subscripts denotes the direction of atomic transition. Their combined effects are included through the net gain G defined as $v_g G = R_{eg} - R_{ge}$, where v_g is the group velocity of laser radiation inside the cavity. This term acts as the gain in the photon rate equation, where the term containing the photon's lifetime τ_p accounts for the loss of photons from the cavity. The last term represents the increase of photons through spontaneous emission. However, only a fraction of the spontaneously emitted photons couple to the cavity mode; β_{sp} represents this fraction and is called the spontaneous emission factor.

The output power P of the laser is related to the photon density N_p as

$$P = (\eta_0 / \tau_p)(N_p V_p \hbar \omega_c), \tag{7.182}$$

where η_0 is the efficiency with which photons of energy $\hbar \omega_c$ leak through a partially transparent mirror of the cavity. Here ω_c is the frequency of the cavity mode that is in resonance with the transition frequency of the gain medium.

Fluctuations in a laser's power occur when N_c, N_p and N_{in} fluctuate around their steady-state values. They can be included by writing these quantities as

$$N_c = \bar{N}_c + \delta N_c, \qquad N_p = \bar{N}_p + \delta N_p, \qquad N_{in} = \bar{N}_{in} + \delta N_{in}, \tag{7.183}$$

where a bar over a quantity denotes its steady state value and a δ in front of it denotes a small fluctuation. Because all fluctuations are assumed to be small, we can linearize the rate equations by expanding all terms up to the first order in a Taylor series to obtain the following set of equations:

$$\frac{d\delta N_c}{dt} = \eta_{cc}\delta N_c + \eta_{cp}\delta N_p + \delta N_{in} \tag{7.184}$$

$$\frac{d\delta N_p}{dt} = \eta_{pc}\delta N_c + \eta_{pp}\delta N_p, \tag{7.185}$$

where the coefficients η_{ij} are defined as

$$\eta_{cc} = -\frac{\partial}{\partial N_c}[R_{sp} + R_{nr} + v_g G(N_c, N_p)N_p], \tag{7.186}$$

$$\eta_{cp} = -\frac{\partial}{\partial N_p}]R_{sp} + R_{nr} + v_g G(N_c, N_p)N_p], \tag{7.187}$$

$$\eta_{pc} = \frac{\partial}{\partial N_c}\left[\left(\Gamma v_g G(N_c, N_p) - \frac{1}{\tau_p}\right)N_p + \Gamma \beta_{sp} R_{sp}\right], \tag{7.188}$$

$$\eta_{pp} = \frac{\partial}{\partial N_p}\left[\left(\Gamma v_g G(N_c, N_p) - \frac{1}{\tau_p}\right)N_p + \Gamma \beta_{sp} R_{sp}\right]. \tag{7.189}$$

All of these derivatives are evaluated using the steady-state quantities: $\bar{N}_c, \bar{N}_p$, and $\bar{N}_{in}$.

7.4.3 Relative Intensity Noise of a Laser

Fluctuations in the output power of a laser are measured through the relative intensity noise (RIN), defined as the noise variance normalized to the average-power level. If we write the instantaneous output power $P(t)$ of the laser as $P(t) = \langle P(t) \rangle + \delta P(t)$, the RIN is defined as

$$\text{RIN} = \frac{\langle \delta P^2(t) \rangle}{\langle P(t) \rangle^2}. \tag{7.190}$$

To calculate the RIN , we use the two linearized rate equations given in in Eqs. (7.184) and (7.185). However, we must add a Langevin noise term to them representing noise sources $F_c(t)$ and $F_p(t)$. The resulting equations are

$$\frac{d\delta N_c}{dt} = \eta_{cc}\delta N_c + \eta_{cp}\delta N_p + \delta N_{in} + F_c(t) \tag{7.191}$$

$$\frac{d\delta N_p}{dt} = \eta_{pc}\delta N_c + \eta_{pp}\delta N_p + F_p(t). \tag{7.192}$$

The correlation functions of Langevin-noise terms can be easily found by following the simple strategy outlined in Aside 7.4. The resulting expressions are

$$\langle F_c(t)F_c(t') \rangle = \frac{1}{V_c}\left[\bar{N}_{in} + (R_{sp} + R_{nr} + R_{eg} + R_{ge})\right]\delta(t - t'), \tag{7.193}$$

$$\langle F_p(t)F_p(t') \rangle = \frac{1}{V_p^2}\left[V_p\bar{N}_p/\tau_p + V_c(R_{eg} + R_{ge} + \beta_{sp}R_{sp})\right]\delta(t - t'), \tag{7.194}$$

$$\langle F_p(t)F_c(t') \rangle = \langle F_c(t)F_p(t') \rangle = -\frac{V_c}{V_p}(R_{eg} + R_{ge} + \beta_{sp}R_{sp})\delta(t - t'). \tag{7.195}$$

It is easier to solve the preceding two equations in the frequency domain. When we take their Fourier transform, the time derivative is replaced with $i\omega$. The resulting equations can be written in a matrix form as

$$\begin{pmatrix} -\eta_{cc} + i\omega & -\eta_{cp} \\ -\eta_{pc} & -\eta_{pp} + i\omega \end{pmatrix}\begin{pmatrix} \mathcal{F}\{\delta N_c(t)\} \\ \mathcal{F}\{\delta N_p(t)\} \end{pmatrix} = \begin{pmatrix} \mathcal{F}\{F_c(t)\} \\ \mathcal{F}\{F_p(t)\} \end{pmatrix}. \tag{7.196}$$

The solution is found by inverting the coefficient matrix and is given by

$$\begin{pmatrix} \mathcal{F}\{\delta N_c(t)\} \\ \mathcal{F}\{\delta N_p(t)\} \end{pmatrix} = \frac{1}{D(\omega)}\begin{pmatrix} -\eta_{pp} + i\omega & \eta_{cp} \\ \eta_{pc} & -\eta_{cc} + i\omega \end{pmatrix}\begin{pmatrix} \mathcal{F}\{F_c(t)\} \\ \mathcal{F}\{F_p(t)\} \end{pmatrix}, \tag{7.197}$$

where $D(\omega) = (\eta_{cc} - i\omega)(\eta_{ppP} - i\omega) - \eta_{pc}\eta_{cp}$.

We can use these expressions to calculate the spectral densities of N_c and N_p. However, we are interested in the spectral density of power fluctuations $\delta P(t)$. Even though the output power is proportional to N_p, its spectral density is not just related to $|\mathcal{F}\{\delta N_p(t)\}|^2$, as pointed out by Yamamoto and Imoto in their 1986 paper [406]. They found that the power spectral density of a laser operating above its threshold is approximately equal to the classical shot-noise level found by Fourier transforming $\langle F_p(t)F_p(t') \rangle$. However, the spectral density of δN_p is Lorentzian in shape, and its variance is equal to the average photon number $\bar{N}_p$, as dictated by its Poisson statistics. The difference stems from the presence of the output mirror, which acts as a random selector that divides the internal photon stream into its reflected and transmitted parts. Even though the average transmission of the mirror is constant, the selection of photons takes place at random times. As a result, over a short time duration, irregularities in the transmission process lead to partition noise, which affects the statistics of output power fluctuations.

We can account for this partition noise at the output mirror by adding a Langevin force $F_m(t)$ to the output power fluctuations. The modified expression for the output power variation $\delta P(t)$ becomes

$$\delta P(t) = (\eta_0/\tau_p)V_p\hbar\omega_{eg}\delta N_P(t) + F_m(t). \tag{7.198}$$

The spectral density of $\delta P(t)$ is given by

$$S_{PP}(\omega) = \frac{1}{2\pi}\int_{-\infty}^{\infty}\langle\mathcal{F}\{\delta P\}(\omega)\mathcal{F}\{\delta P\}^*(\omega')\rangle\,d\omega'. \tag{7.199}$$

By taking the Fourier transform of Eq. (7.198), we obtain the power spectral density in the form

$$S_{PP}(\omega) = [(\eta_0/\tau_p)V_p\hbar\omega_{eg}]^2[S_{\delta N_p\delta N_p}(\omega) + S_{F_mF_m}(\omega)S_{\delta N_pF_m}(\omega) + S_{F_m\delta N_p}(\omega)]. \tag{7.200}$$

The last two terms result from interference between the two noise sources.

The correlation function for $F_m(t)$ can be calculated by following the guidelines in Aside 7.4. As $(\eta_0/\tau_p)\bar{N}_pV_p$ is the rate at which photons escape the output mirror, it can be written as

$$\langle F_m(t)F_m(t')\rangle = (\eta_0/\tau_p)\bar{N}_pV_p(\hbar\omega_{eg})^2\delta(t-t'), \tag{7.201}$$

where the factor $(\hbar\omega_{eg})^2$ was added because $F_m(t)$ has power units, but the correlation strengths are calculated using the photon number N_p. It follows that the spectral density of F_m is given by

$$S_{F_mF_m}(\omega) = (\eta_0/\tau_p)\bar{N}_pV_p(\hbar\omega_{eg})^2. \tag{7.202}$$

The cross-correlation term can be calculated by considering the photon numbers inside and outside the cavity and is found to be

$$\langle F_p(t)F_m(t')\rangle = -(\eta_0/\tau_p)\bar{N}_p\hbar\omega_{eg}\delta(t-t') \to S_{F_pF_m}(\omega) = -(\eta_0/\tau_p)\bar{N}_p\hbar\omega_{eg}. \tag{7.203}$$

Note also that $S_{F_pF_m}(\omega) = S_{F_mF_p}(\omega)$. We can evaluate $S_{N_pF_m}(\omega)$ and $S_{F_mN_p}(\omega)$ by multiplying the left or right side of Eq. (7.197) with $\mathcal{F}\{F_m(t)\}(\omega')$ and carrying out the integration as specified in Eq. (7.199). Note that there is no correlation between $F_N(t)$ and $F_m(t)$ because the two are completely independent processes. The resulting expressions are:

$$S_{\delta N_pF_m}(\omega) = [D(\omega)]^{-1}(-\eta_{NN} + i\omega)S_{F_pF_m}(\omega), \tag{7.204}$$

$$S_{F_m\delta N_p}(\omega) = -[D*(\omega)]^{-1}(\eta_{NN}^* + i\omega)S_{F_mF_p}(\omega). \tag{7.205}$$

At this point we have everything needed to calculate the RIN of a laser operating above its threshold. The underlying details of the derivation are useful for learning the intricate reasoning required to calculate various correlation functions associated with the operation of any quantum device (not just lasers). For example, the RIN is a measure of the performance of a laser that can be measured and validated experimentally. The utility of such a measure is that, even though one may not know the exact origin of all stochastic terms, the derived quantities can be measured experimentally and used to validate the underlying model.

7.4.4 Spectral Bandwidth of a Laser

Even a laser oscillating in a single mode of its cavity has a finite bandwidth, often called the laser's line width. For an intuitive understanding of this line width, we must consider the role of spontaneous emission. When a spontaneously emitted photon couples into a specific mode of the cavity, the laser's phase changes suddenly in a random fashion. Thus, each such spontaneous-emission event acts as a noise source, which we can model using a Langevin force. We stress that events other than spontaneous emission also influence the line width of a real laser. The value we calculate using spontaneous emission represents the smallest value of the laser's line width having its origin in the quantum noise.

We use again the rate-equation model of Section 7.4.2 but add an additional rate equation for the phase of the laser's electric field. We construct this additional equation by considering how the refractive index of the laser's material changes with the carrier density N_c. This refractive index is a complex quantity written as $n = n_r + in_i$. Its real part n_r changes the phase of the electric field, while its imaginary part n_i is related to the g of the medium through the relation $g = 2n_i\omega_{eg}/c$. However, the real and imaginary parts are related through the Kramers–Kronig relations (see Aside 3.2). In the literature on semiconductor lasers, it is common to introduce a parameter called the line-width enhancement factor [409] to account for this relationship. This parameter is defined as

$$\alpha_h = -\frac{dn_r/dN_c}{dn_i/dN_c}. \tag{7.206}$$

Using this parameter, small variations in the optical phase satisfy [410]

$$\frac{d}{dt}\delta\phi(t) = \frac{\alpha_h}{2}\Gamma v_g\left(\frac{dG}{dN_c}\right)\delta N_c(t) + F_\phi(t), \tag{7.207}$$

where the Langevin noise term $F_\phi(t)$ accounts for phase fluctuations induced by spontaneous emission. Note that fluctuations in the laser's frequency are related to the phase through the derivative $d(\delta\phi)/dt$.

Next, we need to establish various correlation functions of $F_\phi(t)$ with itself and other Langevin forces. As this question has been addressed in Ref. [409], we quote the results here:

$$\langle F_\phi(t)F_\phi(t')\rangle = \frac{\beta_{sp}R_{sp}}{2N_{P0}}\delta(t-t'), \tag{7.208}$$

$$\langle F_\phi(t)F_P(t')\rangle = 0, \quad \langle F_\phi(t)F_N(t')\rangle = 0, \quad \langle F_\phi(t)\delta N_c(t')\rangle = 0. \tag{7.209}$$

We use these expressions to find the spectral density of $\delta\omega_{eg}$ using the relation in Eq. (7.199) but with the substitution $\delta P(t) \to \delta\omega_{eg}$:

$$S_{\delta\omega\delta\omega}(\omega) = \left(\frac{\alpha_h}{2}\Gamma v_g\frac{dG}{dN_c}\right)^2 S_{\delta N_c\delta N_c} + S_{F_\phi F_\phi}(\omega), \tag{7.210}$$

where we used the fact that phase fluctuations are not correlated with the carrier-density fluctuations. This expression shows that the spectral density of frequency fluctuations is enhanced because of carrier fluctuations when $\alpha_h \neq 0$.

The line width of a laser, $\delta\omega_l$, is defined as the full width (at half maximum) of its spectral distribution and is related to the preceding spectral density as [411]

$$\delta\omega_l = S_{\delta\omega\delta\omega}(0) = (1 + \alpha_h^2)S_{F_\phi F_\phi}(0), \tag{7.211}$$

where we combined the two terms. This relation shows why α_h is called the line-width enhancement factor. In the case of semiconductor lasers, values of α_h can vary from 2 to 8, depending on the laser's design. Thus, carrier fluctuations broaden considerably the line width of such lasers. In the absence of carrier fluctuations (or $\alpha_h = 0$), a laser's line width acquires its smallest value, dictated by phase fluctuations induced by the coupling of spontaneously emitted photons into the laser mode at random times. This is the fundamental limit set by the quantum noise.

7.5 Squeezed States of Light

One may ask whether it is possible to reduce the noise of an optical device below the fundamental limit set by the quantum noise. Although the answer is clearly no if one wants to reduce the noise at all frequencies, it turns out that the noise level of a device can be reduced below the limit set by the quantum noise in a narrow frequency range. This reduction is realized through the use of special quantum states known as *squeezed states*. In recent years, the use of squeezed states for noise reduction has been proposed for several applications.

An important application is related to the laser interferometer gravitational-wave observatory (LIGO). The LIGO employs a Michelson interferometer with two 4-km-long arms to detect gravitational waves creating relative displacements as small as 10^{-20} meter. This instrument in 2016 succeeded in detecting gravitational waves, resulting in a Nobel Prize for the team. However, its sensitivity is limited by the quantum noise and can be improved by using an instrument known as the quantum vacuum squeezer, which reduces the effects of vacuum fluctuations by squeezing them out. By the end of 2019, all of the world's LIGO detectors have been upgraded to use a quantum vacuum squeezer [412]. The use of such a device has improved the LIGO's range by 15% and has allowed the observation of gravitational waves that would have been unobservable without the use of squeezed states. Other similar applications are emerging. Thus, it is worth looking at the fundamentals behind the squeezed-state concept.

As we have discussed in Section 2.2, quantization imposes fundamental limitations on the accuracy with which we can simultaneously measure certain properties of a quantum device. Heisenberg's uncertainty principle describes the constraint imposed on the variances of two noncommuting Hermitian operators. For two such operators A and B, with the associated variances $\Delta A^2 = \langle (A - \langle A \rangle)^2 \rangle$ and $\Delta B^2 = \langle (B - \langle B \rangle)^2 \rangle$, the most general form of the uncertainty principle is given by [413]

$$\Delta A^2 \Delta B^2 \geq \frac{1}{4}\left[\langle C\rangle^2 + \langle D\rangle^2\right], \tag{7.212}$$

where $[A, B] = iC$ and $\langle D\rangle = \langle AB + BA\rangle - 2\langle A\rangle\langle B\rangle$. Clearly, $\langle D\rangle$ is a measure of the correlation between A and B.

When no correlation exists between the operators A and B such that $\langle D\rangle = 0$, we obtain the standard uncertainty relation (see Aside 2.11)

$$\Delta A^2 \Delta B^2 \geq \frac{1}{4}\langle C\rangle^2. \tag{7.213}$$

For example, the position and momentum operators, for which $[x, p] = i\hbar$ (or $\langle c\rangle = \hbar$), obey Heisenberg's uncertainty relation

$$(\Delta x)(\Delta p) \geq \hbar/2. \tag{7.214}$$

All preceding uncertainty relations show that one can reduce (or squeeze) the variance of one observable, provided the variance of the other observable increases at the same time. For example, we can squeeze Δx to a relatively small value, provided the standard deviation Δp increases such that their product remains larger than $\hbar/2$. This is because quantum mechanics imposes a constraint only on the product of these two quantities.

The best example of a squeezed state comes from quantum optics through the so-called squeezed light [118]. The underline concept is best understood by considering the phasor representation of the electric field of light in one mode of the optical field. We write its complex amplitude as $A = A_r + iA_i = |A|e^{i\phi}$, where the real and imaginary parts of A correspond to two different *quadratures*, often called the in-phase (I) and quadrature (Q) components. The uncertainty principle imposes the constraint $(\Delta A_r)(\Delta A_i) \geq \hbar/2$ when A is expressed in suitable units. This means for a given measurement, the more precisely we know the in-phase part, the more potential error there is in the quadrature part (and vice versa).

The quantum state of radiation emitted by a laser is represented by a coherent state, for which the variances of two quadratures are equal in magnitude. As shown in Figure 7.3, this gives rise to a circularly symmetric region of uncertainty for the two quadratures. By using specialized techniques, the circle can be squeezed into an ellipse, as shown schematically on the right side in Figure 7.3. This deformation implies that uncertainty in the

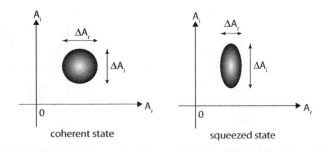

Fig. 7.3   The basics of the squeezed state.

A_r quadrature is reduced, while the uncertainty in the less relevant quadrature A_i has increased. Other kinds of squeezing are possible, depending on the physical parameter that exhibits noise below the shot-noise level. For example, photon-number squeezing occurs if the variance of the photon-number operator becomes less than the shot-noise level. We focus on *quadrature squeezing* in this section.

When an optical field is quantized, its complex amplitude A is related to the expectation value of the annihilation operator $\hat{a}$. The operators for the two quadrature components $\hat{X}$ and $\hat{Y}$ can then be introduced as

$$\hat{X} = \hat{a}^\dagger e^{i\theta} + \hat{a} e^{-i\theta}, \qquad \hat{Y} = i(\hat{a}^\dagger e^{i\theta} - \hat{a} e^{-i\theta}), \qquad (7.215)$$

where θ is an arbitrary angle. Using the commutation relation $[\hat{a}, \hat{a}^\dagger] = 1$, it is easy to show that $[\hat{X}, \hat{Y}] = 2i$ for any value of θ. For any quantum state $|\psi\rangle$, Heisenberg's uncertainty principle imposes the constraint $\sigma_X^2 \sigma_Y^2 \geq 1$, where the noise variance is calculated using $\sigma_X^2 = \langle \psi | (\hat{X} - \bar{X})^2 | \psi \rangle$, where $\bar{X} = \langle \psi | X | \psi \rangle$ is the mean value. In the case of a coherent state $\sigma_X = \sigma_Y$.

Figure 7.4(a) shows a coherent state schematically with the same amount of noise in its two quadratures, resulting in a circular shape of fluctuations around the mean value. It turns out that the nonlinear effects inside a medium can turn a coherent state into a squeezed state similar to that shown in part (b). Such states have the property that quantum fluctuations in one quadrature are reduced below those of a coherent state. In the case of Figure 7.4(b), fluctuations decrease in the Y quadrature but are enhanced in the X quadrature, resulting in an elongated noise ellipse. This is also evident from the temporal trace where amplitude noise is enhanced considerably. A phase-sensitive detection scheme such as heterodyne detection must be employed to observe noise reduction in one quadrature.

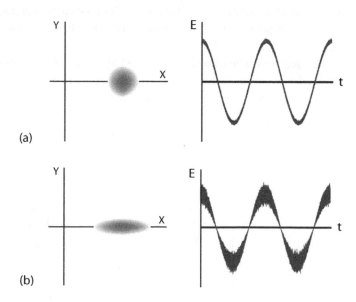

Fig. 7.4   (a) Coherent and (b) squeezed quantum states. Noise distribution along the two quadratures is shown together with the time dependence of the optical field.

A nonlinear process is required to transform a coherent state into a squeezed state. The use of four-wave mixing (FWM) for squeezing was suggested during the 1980s, and a detailed theory was developed in a 1985 paper [414]. In one implementation, a pump beam initiates the FWM process inside an optical fiber (acting as the nonlinear medium), but no signal is launched at the input end [415]. Rather, vacuum fluctuations provide the initial seed for the growth of the signal and idler fields. Such a process is called *spontaneous FWM* to differentiate it from its stimulated version. Squeezing occurs because the noise components at the signal and idler frequencies are coupled through the fiber nonlinearity. Mathematically, it is governed by the following two equations:

$$\frac{d\hat{a}_s}{dz} = i\delta\hat{a}_s + 2i\gamma A_p^2\hat{a}_i^\dagger, \tag{7.216}$$

$$\frac{d\hat{a}_i}{dz} = i\delta\hat{a}_i + 2i\gamma A_p^2\hat{a}_s^\dagger, \tag{7.217}$$

where δ provides a measure of the phase mismatch, γ is the nonlinear parameter, and A_p is the amplitude of the pump (treated classically). Vacuum fluctuations are included through the commutation relation $[\hat{a}_j(z), \hat{a}_k^\dagger(z)] = \delta_{jk}$ (where j, $k = s$ or i), which must be satisfied for all values of z.

Equations (7.216) and (7.217) can be solved easily because of their linear nature. Their general solution is given by [416]

$$\hat{a}_s(z) = \hat{a}_s(0)[\cosh(gz) + (i\delta/g)\sinh(gz)] + i(\gamma/g)A_p^2\hat{a}_i^\dagger(0)\sinh(gz), \tag{7.218}$$

$$\hat{a}_i(z) = \hat{a}_i(0)[\cosh(gz) + (i\delta/g)\sinh(gz)] + i(\gamma/g)A_p^2\hat{a}_s^\dagger(0)\sinh(gz), \tag{7.219}$$

where the parametric gain g is defined as $g = (\gamma^2|A_p|^4 - \delta^2)^{1/2}$. This solution reduces to that given in Ref. [414] in the case of perfect phase matching ($\delta = 0$). Note that the signal amplitude at a distance z inside the fiber evolves as a linear combination of $\hat{a}_s(0)$ and $\hat{a}_i^\dagger(0)$. It is this feature that is responsible for squeezing. The total field at the output end of a fiber of length L is given by

$$\hat{A}_t(t) = A_p(L) + \hat{a}_s(L)\exp(-i\Omega t) + \hat{a}_i(L)\exp(i\Omega t), \tag{7.220}$$

where $\Omega = \omega_s - \omega_p$ is the signal's detuning from the pump frequency ω_p.

From a physical standpoint, squeezing can be understood as deamplification of signal and idler waves for certain values of the relative phase between them. A phase-sensitive detection scheme is employed, and the phase of the local oscillator is adjusted to change the relative phase. In practice, the pump itself is used as a local oscillator with an adjustable phase θ. Its beating with the signal and idler fields at a photodetector generates an electric current whose noise power varies with both Ω and θ. In a 1986 experiment, a 647-nm CW pump beam was launched into a 114-m-long optical fiber [417]. It was necessary to cool the fiber to liquid-helium temperature to overcome the noise produced by spontaneous Brillouin scattering. Cooling also reduced the threshold of stimulated Brillouin scattering (SBS), which was suppressed by modulating the pump beam at a frequency much larger than the bandwidth of the Brillouin gain spectrum. Thermal Brillouin scattering from guided acoustic waves was still the most limiting factor in the experiment; it limited both the frequency range and the amount of noise squeezing. Squeezing was observed in two spectral bands located around 45 and 55 MHz but its magnitude was below 1 dB on the decibel scale. More recently, values as large as 10 dB have been realized.

In a 2013 experiment, squeezing was observed by measuring fluctuations in two quadratures of the electromagnetic field generated by a tunnel junction acting as a quantum conductor [418]. It was necessary to cool the tunnel junction to 10 mK. Even then, the observed squeezing was limited to below 1 dB. It was predicted in a 2019 study that noise can be squeezed by as much as 12 dB through parametric amplification inside a superconducting tunnel junction [419]. Considerable research has been done in recent years for investigating the statistics of photons in the nonclassical radiation emitted by a tunnel junction. Squeezing is expected to continue to play an important role in such studies.

References

[1] M. Boholm, "The use and meaning of nano in American English: Towards a systematic description," *Ampersand*, vol. 3, pp. 163–173, 2016.

[2] N. Taniguchi, "On the basic concept of nanotechnology," in *Proc. Int. Conf. Prod. Eng. Part II*. Japan Society of Precision Engineering, 1974, pp. 18–23.

[3] D. Guo, G. Xie, and J. Luo, "Mechanical properties of nanoparticles: Basics and applications," *J. Physics D: Appl. Phys.*, vol. 47, p. 013001, 2014.

[4] F. G. Shi, "Size dependent thermal vibrations and melting in nanocrystals," *Journal of Materials Research*, vol. 9, no. 5, pp. 1307–1314, 1994.

[5] E. Győry and F. Márkus, "Size dependent thermal conductivity in nano-systems," *Thin Solid Films*, vol. 565, pp. 89–93, 2014.

[6] C. N. R. Rao, G. U. Kulkarni, P. J. Thomas, and P. P. Edwards, "Size dependent chemistry: Properties of nanocrystals," *Chemistry*, vol. 8, pp. 28–35, 2002.

[7] F. Yang, D. Deng, X. Pan, Q. Fu, and X. Bao, "Understanding nano effects in catalysis," *National Science Review*, vol. 2, no. 2, pp. 183–201, 2015.

[8] Y. V. Nazarov and Y. M. Blanter, *Quantum Transport: Introduction to Nanoscience*. Cambridge University Press, 2009.

[9] S. Maekawa and T. Shinjo, *Spin Dependent Transport in Magnetic Nanostructures*, ser. Advances in Condensed Matter Science. CRC Press, 2002.

[10] M. Quinten, *Optical Properties of Nanoparticle Systems: Mie and Beyond*. Wiley, 2010.

[11] P. Pawlow, "The dependency of the melting point on the surface energy of a solid body," *Z. Phys. Chem.*, vol. 65, no. 5, p. 545, 1909.

[12] P. Buffat and J.-P. Borel, "Size effect on the melting temperature of gold particles," *Phys. Rev. A*, vol. 13, pp. 2287–2298, 1976.

[13] C. Q. Sun, Y. Wang, B. K. Tay, S. Li, H. Huang, and Y. B. Zhang, "Correlation between the melting point of a nanosolid and the cohesive energy of a surface atom," *J. Phys. Chem. B*, vol. 106, no. 41, pp. 10701–10705, 2002.

[14] X. Q. Wang and A. S. Mujumdar, "Heat transfer characteristics of nanofluids: a review," *Int. J. Thermal Sciences*, vol. 46, pp. 1–19, 2007.

[15] S. G. Volz and G. Chen, "Molecular dynamics simulation of thermal conductivity of silicon nanowires," *Appl. Phys. Lett.*, vol. 75, no. 14, pp. 2056–2058, 1999.

[16] N. Mingo and D. A. Broido, "Carbon nanotube ballistic thermal conductance and its limits," *Phys. Rev. Lett.*, vol. 95, no. 9, p. 096105, 2005.

[17] G. C. Bond, "Gold: a relatively new catalyst," *Catalysis Today*, vol. 72, pp. 5–9, 2002.

[18] L. Li and Y.-J. Zhu, "High chemical reactivity of silver nanoparticles toward hydrochloric acid," *J. Colloidal Interf. Sci.*, vol. 303, pp. 415–418, 2006.

[19] A. C. Ford, J. C. Ho, Y.-L. Chueh, Y.-C. Tseng, Z. Fan, J. Guo, J. Bokor, and A. Javey, "Diameter-dependent electron mobility of InAs nanowires," *Nano Letters*, vol. 9, no. 1, pp. 360–365, 2008.

[20] L. G. Zhou and H. Huang, "Are surfaces elastically softer or stiffer?" *Appl. Phys. Lett.*, vol. 84, no. 11, pp. 1940–1942, 2004.

[21] H. Liang, M. Upmanyu, and H. Huang, "Size-dependent elasticity of nanowires: Nonlinear effects," *Phys. Rev. B*, vol. 71, no. 24, p. 241403, 2005.

[22] X. Y. Kong and Z. L. Wang, "Spontaneous polarization-induced nanohelixes, nanosprings, and nanorings of piezoelectric nanobelts," *Nano Letters*, vol. 3, no. 12, pp. 1625–1631, 2003.

[23] L. Mandel and E. Wolf, "The measures of bandwidth and coherence time in optics," *Proc. Phys. Soc.*, vol. 80, no. 4, p. 894, 1962.

[24] C. L. Mehta, "Coherence time and effective bandwidth of blackbody radiation," *Nuovo Cimento*, vol. 28, no. 2, pp. 401–408, 1963.

[25] G. Chen, "Size and interface effects on thermal conductivity of superlattices and periodic thin-film structures," *J. Heat Transfer*, vol. 119, no. 2, pp. 220–229, 1997.

[26] E. Stern, R. Wagner, F. J. Sigworth, R. Breaker, T. M. Fahmy, and M. A. Reed, "Importance of the Debye screening length on nanowire field effect transistor sensors," *Nano Letters*, vol. 7, no. 11, pp. 3405–3409, 2007.

[27] L. Guo, E. Leobandung, and S. Y. Chou, "A room-temperature silicon single-electron metal-oxide-semiconductor memory with nanoscale floating-gate and ultranarrow channel," *Appl. Phys. Lett.*, vol. 70, pp. 850–852, 1997.

[28] S. Maier, *Plasmonics: Fundamentals and Applications.* Springer, 2007.

[29] L. Novotny and B. Hecht, *Principles of Nano-Optics.* Cambridge University Press, 2012.

[30] J. D. Jackson, *Classical Electrodynamics, 3rd ed.* Wiley, 2007.

[31] V. V. Pokropivny and V. V. Skorokhod, "Classification of nanostructures by dimensionality and concept of surface forms engineering in nanomaterial science," *Materials Science and Engineering*, vol. 27, no. 5, pp. 990–993, 2007.

[32] M. G. Bawendi, M. L. Steigerwald, and L. E. Brus, "The quantum mechanics of larger semiconductor clusters (quantum dots)," *Annu. Rev. Phys. Chem.*, vol. 41, no. 1, pp. 477–496, 1990.

[33] E. N. Economou, *The Physics of Solids: Essentials and Beyond.* Springer, 2010.

[34] B. Lamprecht, J. R. Krenn, A. Leitner, and F. R. Aussenegg, "Resonant and off-resonant light-driven plasmons in metal nanoparticles studied by femtosecond-resolution third-harmonic generation," *Phys. Rev. Lett.*, vol. 83, pp. 4421–4424, 1999.

[35] R. A. Molina, D. Weinmann, and R. A. Jalabert, "Oscillatory size dependence of the surface plasmon linewidth in metallic nanoparticles," *Phys. Rev. B*, vol. 65, p. 155427, 2002.

[36] D. F. Zaretsky, P. A. Korneev, S. V. Popruzhenko, and W. Becker, "Landau damping in thin films irradiated by a strong laser field," *J. Phys. B: At. Mol. Opt. Phys.*, vol. 37, no. 24, pp. 4817–4830, 2004.

[37] P. P. Edwards, R. L. Johnston, and C. N. R. Rao, *On the Size-Induced Metal-Insulator Transition in Clusters and Small Particles*. Wiley-Blackwell, 2008, ch. 4.8, pp. 1454–1481.

[38] R. Kubo, "Electronic properties of metallic fine particles." *J. Phys. Soc. Japan*, vol. 17, no. 6, pp. 975–986, 1962.

[39] A. Kawabata and R. Kubo, "Electronic properties of fine metallic particles. ii. plasma resonance absorption," *Journal of the Physical Society of Japan*, vol. 21, no. 9, pp. 1765–1772, 1966.

[40] L. D. Landau and E. M. Lifshitz, *Quantum Mechanics: Non-Relativistic Theory*. Elsevier, 1981.

[41] M. Mucha-Kruczynski, *Theory of Bilayer Graphene Spectroscopy*. Springer, 2012.

[42] T. V. Shahbazyan, "Local density of states for nanoplasmonics," *Phys. Rev. Lett.*, vol. 117, p. 207401, 2016.

[43] W. K. Wootters and W. H. Zurek, "A single quantum cannot be cloned," *Nature*, vol. 299, pp. 802–803, 1982.

[44] J. S. Lundeen, B. Sutherland, A. Patel, C. Stewart, and C. Bamber, "Direct measurement of the quantum wavefunction," *Nature*, vol. 474, pp. 188–191, 2011.

[45] N. W. M. Ritchie, J. G. Story, and R. G. Hulet, "Realization of a measurement of a weak value," *Phys. Rev. Lett.*, vol. 66, pp. 1107–1110, 1991.

[46] A. Messiah, *Quantum Mechanics*. Dover Publications, 2014.

[47] Y. Aharonov and D. Bohm, "Significance of electromagnetic potentials in the quantum theory," *Phys. Rev.*, vol. 115, pp. 485–491, 1959.

[48] J. Anandan, "The geometric phase," *Nature*, vol. 360, pp. 307–313, 1992.

[49] J. Anandan and Y. Aharonov, "Geometric quantum phase and angles," *Phys. Rev. D*, vol. 38, pp. 1863–1870, 1988.

[50] Y. Aharonov and J. Anandan, "Phase change during a cyclic quantum evolution," *Phys. Rev. Lett.*, vol. 58, pp. 1593–1596, 1987.

[51] Y. Aharonov and A. Casher, "Topological quantum effects for neutral particles," *Phys. Rev. Lett.*, vol. 53, pp. 319–321, 1984.

[52] R. A. Webb, S. Washburn, C. P. Umbach, and R. B. Laibowitz, "Observation of h/e Aharonov–Bohm oscillations in normal-metal rings," *Phys. Rev. Lett.*, vol. 54, pp. 2696–2699, 1985.

[53] B. L. Al'tshuler, A. G. Aronov, and B. Z. Spivak, "The Aaronov–Bohm effect in disordered conductors," *JETP Lett.*, vol. 33, no. 2, pp. 94–97, 1981.

[54] K. K. Choi, D. C. Tsui, and K. Alavi, "Dephasing time and one-dimensional localization of two-dimensional electrons in $GaAsAl_xGa_{1-x}As$ heterostructures," *Phys. Rev. B*, vol. 36, pp. 7751–7754, 1987.

[55] L. P. Lévy, G. Dolan, J. Dunsmuir, and H. Bouchiat, "Magnetization of mesoscopic copper rings: Evidence for persistent currents," *Phys. Rev. Lett.*, vol. 64, pp. 2074–2077, 1990.

[56] M. Büttiker, Y. Imry, and R. Landauer, "Josephson behavior in small normal one-dimensional rings," *Phys. Lett. A*, vol. 96, no. 7, pp. 365–367, 1983.

[57] H.-F. Cheung, E. K. Riedel, and Y. Gefen, "Persistent currents in mesoscopic rings and cylinders," *Phys. Rev. Lett.*, vol. 62, pp. 587–590, 1989.

[58] F. Schwerer, J. Conroy, and S. Arajs, "Matthiessen's rule and the electrical resistivity of iron-silicon solid solutions," *J. Phys. Chem. Solids*, vol. 30, no. 6, pp. 1513–1525, 1969.

[59] B. J. van Wees, H. van Houten, C. W. J. Beenakker, J. G. Williamson, L. P. Kouwenhoven, D. van der Marel, and C. T. Foxon, "Quantized conductance of point contacts in a two-dimensional electron gas," *Phys. Rev. Lett.*, vol. 60, pp. 848–850, 1988.

[60] D. A. Wharam, T. J. Thornton, R. Newbury, M. Pepper, H. Ahmed, J. E. F. Frost, D. G. Hasko, D. C. Peacock, D. A. Ritchie, and G. A. C. Jones, "One-dimensional transport and the quantisation of the ballistic resistance," *J. Phys. C: Solid State Phys.*, vol. 21, no. 8, pp. L209–L214, 1988.

[61] M. Büttiker, "Voltage fluctuations in small conductors," *Physical Review B*, vol. 35, no. 8, pp. 4123–4126, 1987.

[62] M. Büttiker, "Symmetry of electrical conduction," *IBM J. Res. Develop.*, vol. 32, no. 3, pp. 317–334, 1988.

[63] R. Landauer, "Spatial variation of currents and fields due to localized scatterers in metallic conduction," *IBM Journal of Research and Development*, vol. 1, no. 3, pp. 223–231, 1957.

[64] R. Landauer, "Electrical resistance of disordered one-dimensional lattices," *Philos. Mag.*, vol. 21, no. 172, pp. 863–867, 1970.

[65] C. W. J. Beenakker and H. van Houten, "Quantum transport in semiconductor nanostructures," vol. 44, pp. 1–228, 1991.

[66] K. K. Likharev, "Single-electron devices and their applications," *Proc. IEEE*, vol. 87, no. 4, pp. 606–632, 1999.

[67] P. Lafarge, H. Pothier, E. R. Williams, D. Esteve, C. Urbina, and M. H. Devoret, "Direct observation of macroscopic charge quantization," *Zeit. Physik B*, vol. 85, no. 3, pp. 327–332, 1991.

[68] A. Aassime, D. Gunnarsson, K. Bladh, P. Delsing, and R. Schoelkopf, "Radio-frequency single-electron transistor: Toward the shot-noise limit," *Appl. Phys. Lett.*, vol. 79, no. 24, pp. 4031–4033, 2001.

[69] T. A. Fulton and G. J. Dolan, "Observation of single-electron charging effects in small tunnel junctions," *Phys. Rev. Lett.*, vol. 59, no. 1, pp. 109–112, 1987.

[70] C. Garola and A. Rossi, *The Foundations of Quantum Mechanics: Historical Analysis and Open Questions*. Springer, 2012.

[71] J. Zheng-Johansson and P. Johansson, *Unification of Classical, Quantum, and Relativistic Mechanics and of the Four Forces*. Nova Science Publishers, 2006.

[72] A. Shabana, *Dynamics of Multibody Systems*. Cambridge University Press, 2013.

[73] F. Scheck, *Mechanics: From Newton's Laws to Deterministic Chaos*. Springer, 2010.

[74] A. Tikhonov and A. Samarskii, *Equations of Mathematical Physics*. Dover Publications, 2013.

[75] M. Calkin, *Lagrangian and Hamiltonian Mechanics*. World Scientific, 1996.

[76] H. Muller-Kirsten, *Electrodynamics: An Introduction Including Quantum Effects*. World Scientific, 2004.

[77] A. Zangwill, *Modern Electrodynamics*. Cambridge University Press, 2013.

[78] I. Gelfand, E. Glagoleva, and A. Kirillov, *The Method of Coordinates*. Dover Publications, 2002.

[79] J. Maruskin, *Introduction to Dynamical Systems and Geometric Mechanics*. Solar Crest Publishing, 2012.

[80] P. Hamill, *A Student's Guide to Lagrangians and Hamiltonians*. Cambridge University Press, 2014.

[81] J. Török, *Analytical Mechanics: With an Introduction to Dynamical Systems*. Wiley, 2000.

[82] C. Lanczos, *The Variational Principles of Mechanics*. Dover Publications, 2012.

[83] D. Holm, *Geometric Mechanics: Rotating, Translating and Rolling*. Imperial College Press, 2008.

[84] B. Dacorogna, *Introduction to the Calculus of Variations: Third Edition*. World Scientific, 2014.

[85] M. Frémond, *Virtual Work and Shape Change in Solid Mechanics*. Springer, 2016.

[86] P. Mann, *Lagrangian and Hamiltonian Dynamics*. Oxford University Press, 2018.

[87] J. José and E. Saletan, *Classical Dynamics: A Contemporary Approach*. Cambridge University Press, 1998.

[88] W. Cottingham and D. Greenwood, *An Introduction to the Standard Model of Particle Physics*. Cambridge University Press, 2007.

[89] J. Taylor, *Classical Mechanics*. University Science Books, 2005.

[90] G. Giachetta, L. Mangiarotti, and G. Sardanashvili, *New Lagrangian and Hamiltonian Methods in Field Theory*. World Scientific, 1997.

[91] T. Lee, M. Leok, and N. McClamroch, *Global Formulations of Lagrangian and Hamiltonian Dynamics on Manifolds*. Springer, 2017.

[92] K. Vogtmann, A. Weinstein, and V. Arnold, *Mathematical Methods of Classical Mechanics*. Springer, 1997.

[93] O. Johns, *Analytical Mechanics for Relativity and Quantum Mechanics*. Oxford University Press, 2011.

[94] H. Jong-Ping and Z. Yuan-Zhong, *Lorentz and Poincare Invariance: 100 Years Of Relativity*. World Scientific, 2001.

[95] R. Shankar, *Principles of Quantum Mechanics*. Springer, 2012.

[96] J. Sakurai and J. Napolitano, *Modern Quantum Mechanics*. Cambridge University Press, 2017.

[97] H. Lüth, *Quantum Physics in the Nanoworld: Schrödinger's Cat and the Dwarfs*. Springer, 2015.

[98] W. Greiner and J. Reinhardt, *Quantum Electrodynamics*. Springer, 2013.

[99] N. Young, *An Introduction to Hilbert Space*. Cambridge University Press, 1988.

[100] N. N. Bogolubov, *Introduction to Quantum Statistical Mechanics*. World Scientific, 2010.

[101] Y. Ohnuki and S. Kamefuchi, *Quantum Field Theory and Parastatistics*. Springer, 2011.

[102] F. Gross, *Relativistic Quantum Mechanics and Field Theory*. Wiley, 2008.

[103] E. Prugovecki, *Quantum Mechanics in Hilbert Space*. Academic Press, 1971.

[104] P. Bongaarts, *Quantum Theory: A Mathematical Approach*. Springer, 2014.

[105] G. Jaeger, *Entanglement, Information, and the Interpretation of Quantum Mechanics*. Springer, 2009.

[106] A. Aczel, *Entanglement: The Greatest Mystery in Physics*. Wiley, 2003.

[107] B. Thaller, *Advanced Visual Quantum Mechanics*. Springer, 2005.

[108] J. S. Bell, M. Bell, K. Gottfried, and M. Veltman, *On the Foundations of Quantum Mechanics*. World Scientific, 2001.

[109] F. Selleri, *Quantum Mechanics Versus Local Realism: The Einstein-Podolsky-Rosen Paradox*. Springer, 2013.

[110] D. Hemmick and A. Shakur, *Bell's Theorem and Quantum Realism: Reassessment in Light of the Schrödinger Paradox*. Springer, 2011.

[111] F. Levi, "Nonlocal legacy," *Nature Physics*, vol. 11, p. 384, 2015.

[112] M. Fadel, T. Zibold, B. Décamps, and P. Treutlein, "Spatial entanglement patterns and Einstein-Podolsky-Rosen steering in Bose-Einstein condensates," *Science*, vol. 360, no. 6387, pp. 409–413, 2018.

[113] P. Busch, P. Lahti, J. Pellonpää, and K. Ylinen, *Quantum Measurement*. Springer, 2016.

[114] J. Audretsch, *Entangled Systems: New Directions in Quantum Physics*. Wiley, 2008.

[115] M. Nielsen, M. Nielsen, and I. Chuang, *Quantum Computation and Quantum Information*. Cambridge University Press, 2000.

[116] A. N. Whitehead and B. Russell, *Principia Mathematica to *56*. Cambridge University Press, 1997.

[117] B. Venkatachala, *Inequalities: An Approach through Problems*. Springer, 2018.

[118] M. Scully and M. Zubairy, *Quantum Optics*. Cambridge University Press, 1997.

[119] J. S. Townsend, *A Modern Approach to Quantum Mechanics*. University Science Books, 2000.

[120] M. Ohya and D. Petz, *Quantum Entropy and Its Use*. Springer, 2004.

[121] T. Heinosaari and M. Ziman, *The Mathematical Language of Quantum Theory: From Uncertainty to Entanglement*. Cambridge University Press, 2011.

[122] G. Sterman, *An Introduction to Quantum Field Theory*. Cambridge University Press, 1993.

[123] D. Finkelstein, *Quantum Relativity: A Synthesis of the Ideas of Einstein and Heisenberg*. Springer, 2012.

[124] D. A. Dubin, M. A. Hennings, and T. B. Smith, *Mathematical Aspects of Weyl Quantization and Phase*. World Scientific Publishing Company, 2000.

[125] M. Kachelriess, *Quantum Fields: From the Hubble to the Planck Scale*. Oxford University Press, 2017.

[126] P. Francesco, P. Di Francesco, P. Mathieu, D. Sénéchal, and D. Senechal, *Conformal Field Theory*. Springer, 1997.

[127] C. Enz, *A Course on Many-Body Theory Applied to Solid-State Physics*. World Scientific, 1992.

[128] R. Newton, *Scattering Theory of Waves and Particles*. Springer, 2013.

[129] G. J. Kalman, J. M. Rommel, and K. Blagoev, *Strongly Coupled Coulomb Systems*. Springer, 2007.

[130] T. Wu and T. Ohmura, *Quantum Theory of Scattering*. Dover Publications, 2014.

[131] R. Jishi, *Feynman Diagram Techniques in Condensed Matter Physics*. Cambridge University Press, 2013.

[132] H. Bruus and K. Flensberg, *Many-Body Quantum Theory in Condensed Matter Physics*. Oxford University Press, 2004.

[133] J. A. Heras, "Can Maxwell's equations be obtained from the continuity equation?" *Am. J. Phys.*, vol. 75, no. 7, pp. 652–657, 2007.

[134] F. Miller, A. Vandome, and J. McBrewster, *Covariant Formulation of Classical Electromagnetism*. VDM Publishing, 2010.

[135] A. Garg, *Classical Electromagnetism in a Nutshell*. Princeton University Press, 2012.

[136] W. Rosser, *Introductory Special Relativity*. Taylor & Francis, 1992.

[137] M. Dalarsson and N. Dalarsson, *Tensors, Relativity, and Cosmology*. Elsevier Science, 2015.

[138] J. Schröter and C. Pfeifer, *Minkowski Space: The Spacetime of Special Relativity*. De Gruyter, 2017.

[139] C. Quigg, *Gauge Theories of the Strong, Weak, and Electromagnetic Interactions*, 2nd ed. Princeton University Press, 2013.

[140] A. Babin and A. Figotin, *Neoclassical Theory of Electromagnetic Interactions: A Single Theory for Macroscopic and Microscopic Scales*. Springer, 2016.

[141] K.-H. Yang, "The physics of gauge transformations," *Am. J. Phys.*, vol. 73, no. 8, pp. 742–751, 2005.

[142] J. D. Jackson, "From Lorenz to Coulomb and other explicit gauge transformations," *Am. J. Phys.*, vol. 70, no. 9, pp. 917–928, 2002.

[143] J. A. Heras, "The Kirchhoff gauge," *Annals of Physics*, vol. 321, no. 5, pp. 1265–1273, 2006.

[144] J. A. Heras, "The Coulomb static gauge," *Am. J. Phys.*, vol. 75, no. 5, pp. 459–462, 2007.

[145] R. Nevels and C.-S. Shin, "Lorenz, Lorentz, and the gauge," *IEEE Anten. Propag. Mag.*, vol. 43, no. 3, pp. 70–71, 2001.

[146] T. Barrett, *Topological Foundations of Electromagnetism*. World Scientific, 2008.

[147] H. Kleinert, *Multivalued Fields in Condensed Matter, Electromagnetism, and Gravitation*. World Scientific, 2008.

[148] M. Schwartz, *Quantum Field Theory and the Standard Model*. Cambridge University Press, 2014.

[149] N. V. Mitskievich, *Relativistic Physics in Arbitrary Reference Frames*. Nova Science Publishers, 2006.

[150] M. Maggiore, *A Modern Introduction to Quantum Field Theory*. Oxford University Press, 2005.

[151] F. Schwabl, R. Hilton, and A. Lahee, *Advanced Quantum Mechanics.*   Springer, 2008.

[152] M. Srednicki, *Quantum Field Theory.* Cambridge University Press, 2007.

[153] G. Schatz and M. Ratner, *Quantum Mechanics in Chemistry.* Dover Publications, 2012.

[154] R. Prasad, *Electronic Structure of Materials.* CRC Press, 2013.

[155] K. Ogata, *Modern Control Engineering.* Prentice Hall, 2010.

[156] B. P. Lathi, R. A. Green, and R. Green, *Linear Systems and Signals.* Oxford University Press, 2017.

[157] R. H. Bishop, *Mechatronic Systems, Sensors, and Actuators: Fundamentals and Modeling.* CRC Press, 2017.

[158] R. K. R. Yarlagadda, *Analog and Digital Signals and Systems.* Springer, 2010.

[159] C. M. Van Vliet, *Equilibrium and Non-equilibrium Statistical Mechanics.* World Scientific, 2008.

[160] D. Pines and P. Nozieres, *The Theory of Quantum Liquids.* CRC Press, 1966.

[161] J. P. Hespanha, *Linear Systems Theory: Second Edition.* Princeton University Press, 2018.

[162] A. V. Oppenheim and A. S. Willsky, *Signals and Systems.* Pearson, 2013.

[163] R. M. Gray, *Entropy and Information Theory.* Springer, 2013.

[164] W. C. Schieve and L. P. Horwitz, *Quantum Statistical Mechanics.*   Cambridge University Press, 2009.

[165] L. D. Landau and E. M. Lifshitz, *Statistical Physics.* Elsevier Science, 2013.

[166] R. Kubo, "Lectures in theoretical physics," *Interscience*, pp. 120–203, 1959.

[167] R. Mazo, *Statistical mechanical theories of transport processes.*   Pergamon Press, 1967.

[168] K. Van Vliet, "Linear response theory revisited. The many-body van Hove limit," *J. Math. Phys.*, vol. 19, no. 6, pp. 1345–1370, 1978.

[169] M. Abramowitz and I. A. Stegun, *Handbook of Mathematical Functions: With Formulas, Graphs, and Mathematical Tables.* Dover Publications, 1965.

[170] V. Lucarini, J. Saarinen, K. Peiponen, and E. Vartiainen, *Kramers-Kronig Relations in Optical Materials Research.* Springer, 2005.

[171] B. Nistad and J. Skaar, "Causality and electromagnetic properties of active media," *Phys. Rev. E*, vol. 78, p. 036603, 2008.

[172] M. V. S. Bonança, "Fluctuation-dissipation theorem for the microcanonical ensemble," *Phys. Rev. E*, vol. 78, p. 031107, 2008.

[173] R. K. Pathria and P. D. Beale, *Statistical Mechanics*, 3rd ed. Elsevier Science, 2011.

[174] R. Buccheri, M. Saniga, and W. M. Stuckey, *The Nature of Time: Geometry, Physics and Perception.* Springer, 2012.

[175] H. B. Callen and T. A. Welton, "Irreversibility and generalized noise," *Phys. Rev.*, vol. 83, pp. 34–40, 1951.

[176] H. Hellmann, *Einführung in die quantenchemie.* Deuticke, 1937, 350 pages.

[177] R. P. Feynman, "Forces in molecules," *Phys. Rev.*, vol. 56, pp. 340–343, 1939.

[178] P. Güttinger, "Das verhalten von atomen im magnetischen drehfeld," *Zeitschrift für Physik*, vol. 73, no. 3, pp. 169–184, 1932.

[179] W. Pauli, P. Achuthan, and K. Venkatesan, *General Principles of Quantum Mechanics*. Springer, 2012.

[180] J. B. Johnson, "Thermal agitation of electricity in conductors," *Phys. Rev.*, vol. 32, pp. 97–109, 1928.

[181] H. Nyquist, "Thermal agitation of electric charge in conductors," *Phys. Rev.*, vol. 32, pp. 110–113, 1928.

[182] H. Hapuarachchi, M. Premaratne, Q. Bao, W. Cheng, S. D. Gunapala, and G. P. Agrawal, "Cavity QED analysis of an exciton-plasmon hybrid molecule via the generalized nonlocal optical response method," *Phys. Rev. B*, vol. 95, p. 245419, 2017.

[183] H. Hapuarachchi, S. D. Gunapala, Q. Bao, M. I. Stockman, and M. Premaratne, "Exciton behavior under the influence of metal nanoparticle near fields: Significance of nonlocal effects," *Phys. Rev. B*, vol. 98, p. 115430, 2018.

[184] M. Premaratne and M. I. Stockman, "Theory and technology of SPASERs," *Adv. Opt. Photon.*, vol. 9, no. 1, pp. 79–128, 2017.

[185] N. Ashcroft and N. Mermin, *Solid State Physics*. Holt, Rinehart and Winston, 1976.

[186] A. A. Govyadinov, G. Y. Panasyuk, J. C. Schotland, and V. A. Markel, "Theoretical and numerical investigation of the size dependent optical effects in metal nanoparticles," *Phys. Rev. B*, vol. 84, p. 155461, 2011.

[187] W. Götze and P. Wölfle, "Homogeneous dynamical conductivity of simple metals," *Phys. Rev. B*, vol. 6, pp. 1226–1238, 1972.

[188] C. S. Kumarasinghe, M. Premaratne, S. D. Gunapala, and G. P. Agrawal, "Design of all-optical, hot-electron current-direction-switching device based on geometrical asymmetry," *Scientific Reports*, vol. 6, p. 21470, 2016.

[189] C. S. Kumarasinghe, M. Premaratne, S. D. Gunapala, and G. P. Agrawal, "Theoretical analysis of hot electron injection from metallic nanotubes into a semiconductor interface," *Phys. Chem. Chem. Phys.*, vol. 18, pp. 18 227–18 236, 2016.

[190] C. S. Kumarasinghe, M. Premaratne, Q. Bao, and G. P. Agrawal, "Theoretical analysis of hot electron dynamics in nanorods," *Scientific Reports*, vol. 5, p. 12140, 2015.

[191] P. B. Allen, "Electron-phonon effects in the infrared properties of metals," *Phys. Rev. B*, vol. 3, pp. 305–320, 1971.

[192] J. B. Smith and H. Ehrenreich, "Frequency dependence of the optical relaxation time in metals," *Phys. Rev. B*, vol. 25, pp. 923–930, 1982.

[193] D. F. Escande, F. Doveil, and Y. Elskens, "N-body description of Debye shielding and Landau damping," *Plasma Phys. Control. Fusion*, vol. 58, no. 1, p. 014040, 2016.

[194] G. Ghosh, "Sellmeier coefficients and dispersion of thermo-optic coefficients for some optical glasses," *Appl. Opt.*, vol. 36, no. 7, pp. 1540–1546, 1997.

[195] A. D. Rakic, A. B. Djurišic, J. M. Elazar, and M. L. Majewski, "Optical properties of metallic films for vertical-cavity optoelectronic devices," *Appl. Opt.*, vol. 37, no. 22, pp. 5271–5283, Aug 1998.

[196] G. E. Jellison and F. A. Modine, "Parameterization of the optical functions of amorphous materials in the interband region," *Appl. Phys. Lett.*, vol. 69, no. 3, pp. 371–373, 1996.

[197] C. Rupasinghe, I. D. Rukhlenko, and M. Premaratne, "Spaser made of graphene and carbon nanotubes," *ACS Nano*, vol. 8, no. 3, pp. 2431–2438, 2014.

[198] V. Apalkov and M. I. Stockman, "Proposed graphene nanospaser," *Light: Science & Appl.*, vol. 3, p. e191, 2014.

[199] C. Jayasekara, M. Premaratne, S. D. Gunapala, and M. I. Stockman, "MoS_2 spaser," *J. Appl. Phys.*, vol. 119, no. 13, p. 133101, 2016.

[200] W. Zhu, I. D. Rukhlenko, and M. Premaratne, "Graphene metamaterial for optical reflection modulation," *Appl. Phys. Lett.*, vol. 102, no. 24, p. 241914, 2013.

[201] W. Zhu, I. D. Rukhlenko, L.-M. Si, and M. Premaratne, "Graphene-enabled tunability of optical fishnet metamaterial," *Appl. Phys. Lett.*, vol. 102, no. 12, p. 121911, 2013.

[202] V. P. Gusynin and S. G. Sharapov, "Transport of Dirac quasiparticles in graphene: Hall and optical conductivities," *Phys. Rev. B*, vol. 73, p. 245411, 2006.

[203] V. P. Gusynin, S. G. Sharapov, and J. P. Carbotte, "Unusual microwave response of Dirac quasiparticles in graphene," *Phys. Rev. Lett.*, vol. 96, p. 256802, 2006.

[204] V. P. Gusynin, S. G. Sharapov, and J. P. Carbotte, "Anomalous absorption line in the magneto-optical response of graphene," *Phys. Rev. Lett.*, vol. 98, p. 157402, 2007.

[205] N. M. R. Peres, F. Guinea, and A. H. Castro Neto, "Electronic properties of disordered two-dimensional carbon," *Phys. Rev. B*, vol. 73, p. 125411, 2006.

[206] I. F. Herbut, V. Juricic, and O. Vafek, "Coulomb interaction, ripples, and the minimal conductivity of graphene," *Phys. Rev. Lett.*, vol. 100, p. 046403, 2008.

[207] L. Fritz, J. Schmalian, M. Müller, and S. Sachdev, "Quantum critical transport in clean graphene," *Phys. Rev. B*, vol. 78, p. 085416, 2008.

[208] B. Wunsch, T. Stauber, F. Sols, and F. Guinea, "Dynamical polarization of graphene at finite doping," *New J. Phys.*, vol. 8, no. 12, p. 318, 2006.

[209] E. H. Hwang and S. Das Sarma, "Dielectric function, screening, and plasmons in two-dimensional graphene," *Phys. Rev. B*, vol. 75, p. 205418, 2007.

[210] M. Koshino and T. Ando, "Transport in bilayer graphene: Calculations within a self-consistent Born approximation," *Phys. Rev. B*, vol. 73, p. 245403, 2006.

[211] J. Nilsson, A. H. Castro Neto, F. Guinea, and N. M. R. Peres, "Electronic properties of bilayer and multilayer graphene," *Phys. Rev. B*, vol. 78, p. 045405, 2008.

[212] D. S. L. Abergel and V. I. Fal'ko, "Optical and magneto-optical far-infrared properties of bilayer graphene," *Phys. Rev. B*, vol. 75, p. 155430, 2007.

[213] E. J. Nicol and J. P. Carbotte, "Optical conductivity of bilayer graphene with and without an asymmetry gap," *Phys. Rev. B*, vol. 77, p. 155409, 2008.

[214] N. M. R. Peres, J. M. B. Lopes dos Santos, and T. Stauber, "Phenomenological study of the electronic transport coefficients of graphene," *Phys. Rev. B*, vol. 76, p. 073412, 2007.

[215] T. Stauber, N. M. R. Peres, and F. Guinea, "Electronic transport in graphene: A semiclassical approach including midgap states," *Phys. Rev. B*, vol. 76, p. 205423, 2007.

[216] L. A. Falkovsky and A. A. Varlamov, "Space-time dispersion of graphene conductivity," *European Phys. J. B*, vol. 56, no. 4, pp. 281–284, 2007.

[217] L. A. Falkovsky and S. S. Pershoguba, "Optical far-infrared properties of a graphene monolayer and multilayer," *Phys. Rev. B*, vol. 76, p. 153410, 2007.

[218] T. Stauber, N. M. R. Peres, and A. K. Geim, "Optical conductivity of graphene in the visible region of the spectrum," *Phys. Rev. B*, vol. 78, p. 085432, 2008.

[219] A. Vakil and N. Engheta, "Transformation optics using graphene," *Science*, vol. 332, no. 6035, pp. 1291–1294, 2011.

[220] S. A. Mikhailov and K. Ziegler, "New electromagnetic mode in graphene," *Phys. Rev. Lett.*, vol. 99, p. 016803, 2007.

[221] G. F. Bertsch, R. A. Broglia, P. V. Landshoff, D. R. Nelson, D. W. Sciama, and S. Weinberg, *Oscillations in Finite Quantum Systems*. Cambridge University Press, 1994.

[222] M. Baer, *Beyond Born-Oppenheimer: Electronic Nonadiabatic Coupling Terms and Conical Intersections*. Wiley, 2006.

[223] F. Wang and Y. R. Shen, "General properties of local plasmons in metal nanostructures," *Phys. Rev. Lett.*, vol. 97, p. 206806, 2006.

[224] M. I. Stockman, "Nanoplasmonics: past, present, and glimpse into future," *Opt. Express*, vol. 19, no. 22, pp. 22 029–22 106, 2011.

[225] W. L. Barnes, "Surface plasmon-polariton length scales: a route to sub-wavelength optics," *J. Opt. A: Pure Appl. Opt.*, vol. 8, no. 4, p. S87, 2006.

[226] J. B. Khurgin, "How to deal with the loss in plasmonics and metamaterials," *Nature Nanotechnology*, vol. 10, pp. 2–6, 2015.

[227] V. P. Gusynin, S. G. Sharapov, and J. P. Carbotte, "Magneto-optical conductivity in graphene," *J. Phys.: Condens. Matter*, vol. 19, no. 2, p. 026222, 2007.

[228] Z. Q. Li, E. A. Henriksen, Z. Jiang, Z. Hao, M. C. Martin, P. Kim, H. L. Stormer, and D. N. Basov, "Dirac charge dynamics in graphene by infrared spectroscopy," *Nature Phys.*, vol. 4, pp. 532–535, 2008.

[229] A. N. Grigorenko, M. Polini, and K. S. Novoselov, "Graphene plasmonics," *Nature Photon.*, vol. 6, pp. 749–758, 2012.

[230] F. J. Garciía de Abajo, "Graphene plasmonics: Challenges and opportunities," *ACS Photonics*, vol. 1, no. 3, pp. 135–152, 2014.

[231] M. Jablan, H. Buljan, and M. Soljacic, "Plasmonics in graphene at infrared frequencies," *Phys. Rev. B*, vol. 80, p. 245435, 2009.

[232] G. W. Hanson, "Dyadic Green's functions and guided surface waves for a surface conductivity model of graphene," *J. Appl. Phys.*, vol. 103, no. 6, p. 064302, 2008.

[233] M. Jablan, M. Soljacic, and H. Buljan, "Plasmons in graphene: Fundamental properties and potential applications," *Proc. IEEE*, vol. 101, no. 7, pp. 1689–1704, 2013.

[234] B. Dastmalchi, P. Tassin, T. Koschny, and C. M. Soukoulis, "A new perspective on plasmonics: Confinement and propagation length of surface plasmons for different materials and geometries," *Adv. Opt. Materials*, vol. 4, pp. 177–184, 2016.

[235] Y. He and T. Zeng, "First-principles study and model of dielectric functions of silver nanoparticles," *J. Phys. Chem. C*, vol. 114, no. 42, pp. 18023–18030, 2010.

[236] J. A. Scholl, A. L. Koh, and J. A. Dionne, "Quantum plasmon resonances of individual metallic nanoparticles," *Nature*, vol. 483, pp. 421–427, 2012.

[237] H. D. Zeh, *The Physical Basis of The Direction of Time*. Springer, 2007.

[238] F. Weinert, *The Scientist as Philosopher: Philosophical Consequences of Great Scientific Discoveries*. Springer, 2004.

[239] E. Wigner, "Über die operation der zeitumkehr in der quantenmechanik," *Nachrichten von der Gesellschaft der Wissenschaften zu Göttingen, Mathematisch-Physikalische Klasse*, vol. 1932, pp. 546–559, 1932.

[240] E. U. Condon and G. H. Shortley, *The Theory of Atomic Spectra*. Cambridge University Press, 1951.

[241] P. W. Milonni, "Semiclassical and quantum-electrodynamical approaches in nonrelativistic radiation theory," *Phys. Rep.*, vol. 25, no. 1, pp. 1–81, 1976.

[242] P. W. Milonni, "Why spontaneous emission?" *Am. J. Phys.*, vol. 52, no. 4, pp. 340–343, 1984.

[243] E. M. Purcell, "Spontaneous emission probabilities at radio frequencies," *Phys. Rev.*, vol. 69, p. 681, 1946.

[244] P. W. Milonni and P. L. Knight, "Spontaneous emission between mirrors," *Opt. Commun.*, vol. 9, no. 2, pp. 119–122, 1973.

[245] J.-K. Hwang, H.-Y. Ryu, and Y.-H. Lee, "Spontaneous emission rate of an electric dipole in a general microcavity," *Phys. Rev. B*, vol. 60, no. 7, pp. 4688–4695, 1999.

[246] F. D. Martini, G. Innocenti, G. R. Jacobovitz, and P. Mataloni, "Anomalous spontaneous emission time in a microscopic optical cavity," *Phys. Rev. Lett.*, vol. 59, pp. 2955–2958, 1987.

[247] D. Kleppner, "Inhibited spontaneous emission," *Phys. Rev. Lett.*, vol. 47, pp. 233–236, 1981.

[248] E. Yablonovitch, "Inhibited spontaneous emission in solid-state physics and electronics," *Phys. Rev. Lett.*, vol. 58, pp. 2059–2062, 1987.

[249] P. A. M. Dirac, "The quantum theory of the emission and absorption of radiation," *Proc. Roy. Soc. London*, vol. A114, p. 243, 1927.

[250] V. Weisskopf and E. Wigner, "Berechnung der natürlichen linienbreite auf grund der diracschen lichttheorie," *Zeitschrift für Physik*, vol. 63, no. 1, pp. 54–73, 1930.

[251] L. J. Curtis, *Atomic Structure and Lifetimes: A Conceptual Approach*. Cambridge University Press, 2003.

[252] W. Pfeifer, *The Lie Algebras su(N): An Introduction*. Birkhäuser, 2003.

[253] P. Meystre and M. Sargent, *Elements of Quantum Optics*. Springer, 2007.

[254] H. I. Yoo and J. H. Eberly, "Dynamical theory of an atom with two or three levels interacting with quantized cavity fields," *Phys. Rep.*, vol. 118, no. 5, pp. 239–337, 1985.

[255] B. W. Shore and P. L. Knight, "The Jaynes-Cummings model," *J. Mod. Opt.*, vol. 40, no. 7, pp. 1195–1238, 1993.

[256] P. L. Knight and P. W. Milonni, "The Rabi frequency in optical spectra," *Phys. Rep.*, vol. 66, no. 2, pp. 21–107, 1980.

[257] S. Stenholm, "Quantum theory of electromagnetic fields interacting with atoms and molecules," *Phys. Rep.*, vol. 6, no. 1, pp. 1–121, 1973.

[258] W. H. Louisell, *Quantum Statistical Properties of Radiation*. Wiley, 1973.

[259] M. Sargent, M. O. Scully, and W. E. Lamb, *Laser Physics*. Addison-Wesley, 1974.

[260] I. R. Senitzky, "Dissipation in quantum mechanics. The harmonic oscillator," *Phys. Rev.*, vol. 119, pp. 670–679, 1960.

[261] I. R. Senitzky, "Dissipation in quantum mechanics. the harmonic oscillator. II," *Phys. Rev.*, vol. 124, pp. 642–648, 1961.

[262] J. R. Ray, "Dissipation and quantum theory," *Lett. Nuovo Cimento*, vol. 25, no. 2, pp. 47–50, 1979.

[263] A. O. Caldeira and A. J. Leggett, "Quantum tunnelling in a dissipative system," *Annal. Phys.*, vol. 149, no. 2, pp. 374–456, 1983.

[264] S. Nakajima, "On quantum theory of transport phenomena: steady diffusion," *Prog. Theor. Phys.*, vol. 20, no. 6, pp. 948–959, 1958.

[265] R. Zwanzig, "Ensemble method in the theory of irreversibility," *J. Chem. Phys.*, vol. 33, no. 5, pp. 1338–1341, 1960.

[266] I. Prigonine and P. Resibois, "On the kinetics of the approach to equilibrium," *Physica*, vol. 27, no. 7, pp. 629–646, 1961.

[267] G. W. Ford, M. Kac, and P. Mazur, "Statistical mechanics of assemblies of coupled oscillators," *J. Math. Phys.*, vol. 6, no. 4, pp. 504–515, 1965.

[268] H. Mori, "Transport, collective motion, and Brownian motion," *Prog. Theor. Phys.*, vol. 33, no. 3, pp. 423–455, 1965.

[269] L. Van Hove, "Quantum-mechanical perturbations giving rise to a statistical transport equation," *Physica*, vol. 21, no. 1-5, pp. 517–540, 1954.

[270] I. Prigogine, *Non-Equilibrium Statistical Mechanics*. Dover Publications, 2017.

[271] F. Haake and R. Reibold, "Strong damping and low-temperature anomalies for the harmonic oscillator," *Phys. Rev. A*, vol. 32, pp. 2462–2475, 1985.

[272] H. P. Breuer and F. Petruccione, *The Theory of Open Quantum Systems*. Oxford University Press, 2002.

[273] U. Weiss, *Quantum Dissipative Systems*. World Scientific, 2012.

[274] M. A. Schlosshauer, *Decoherence and the Quantum-to-Classical Transition*. Springer, 2007.

[275] V. Gorini, A. Kossakowski, and E. C. G. Sudarshan, "Completely positive dynamical semigroups of N-level systems," *J. Math. Phys.*, vol. 17, no. 5, pp. 821–825, 1976.

[276] G. Lindblad, "On the generators of quantum dynamical semigroups," *Commun. Math. Phys.*, vol. 48, no. 2, pp. 119–130, 1976.

[277] R. Bausch, "Bewegungsgesetze nicht abgeschlossener quantensysteme," *Zeitschrift für Physik*, vol. 193, no. 2, pp. 246–265, 1966.

[278] F. Haake, "Statistical treatment of open systems by generalized master equations," in *Quantum Statistics in Optics and Solid-State Physics, Vol. 66*. Springer, 1973.

[279] H. Haken, *Laser Theory*. Springer, 2012.

[280] E. C. G. Sudarshan, "Quantum dynamics, metastable states, and contractive semi-groups," *Phys. Rev. A*, vol. 46, pp. 37–48, 1992.

[281] W. F. Stinespring, "Positive functions on C*-algebras," *Proc. Am. Math. Soc.*, vol. 6, no. 2, pp. 211–216, 1955.

[282] K. Kraus, "General state changes in quantum theory," *Annals of Physics*, vol. 64, no. 2, pp. 311–335, 1971.

[283] K. Kraus, W. H. Wootters, J. D. Dollard, and A. Böhn, *States, Effects, and Operations: Fundamental Notions of Quantum Theory*. Springer, 1983.

[284] I. Burghardt, "Dynamics of predissociation in the condensed phase: Markovian master equation," *J. Phys. Chem. A*, vol. 102, no. 23, pp. 4192–4206, 1998.

[285] P. Saalfrank and R. Kosloff, "Quantum dynamics of bond breaking in a dissipative environment: Indirect and direct photodesorption of neutrals from metals," *J. Chem. Phys.*, vol. 105, no. 6, pp. 2441–2455, 1996.

[286] E. Stefanescu, R. J. Liotta, and A. Sandulescu, "Giant resonances as collective states with dissipative coupling," *Phys. Rev. C*, vol. 57, no. 2, p. 798, 1998.

[287] R. A. Horn and C. R. Johnson, *Matrix Analysis*. Cambridge University Press, 2012.

[288] G. Schaller, *Open Quantum Systems Far from Equilibrium*. Springer, 2014.

[289] A. G. Redfield, "The theory of relaxation processes," in *Advances in Magnetic Resonance*, J. S. Waugh, Ed. Academic Press, 1965, pp. 1–32.

[290] D. Gamliel and H. Levanon, *Stochastic Processes in Magnetic Resonance*. World Scientific, 1995.

[291] P. A. Abragam and A. Abragam, *The Principles of Nuclear Magnetism*. Clarendon Press, 1961.

[292] D. A. Wiersma, *Coherent Optical Transient Studies of Dephasing and Relaxation in Electronic Transitions of Large Molecules in the Condensed Phase*. Wiley, 2007, pp. 421–485.

[293] E. B. Davies, "Markovian master equations," *Commun. Math. Phys.*, vol. 39, no. 2, pp. 91–110, 1974.

[294] R. Dümcke and H. Spohn, "The proper form of the generator in the weak coupling limit," *Zeitschrift für Physik B*, vol. 34, no. 4, pp. 419–422, 1979.

[295] C. Cohen-Tannoudji, J. Dupont-Roc, and G. Grynberg, *Atom-Photon Interactions: Basic Processes and Applications*. Wiley, 1998.

[296] C. Gardiner, P. Zoller, and P. Zoller, *Quantum Noise*, 3rd ed. Springer, 2004.

[297] D. F. Walls and G. J. Milburn, *Quantum Optics*. Springer, 2008.

[298] B. Weber, S. Mahapatra, H. Ryu, S. Lee, A. Fuhrer, T. C. G. Reusch, D. L. Thompson, W. C. T. Lee, G. Klimeck, L. C. L. Hollenberg, and M. Y. Simmons, "Ohm's law survives to the atomic scale," *Science*, vol. 335, no. 6064, pp. 64–67, 2012.

[299] J. P. Pekola, O.-P. Saira, V. F. Maisi, A. Kemppinen, M. Möttönen, Y. A. Pashkin, and D. V. Averin, "Single-electron current sources: Toward a refined definition of the Ampere," *Rev. Mod. Phys.*, vol. 85, pp. 1421–1472, 2013.

[300] M. Büttiker, "Four-terminal phase-coherent conductance," *Phys. Rev. Lett.*, vol. 57, pp. 1761–1764, 1986.

[301] J. Verspecht, "Large-signal network analysis," *IEEE Microwave Magazine*, vol. 6, no. 4, pp. 82–92, 2005.

[302] S. Datta, H. Ahmad, and M. Pepper, *Electronic Transport in Mesoscopic Systems*. Cambridge University Press, 1997.

[303] A. Szafer and A. D. Stone, "Theory of quantum conduction through a constriction," *Phys. Rev. Lett.*, vol. 62, pp. 300–303, 1989.

[304] H. L. Engquist and P. W. Anderson, "Definition and measurement of the electrical and thermal resistances," *Phys. Rev. B*, vol. 24, pp. 1151–1154, 1981.

[305] D. K. Ferry, S. M. Goodnick, and J. Bird, *Transport in Nanostructures*, 2nd ed. Cambridge University Press, 2009.

[306] P. C. Martin and J. Schwinger, "Theory of many-particle systems. I," *Phys. Rev.*, vol. 115, pp. 1342–1373, 1959.

[307] L. P. Kadanoff and G. Baym, *Quantum Statistical Mechanics*. Benjamin, 1962.

[308] L. V. Keldysh, "Diagram technique for nonequilibrium processes," *Sov. Phys. JETP*, vol. 20, no. 4, pp. 1018–1026, 1965.

[309] D. C. Langreth, *Linear and nonlinear electron transport in solids*. Plenum, 1975, pp. 3–32.

[310] O. V. Konstantinov and V. I. Perel, "A diagram technique for evaluating transport quantities," *Sov. Phys. JETP*, vol. 12, pp. 142–149, 1961.

[311] I. E. Dzyaloshinskii, "A diagram technique for evaluating transport coefficients in statistical physics at finite temperatures," *Sov. Phys. JETP*, vol. 15, pp. 778–783, 1962.

[312] A. A. Abrikosov, L. P. Gor'kov, and I. Y. Dzyaloshinskii, *Quantum Field Theoretical Methods in Statistical Physics*. Pergamon Press, 1965.

[313] Y. Meir and N. S. Wingreen, "Landauer formula for the current through an interacting electron region," *Phys. Rev. Lett.*, vol. 68, pp. 2512–2515, 1992.

[314] E. Runge and H. Ehrenreich, "Response and transit times in quantum-well structures," *Phys. Rev. B*, vol. 45, pp. 9145–9148, 1992.

[315] A.-P. Jauho, N. S. Wingreen, and Y. Meir, "Time-dependent transport in interacting and noninteracting resonant-tunneling systems," *Phys. Rev. B*, vol. 50, pp. 5528–5544, 1994.

[316] S. Datta, "Steady-state quantum kinetic equation," *Phys. Rev. B*, vol. 40, pp. 5830–5833, 1989.

[317] S. Datta, "A simple kinetic equation for steady-state quantum transport," *J. Physics: Cond. Matter*, vol. 2, no. 40, p. 8023, 1990.

[318] S. Datta, "Nanoscale device modeling: The Green's function method," *Superlattices and Microstructures*, vol. 28, no. 4, pp. 253–278, 2000.

[319] S. Datta, "Electrical resistance: an atomistic view," *Nanotechnology*, vol. 15, no. 7, pp. S433–S451, 2004.

[320] S. Datta, *Quantum Transport: Atom to Transistor*. Cambridge University Press, 2005.

[321] "nanoHub.org," 2011. [Online]. Available: https://nanoHub.org

[322] J. Rammer and H. Smith, "Quantum field-theoretical methods in transport theory of metals," *Rev. Mod. Phys.*, vol. 58, pp. 323–359, 1986.

[323] C. Caroli, R. Combescot, P. Nozieres, and D. Saint-James, "Direct calculation of the tunneling current," *J. Physics C: Solid State Phys.*, vol. 4, no. 8, pp. 916–929, 1971.

[324] H. Haug and A. P. Jauho, *Quantum Kinetics in Transport and Optics of Semiconductors*. Springer, 2007.

[325] N. S. Wingreen, A.-P. Jauho, and Y. Meir, "Time-dependent transport through a mesoscopic structure," *Phys. Rev. B*, vol. 48, pp. 8487–8490, 1993.

[326] H. Nakamura and G. Mil'nikov, *Quantum Mechanical Tunneling in Chemical Physics*. CRC Press, 2016.

[327] P. Busch, *Time in Quantum Mechanics*. Springer, 2008, pp. 73–105.

[328] T. E. Hartman, "Tunneling of a wave packet," *J. Appl. Phys.*, vol. 33, no. 12, pp. 3427–3433, 1962.

[329] U. S. Sainadh, H. Xu, X. Wang, A. Atia-Tul-Noor, W. C. Wallace, N. Douguet, A. Bray, I. Ivanov, K. Bartschat, A. Kheifets, R. T. Sang, and I. V. Litvinyuk, "Attosecond angular streaking and tunnelling time in atomic hydrogen," *Nature*, vol. 568, no. 7750, pp. 75–77, 2019.

[330] T. Lucatorto, M. De Graef, J. A. Stroscio, and W. J. Kaiser, *Scanning Tunneling Microscopy*. Elsevier Science, 2013.

[331] J. Hoekstra, *Introduction to Nanoelectronic Single-Electron Circuit Design*. Jenny Stanford Publishing, 2009.

[332] P. Hommelhoff and M. Kling, *Attosecond Nanophysics: From Basic Science to Applications*. Wiley, 2015.

[333] D. A. Bennett, L. Longobardi, V. Patel, W. J. Chen, D. V. Averin, and J. E. Lukens, "Decoherence in rf SQUID qubits," *Quantum Information Processing*, vol. 8, pp. 217–243, 2009.

[334] G. Gamow, "Zur quantentheorie des atomkernes," *Zeitschrift für Physik*, vol. 51, no. 3, pp. 204–212, 1928.

[335] J.-G. J. Zhu and C. Park, "Magnetic tunnel junctions," *Materials Today*, vol. 9, no. 11, pp. 36–45, 2006.

[336] J. E. House, *Fundamentals of Quantum Mechanics*. Elsevier Science, 2017.

[337] F. Hund, "Progress in the classification and theory of molecular spectra, part I," *Z. Phys.*, vol. 42, p. 93, 1927.

[338] F. Hund, "Progress in the classification and theory of molecular spectra, part II," *Z. Phys.*, vol. 43, p. 805, 1927.

[339] H. Hund, "Progress in the classification and theory of molecular spectra, part III," *Z. Phys.*, vol. 44, p. 742, 1927.

[340] R. H. Fowler and L. Nordheim, "Electron emission in intense electric fields," *Proc. Royal Soc. London, Series A*, vol. 119, no. 781, pp. 173–181, 1928.

[341] M. Razavy, *Quantum Theory of Tunneling*. World Scientific, 2003.

[342] L. Esaki, "New phenomenon in narrow germanium p-n junctions," *Phys. Rev.*, vol. 109, pp. 603–604, 1959.

[343] H. Grabert and H. Horner, "Special issue on single charge tunneling," *Z. Phys. B: Cond. Matter*, vol. 85, no. 3, p. 317, 1991.

[344] H. Grabert and M. H. Devoret, Eds., *Single Charge Tunneling*, ser. NATO Science, vol. 294. New York: Plenum Press, 1992.

[345] I. Giaever, "Electron tunneling and superconductivity," *Rev. Mod. Phys.*, vol. 46, pp. 245–250, Apr. 1974.

[346] B. D. Josephson, "The discovery of tunneling supercurrents," *Science*, vol. 184, no. 4136, pp. 527–530, 1974.

[347] J. Bardeen, L. N. Cooper, and J. R. Schrieffer, "Microscopic theory of superconductivity," *Phys. Rev.*, vol. 106, pp. 162–164, 1957.

[348] J. Bardeen, L. N. Cooper, and J. R. Schrieffer, "Theory of superconductivity," *Phys. Rev.*, vol. 108, pp. 1175–1204, 1957.

[349] B. D. Josephson, "The discovery of tunnelling supercurrents," *Rev. Mod. Phys.*, vol. 46, pp. 251–254, 1974.

[350] D. Twerenbold, "Giaever-type superconducting tunnelling junctions as high-resolution X-ray detectors," *Europhys. Lett.*, vol. 1, no. 5, pp. 209–214, 1986.

[351] J. Clarke and A. I. Braginski, *The SQUID Handbook*. Wiley, 2006.

[352] A. Yazdani, "Watching an atom tunnel," *Nature*, vol. 409, no. 6819, pp. 471–472, 2001.

[353] B. M. Karnakov and V. P. Krainov, *WKB Approximation in Atomic Physics*. Springer, 2012.

[354] N. Zettili, *Quantum Mechanics: Concepts and Applications*. Wiley, 2009.

[355] J. Bardeen, "Tunnelling from a many-particle point of view," *Phys. Rev. Lett.*, vol. 6, pp. 57–59, Jan. 1961.

[356] W. A. Harrison, "Tunneling from an independent-particle point of view," *Phys. Rev.*, vol. 123, pp. 85–89, 1961.

[357] M. H. Cohen, L. M. Falicov, and J. C. Phillips, "Superconductive tunneling," *Phys. Rev. Lett.*, vol. 8, pp. 316–318, 1962.

[358] C. B. Duke, "Tunneling in solids," in *Solid State Physics*, J. A. Appelbaum, Ed. Academic Press, 1970.

[359] D. V. Averin, A. N. Korotkov, and K. K. Likharev, "Theory of single-electron charging of quantum wells and dots," *Phys. Rev. B*, vol. 44, pp. 6199–6211, 1991.

[360] D. Ryndyk, *Theory of Quantum Transport at Nanoscale: An Introduction*. Springer, 2015.

[361] L. J. Geerligs, D. V. Averin, and J. E. Mooij, "Observation of macroscopic quantum tunneling through the Coulomb energy barrier," *Phys. Rev. Lett.*, vol. 65, pp. 3037–3040, 1990.

[362] H. Matsuoka and S. Kimura, "Thermally enhanced co-tunneling of single electrons in a Si quantum dot at 4.2K," *Jap. J. Appl. Phys.*, vol. 34, pp. 1326–1328, 1995.

[363] D. V. Averin and Y. V. Nazarov, "Virtual electron diffusion during quantum tunneling of the electric charge," *Phys. Rev. Lett.*, vol. 65, pp. 2446–2449, 1990.

[364] A. E. Hanna, M. T. Tuominen, and M. Tinkham, "Observation of elastic macroscopic quantum tunneling of the charge variable," *Phys. Rev. Lett.*, vol. 68, pp. 3228–3231, 1992.

[365] K. K. Likharev, N. S. Bakhvalov, G. S. Kazacha, and S. I. Serdyukova, "Single-electron tunnel junction array: an electrostatic analog of the Josephson transmission line," *IEEE Transactions on Magnetics*, vol. 25, no. 2, pp. 1436–1439, 1989.

[366] G. Binnig, H. Rohrer, C. Gerber, and E. Weibel, "7×7 reconstruction on Si(111) resolved in real space," *Phys. Rev. Lett.*, vol. 50, pp. 120–123, 1983.

[367] W. A. Hofer, A. S. Foster, and A. L. Shluger, "Theories of scanning probe microscopes at the atomic scale," *Rev. Mod. Phys.*, vol. 75, pp. 1287–1331, 2003.

[368] B. B. Zhou, S. Misra, E. H. da Silva Neto, P. Aynajian, R. E. Baumbach, J. D. Thompson, E. D. Bauer, and A. Yazdani, "Visualizing nodal heavy fermion superconductivity in CeCoIn$_5$," *Nature Phys.*, vol. 9, pp. 474–479, 2013.

[369] R. Wong, W. Rheinboldt, and D. Siewiorek, *Asymptotic Approximations of Integrals*. Elsevier Science, 2014.

[370] H. Lüth, *Surfaces and Interfaces of Solid Materials*. Springer, 2013.

[371] L. L. Chang, L. Esaki, and R. Tsu, "Resonant tunneling in semiconductor double barriers," *Appl. Phys. Lett.*, vol. 24, no. 12, pp. 593–595, 1974.

[372] U. Mizutani, *Introduction to the Electron Theory of Metals*. Cambridge University Press, 2001.

[373] R. Tsu and L. Esaki, "Tunneling in a finite superlattice," *Applied Physics Letters*, vol. 22, no. 11, pp. 562–564, 1973.

[374] J. N. Schulman, "Extension of Tsu–Esaki model for effective mass effects in resonant tunneling," *Appl. Phys. Lett.*, vol. 72, no. 22, pp. 2829–2831, 1998.

[375] E. Croke and J. N. Schulman, "Resonant tunneling diodes," in *Encyclopedia of Materials: Science and Technology*, K. H. J. Buschow, Ed. Elsevier, 2001, pp. 8185–8192.

[376] H. A. Haus and J. A. Mullen, "Quantum noise in linear amplifiers," *Phys. Rev.*, vol. 128, pp. 2407–2413, 1962.

[377] H. Heffner, "The fundamental noise limit of linear amplifiers," *Proc. IRE*, vol. 50, no. 7, pp. 1604–1608, 1962.

[378] D. Marcuse, *Principles of Quantum Electronics*. Elsevier Science, 2012.

[379] J. L. Gimlett, M. Z. Iqbal, L. Curtis, N. K. Cheung, A. Righetti, F. Fontana, and G. Grasso, "Impact of multiple reflection noise in Gbit/s lightwave systems with optical fibre amplifiers," *Electron. Lett.*, vol. 25, pp. 1393–1394, 1989.

[380] N. Bohr, "Über die serienspektra der elemente," *Z. Phys.*, vol. 2, no. 5, pp. 423–469, 1920.

[381] N. G. van Kampen, *Stochastic Processes in Physics and Chemistry*. North-Holland, 1981.

[382] E. Purcell and D. Morin, *Electricity and Magnetism*. Cambridge University Press, 2013.

[383] R. W. Henry, "Random-walk model of thermal noise for students in elementary physics," *Am. J. Phys.*, vol. 41, no. 12, pp. 1361–1363, 1973.

[384] L. Callegaro, "Unified derivation of Johnson and shot noise expressions," *Am. J. Phys.*, vol. 74, no. 5, pp. 438–440, 2006.

[385] C. Kittel, *Elementary Statistical Physics*. Dover, 2004.

[386] D. Kondepudi and I. Prigogine, *Modern Thermodynamics: From Heat Engines to Dissipative Structures*. Wiley, 2014.

[387] W. Schottky, "Uber spontane stromschwankungen in verschiedenen elektrizitätsleitern," *Annalen der Physik*, vol. 362, no. 23, pp. 541–567, 1918.

[388] Y. M. Blanter and M. Büttiker, "Shot noise in mesoscopic conductors," *Phys. Rep.*, vol. 336, pp. 1–166, 2000.

[389] P. Mazur and D. Bedeaux, "When and why is the random force in Brownian motion a Gaussian process," *Biophys. Chem.*, vol. 41, no. 1, pp. 41–49, 1991.

[390] W. A. Hofer, A. S. Foster, and A. L. Shluger, "Causality, time-reversal invariance and the Langevin equation," *Physica A*, vol. 173, no. 1, pp. 155–174, 1991.

[391] S. Swain, "Master equation derivation of quantum regression theorem," *J. Phys. A: Math. General*, vol. 14, no. 10, pp. 2577–2580, 1981.

[392] M. Lax, "Formal theory of quantum fluctuations from a driven state," *Phys. Rev.*, vol. 129, pp. 2342–2348, 1963.

[393] M. Lax, "Quantum noise X. Density-matrix treatment of field and population-difference fluctuations," *Phys. Rev.*, vol. 157, pp. 213–231, 1967.

[394] R. J. Schoelkopf, A. A. Clerk, S. M. Girvin, K. W. Lehnert, and M. H. Devoret, *Quantum Noise in Mesoscopic Physics*. Springer, 2003, pp. 175–203.

[395] A. A. Clerk, M. H. Devoret, S. M. Girvin, F. Marquardt, and R. J. Schoelkopf, "Introduction to quantum noise, measurement, and amplification," *Rev. Mod. Phys.*, vol. 82, pp. 1155–1208, 2010.

[396] J. Schwinger, "Brownian motion of a quantum oscillator," *J. Math. Phys.*, vol. 2, no. 3, pp. 407–432, 1961.

[397] M. I. Dykman and M. A. Krivoglaz, "Quantum theory of nonlinear oscillators interacting with the medium," *Sov. Phys. JETP*, vol. 37, p. 506, 1973.

[398] H. B. G. Casimir and D. Polder, "The influence of retardation on the London–van der Waals forces," *Phys. Rev.*, vol. 73, pp. 360–372, 1948.

[399] M. I. Kolobov, L. Davidovich, E. Giacobino, and C. Fabre, "Role of pumping statistics and dynamics of atomic polarization in quantum fluctuations of laser sources," *Phys. Rev. A*, vol. 47, pp. 1431–1446, 1993.

[400] C. Benkert, M. O. Scully, J. Bergou, L. Davidovich, M. Hillery, and M. Orszag, "Role of pumping statistics in laser dynamics: Quantum Langevin approach," *Phys. Rev. A*, vol. 41, pp. 2756–2765, 1990.

[401] J. Bergou, L. Davidovich, M. Orszag, C. Benkert, M. Hillery, and M. O. Scully, "Role of pumping statistics in maser and laser dynamics: Density-matrix approach," *Phys. Rev. A*, vol. 40, pp. 5073–5080, 1989.

[402] M. Orszag, *Quantum Optics: Including Noise Reduction, Trapped Ions, Quantum Trajectories, and Decoherence*. Springer, 2016.

[403] D. E. McCumber, "Intensity fluctuations in the output of cw laser oscillators. I," *Phys. Rev.*, vol. 141, pp. 306–322, 1966.

[404] C. Harder, J. Katz, S. Margalit, J. Shacham, and A. Yariv, "Noise equivalent circuit of a semiconductor laser diode," *IEEE J. Quantum Electron.*, vol. 18, no. 3, pp. 333–337, 1982.

[405] M. Lax, "Classical noise V. Noise in self-sustained oscillators," *Phys. Rev.*, vol. 160, pp. 290–307, 1967.

[406] Y. Yamamoto and N. Imoto, "Internal and external field fluctuations of a laser oscillator: Part I," *IEEE J. Quantum Electron.*, vol. 22, no. 10, pp. 2032–2042, 1986.

[407] T. Suhara, *Semiconductor Laser Fundamentals*. CRC Press, 2004.

[408] M. Premaratne and G. P. Agrawal, *Light Propagation in Gain Media: Optical Amplifiers*. Cambridge University Press, 2011.

[409] C. H. Henry, "Theory of spontaneous emission noise in open resonators and its application to lasers and optical amplifiers," *J. Lightw. Technol.*, vol. 4, pp. 288–297, 1986.

[410] W. W. Chow, S. W. Koch, and M. Sargent, *Semiconductor Laser Physics*. Springer, 2012.

[411] B. Tromborg, H. Olesen, and X. Pan, "Theory of linewidth for multielectrode laser diodes with spatially distributed noise sources," *IEEE J. Quantum Electron.*, vol. 27, no. 2, pp. 178–192, 1991.

[412] M. Tse et al., "Quantum-enhanced advanced LIGO detectors in the era of gravitational-wave astronomy," *Phys. Rev. Lett.*, vol. 123, p. 231107, 2019.

[413] H. P. Robertson, "The uncertainty principle," *Phys. Rev.*, vol. 34, pp. 163–164, 1929.

[414] M. D. Levenson, R. M. Shelby, A. Aspect, M. Reid, and D. F. Walls, "Generation and detection of squeezed states of light by nondegenerate four-wave mixing in an optical fiber," *Phys. Rev. A*, vol. 32, pp. 1550–1562, 1985.

[415] G. P. Agrawal, *Nonlinear Fiber Optics*, 6th ed. Academic Press, 2019.

[416] G. P. Agrawal, *Applications of Nonlinear Fiber Optics*, 3rd ed. Academic Press, 2020.

[417] R. M. Shelby, M. D. Levenson, S. H. Perlmutter, R. G. D. Voe, and D. F. Walls, "Broad-band parametric deamplification of quantum noise in an optical fiber," *Phys. Rev. Lett.*, vol. 57, pp. 691–694, 1986.

[418] G. Gasse, C. Lupien, and B. Reulet, "Observation of squeezing in the electron quantum shot noise of a tunnel junction," *Phys. Rev. Lett.*, vol. 111, p. 136601, 2013.

[419] U. C. Mendes, S. Jezouin, P. Joyez, B. Reulet, A. Blais, F. Portier, C. Mora, and C. Altimiras, "Parametric amplification and squeezing with an ac- and dc-voltage biased superconducting junction," *Phys. Rev. Applied*, vol. 11, p. 034035, 2019.

Index

acoustic wave, 259
action integral, 45
action principle, 45
Aharonov–Bohm effect, 26, 28, 70
Aharonov–Casher effect, 28
Airy equation, 201
Airy function, 201
alpha particle, 191
Altshuler–Aronov–Spivak effect, 29
amplifier
   linear, 220
   noise-free, 221
   optical, 220
Anderson localization, 29
antiunitary operator, 122
arrow of time, 122
autocorrelation function, 225, 228
   quantum, 233

Baker–Hausdorff formula, 152
ballistic transport, 159
band-bending effect, 218
Bell state, 54
Berry phase, 28
bioluminescence, 125
blackbody radiation, 9, 11, 21, 93, 124
Bloch frequency, 196
Bloch theorem, 9
Bogoliubov hierarchy, 180
Bohr model, 13
Bohr radius, 13
Boltzmann constant, 9, 13
Boltzmann distribution, 12, 113, 124, 147
Born approximation, 68, 137, 138, 150
Born rule, 23
Born series, 68
Born–Oppenheimer approximation, 116, 120, 193
Bose–Einstein distribution, 91, 92
Bose–Einstein statistics, 53, 154
bosons, 79, 91
Brillouin scattering

spontaneous, 259
stimulated, 259
thermal, 259
Brownian motion, 228

Campbell's theorem, 226
canonical quantization, 61
carbon nanotube, 6
Casimir force, 238
Cauchy–Schwarz inequality, 58
causality, 86, 95, 233
cavity QED, 126
central limit theorem, 224
charge transport, 31
   ballistic, 33
   diffusive, 34
   quasi-ballistic, 34
charging energy, 35, 158
chemical potential, 9, 89, 119
chemoluminescence, 125
CMOS technology, 217
co-tunneling
   elastic, 210, 211
   inelastic, 210, 211
coherence
   degree of, 11
   phase, 11
   spatial, 11
   temporal, 11
coherence length, 11, 33
coherence time, 11
coherent state, 258
collision
   elastic, 10, 33
collision time, 10, 31, 156, 226
commutation relation, 258
commutator, 58, 63
   anti-, 58
conductance, 31, 168, 172
   generalized, 98
conductance fluctuations, 28